智能网格预报技术论文集

中国气象局预报与网络司　组织编写
国　家　气　象　中　心

金荣花　代　刊　张志刚　王　云　主　编

内容简介

为了总结智能网格预报技术研究进展，促进智能网格预报技术交流和成果共享，推进新技术新方法的业务应用，本书精选了近年来有关智能网格预报技术的相关论文编纂成册。其内容不仅包括智能网格预报技术的综述，还涉及具体的预报技术和预报方法，以及天气预报预警技术应用和平台建设等，既有新方法新理论，又结合实际应用，达到科研致用的目的。

本书可供全国气象部门从事天气气候预报预测的业务、科研人员和管理人员阅读，也可供其他气象相关专业人员参考。

图书在版编目（CIP）数据

智能网格预报技术论文集 / 金荣花等主编. -- 北京：气象出版社，2022.8
ISBN 978-7-5029-7778-8

Ⅰ．①智… Ⅱ．①金… Ⅲ．①网格－应用－天气预报－文集 Ⅳ．①P45-39

中国版本图书馆CIP数据核字(2022)第148171号

Zhineng Wangge Yubao Jishu Lunwenji
智能网格预报技术论文集

出版发行：气象出版社	
地　　址：北京市海淀区中关村南大街46号	邮政编码：100081
电　　话：010-68407112(总编室)　010-68408042(发行部)	
网　　址：http://www.qxcbs.com	E-mail：qxcbs@cma.gov.cn
责任编辑：张盼娟	终　　审：吴晓鹏
责任校对：张硕杰	责任技编：赵相宁
封面设计：艺点设计	
印　　刷：北京建宏印刷有限公司	
开　　本：889 mm×1194 mm　1/16	印　　张：16.25
字　　数：492千字	
版　　次：2022年8月第1版	印　　次：2022年8月第1次印刷
定　　价：120.00元	

本书如存在文字不清、漏印以及缺页、倒页、脱页等，请与本社发行部联系调换。

目 录

定量降水预报技术进展 ………………………………… 毕宝贵　代　刊　王　毅　等（001）

中国无缝隙精细化网格天气预报技术进展与挑战 ………… 金荣花　代　刊　赵瑞霞　等（019）

国家级格点化定量降水预报系统 ……………………… 曹　勇　刘凑华　宗志平　等（033）

OTS、MOS和OMOS方法及其优化组合应用于72 h内逐3 h降水预报的试验分析研究
…………………………………………………………… 赵瑞霞　代　刊　金荣花　等（042）

基于集合预报系统的日最高和最低温度预报 ………………………………… 熊敏诠（054）

A Tropical Cyclone Similarity Search Algorithm Based on Deep Learning Method
………………………… WANG Yu（王玉）　HAN Lei（韩雷）　LIN Yinjing（林隐静）　et al.（067）

利用高原积雪信号改进中国南方夏季降水预测的新方法及其在2014年降水预测中的应用试验
………………………………………………………………… 刘　颖　任宏利　张培群　等（090）

多源气象数据融合格点实况产品研制进展 …………… 师春香　潘　旸　谷军霞　等（098）

Roles of Synoptic to Quasi-Monthly Disturbances in Generating Two Pre-Summer Heavy Rainfall
　Episodes over South China ………… JIANG Zhina　ZHANG Da-Lin　LIU Hongbo（109）

最优集合预报订正方法在客观温度预报中的应用 ……… 郝　翠　张迎新　王在文　等（130）

上海市无缝隙天气预报技术 …………………………… 储　海　陈　雷　戴建华　等（140）

基于数值模式误差分析的气温预报方法 ……………………………… 蔡凝昊　俞剑蔚（151）

基于数值预报和随机森林算法的强对流天气分类预报技术 ……… 李文娟　赵　放　郦敏杰　等（162）

基于评分最优化的模式降水预报订正算法对比 ……… 吴启树　韩　美　刘　铭　等（174）

A Convection Nowcasting Method Based on Machine Learning
………………………………………………… SU Aifang　LI Han　CUI Liman　et al.（186）

日极端气温的主客观预报能力评估及多模式集成网格释用 ……… 吴乃庚　曾　沁　刘段灵　等（202）

基于SWAN雷达拼图产品在暴雨过程中的对流云降水识别及效果检验
……………………………………………………… 张　勇　吴胜刚　张亚萍　等（214）

基于SCMOC的贵州最高气温预报方法研究 …………………… 李　刚　杨秀庄　刘彦华　等（227）

陕西省精细化网格预报业务系统技术方法 ……………………… 王建鹏　薛春芳　潘留杰　等（237）

基于小波分析的客观预报方法在智能网格高低温预报中的应用 … 刘新伟　段伯隆　黄武斌　等（247）

定量降水预报技术进展

毕宝贵　代刊　王毅　符娇兰　曹勇　刘凑华

(国家气象中心,北京,100081)

摘　要:本文对21世纪以来定量降水预报技术流程中的数值模式预报、统计后处理、检验评估和预报员作用4个方面的研究工作进行了归纳,主要进展包括:业务全球模式对于降水的预报能力持续提升,而发展高分辨率模式(尤其是对流尺度模式)和集合预报是提高定量降水预报精准化水平的主要途径,且将两者相结合以促进短期降水预报是发展趋势;统计后处理技术已经发展到应用数据挖掘方法对海量预报数据中有效信息进行提取和集成,而再预报资料的出现将进一步促进统计后处理技术的发展;为解决评估精细化定量降水预报面临的新问题,多种新的检验技术得到发展和应用,如极端降水检验评分、空间检验技术及概率检验方法等;预报员在模式和后处理方法上能够提供的附加值越来越有限,但在预报流程中仍将处于核心地位,其角色将逐渐向帮助用户进行决策方向转变。最后本文指出,定量降水预报技术的发展所面临的挑战包括大气水汽观测及同化技术改进、暖区和复杂地形下暴雨预报等科学问题的解决。

关键词:定量降水预报;数值模式;统计后处理;检验评估;预报员作用

Advances in Techniques of Quantitative Precipitation Forecast

BI Baogui　DAI Kan　WANG Yi　FU Jiaolan　CAO Yong　LIU Couhua

(National Meteorological Center, Beijing, 100081)

Abstract: The quantitative precipitation forecast (QPF) is a core operation of weather forecast. Modern technological processes of the QPF include numeric weather forecast, verification and evaluation, objective calibration and integration, forecaster's subjective modification and gridding post-processing. Domestic and international research work covering these five aspects are investigated and summarized, to provide reference for development of the quantitative precipitation forecast.

In the aspect of numerical weather forecast, the forecast skill of the operational global model for precipitation has been improving continuously (a gain of about 1 forecast day per decade), and developments of the high resolution model (especially the convection-permitting model) contribute to describing characteristics of the convective precipitation, while the ensemble models provide uncertainty information and the most possible outcome of the forecast. These two techniques are the main way to improve the fine level and accurate of the QPF, and improvement of short-term precipitation forecast by developing operational high-resolution model ensembles is the international tendency. Objective calibration and integration as well as gridding post-processing make up the statistical post-processing technique of the

---

① 本文发表于《应用气象学报》2016年第5期。
资助项目:公益性行业(气象)专项(GYHY201306002);气象关键技术集成与应用项目(CMAGJ2015Z06)。
第一作者:毕宝贵。E-mail:bibg@cma.gov.cn。

QPF, which now reached a level that applies multiple approaches of data mining to extract and integrate more useful information from massive data, and the emergence of reforecast dataset will further promote the development of statistical post-processing. In terms of verification and evaluation, to solve the new problems in assessing the fine level and accurate of the QPF, a variety of new verification approaches are developed and applied, such as the new score for verifying the precipitation forecast of different climate backgrounds, extremes and multiple types, the spatial verification methods for avoiding dual punishments of traditional methods, as well as the probability verification methods for verifying the stability, sharpness, resolution of the probability forecast. In the aspect of forecaster's subjective modification, although the value added to model and post-processing methods become more and more limited, forecasters still play a core role gradually changing to help users make decision. The development of the QPF techniques still face the challenges of solving scientific problems such as the observation of atmosphere moisture and data assimilation methods, as well as heavy rain forecast in warm section and complex topography.

Key words: quantitative precipitation forecast; numerical weather model; statistical post-processing; verification and evaluation; forecaster's role

引 言

降水是公众最为关心的气象要素，人们根据未来降水情况安排自己的工作生活。此外，政府、工业、农业、水文及地质灾害预警等各个领域对降水预报的要求也越来越高。如在水文应用中，降水预报的空间和时间分辨率分别达到10 km和1 h才能满足洪水预报的要求[1]。传统等级降水预报已经无法满足需求，需要提供高时空分辨率、定量化、准确的降水预报产品，并且部分用户还需概率化预报帮助其进行科学决策。

为应对需求，各国的气象业务中心都建立了定量降水预报（Quantitative Precipitation Forecasts，QPF）业务。美国国家环境预报中心（National Centers for Environmental Prediction，NCEP）下属的天气预报中心（Weather Prediction Center，WPC）从1960年开始率先实施QPF业务。经过多年的努力，WPC制作QPF产品的过程已经从依靠预报经验的人工方法演变到更多依靠对数值模式的解释和订正，以及对集合预报产品应用的过程[2]。中国国家气象中心的QPF业务始于20世纪60年代后期，到现在已经扩展为每天两次发布未来24 h时效内逐6 h累积QPF、未来168 h时效内逐24 h累积QPF，以及中期过程降水量预报等特色QPF产品，格点空间分辨率达到5 km；目标到2020年，预报时效延伸到10 d，空间分辨率达到1～3 km，且增加概率定量降水预报（Probabilistic Quantitative Precipitation Forecast，PQPF）预报的发布。

支撑QPF业务的发展，需要现代化QPF技术流程支撑。如图1所示，目前QPF技术流程包括五个部分：数值模式预报技术是基础，主导整个QPF业务的发展；客观订正集成采用数据挖掘方法从海量预报数据中获得最优的客观预报；预报员在模式和客观预报的基础上发挥人的作用，依靠对天气概念模型的构建、模式预报的理解进行可能的主观集成或订正；降尺度技术通过统计方法加入降水的气候、地形分布等信息，获得更高分辨率的产品；实时检验评估技术贯穿整个技术流程，提供模式及产品的误差和质量信息，用于数值模式的改善，支持预报员主观订正以及对预报质量的管理。五个部分当中，客观订正集成和降尺度都属于统计后处理技术。依据上述技术环节，本文总结了QPF技术的数值模式预报、统计后处理、检验评估以及人的作用发挥四个方面的国内外研究进展，并给出未来工作展望，以期能够对QPF技术的发展提供参考。

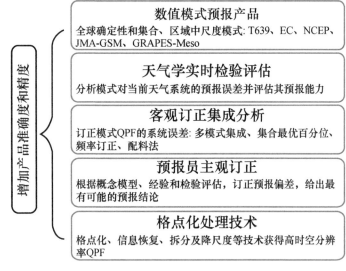

图1 现代化的定量降水预报技术流程

1 定量降水预报的数值预报进展

由于影响时间变化和空间分布的变量太多,定量降水预报被认为是数值模式预报(Numerical Weather Prediction, NWP)最困难的挑战之一[3,4],其预报技巧提升也相对缓慢[5,6]。这主要是因为QPF提供的是一段时间内的降水总量,其准确度受到降水发生位置、移动方向、持续时间、降水效率以及气候背景和天气类型等诸多因素的影响。为了弥补模式对大气环流(高预报技巧)和QPF(低预报技巧)之间愈加明显的发展差异,2004年Roebber[7]指出三种解决途径:发展高分辨率数值模式系统直接预报降水精细特征;发展集合预报系统提供降水预报不确定性信息;将高分辨率模式与集合预报结合起来优势互补。

1.1 业务模式降水预报技巧的提升

国内外许多气象业务中心将QPF预报技巧作为反映模式优劣的关键因子。自从NWP在QPF业务中广泛应用以来,模式开发和预报业务人员对模式QPF性能进行了深入的检验评估[8-12]。1995年,Olson等[9]评估美国国家气象中心的33年QPF表明,相对于对流系统产生的降水,NWP对大型气旋系统或大尺度锋面系统产生的降水具有更好的预报能力。2003年,Ebert等[4]对8个全球和3个区域业务NWP进行了QPF检验评估,得到普遍误差特征:模式QPF预报技巧在冬季显著高于夏季;中纬度地区明显好于热带地区,而在热带地区模式QPF能力仅略微好于"持续"预报(即用前一天的实况作为后一天的预报);模式对小雨量级的预报范围偏大,而对大于20 mm/d的降水模式的预报量级偏低,反映了对强降水预报能力很有限;对降水系统的位置,24 h时效的典型误差在100 km左右。Ebert等[4]同时指出:在短时间内(4~5年)模式的升级并不能促使模式QPF能力的快速提高;模式QPF技巧提升是一个长期过程,除非将准确的降水预报作为其发展的首要目标。正因为如此,许多研究人员不得不发展多种统计后处理方法来进一步提升模式QPF的准确性和可靠性。

尽管模式QPF技巧的提升困难而长期,但近年来通过持续不断地提升模式空间分辨率、改进数据同化能力、完善数值计算方法和次网格物理过程的参数化方案,其对降水的预报技巧不断得到提高。欧洲中期天气预报中心(European Centre for Medium-Range Weather Forecasts, ECMWF)的全球确定性模式作为最先进的NWP代表之一,其降水预报能力持续不断上升,如对强降水的预报能力明显加强,尤其是4～10 d的中长期时效。Forbes等[13]对ECMWF全球模式在2000—2015年的降水性能进行评估表明：在过去10年间模式降水的预报可用时效相当于提高了一天左右，每次QPF技巧的提高对应模式版本的重要升级，如2000年将分辨率从TL399提升到TL511；2003和2006改进了数据同化和云的微物理参数化方案，2010—2013年进一步完善预测雨和雪变量的方案[14]；2015年在显式降水微物理过程参数化方案中引入暖雨和混合相态降水过程的机制，进一步减轻了小雨降水过报的问题并增强了强降水的预报量级，地形强迫产生的极端降水预报也得到改善；2016年3月，ECMWF再次升级模式的空间分辨率至9 km/137层，并引入新网格计算方案[15]，解决了模式"格点暴雨"的问题，个例分析表明对局地强降水的预报能力更强。除ECMWF之外，其他气象业务中心都在通过提升模式的分辨率和改善模式的次网格物理过程参数化来改进QPF预报能力，如2015年初，美国环境模式中心的全球模式GFS(Global Forecast System)分辨率达到13 km/64层；英国气象局的全球模式分辨率达到17 km/70层；日本气象厅全球模式达到20 km/100层；中国自主研发的GRAPES(Global/Regional Assimilation Prediction System)全球模式系统[16]在2016年进入业务流程，检验评估表明其对中国区域的降水预报已经超过原有的T639模式系统。此外，随着全球模式分辨率的快速提高，区域模式系统将会逐步被取代，更高分辨率的模式系统将会得到进一步发展和应用。

1.2 发展高分辨率模式提供降水的精细特征

21世纪以来，全球业务模式的分辨率不断得到提高，使中长期QPF技巧提升显著，但其准确预报短期时效(0～3 d)内局地-区域对流系统以及复杂地形引起的强降水仍然非常困难[17,18]，而该类型强降水正是引起暖季(尤其是远离锋面的暖区)暴雨灾害的主要因素。因此，同时发展能够预报降水精细特征的更高分辨率模式(10 km及其以下)是提升QPF能力的关键。

发展高分辨率模式有利于模拟和预报一些关键的大气现象，例如快速加强的温带气旋[19]、地形引起的风和降水[20]、海陆风环流[21]和对流系统[22]，但同时受到诸多条件的限制。首先是初始数据条件，高分辨率模式对观测数据更加敏感，要求更加严格，例如Zhang等[23]的研究显示，其至单个探空数据将会显著改变一个气旋生成过程的中尺度降水分布特征；Weygandt等[24]建议，通过同化天气雷达数据来加强中尺度系统的模拟能力；然而仅这样还远远不够，需要加入新的高时空分辨率观测资料来提高中尺度系统的模拟[25]。其次是模式次网格物理过程的精确描述，随着时空分辨率的提高，要求更为复杂的模式次网格物理参数化方案，例如模拟大尺度强迫下的对流发生发展，需要准确描述行星边界层中辐射传输、地面热量通量、感热及潜热通量、湍流混合、凝结潜热以及表现云的微物理过程[26-28]等。2003年，Marshall等[29]阐述了准确预报行星边界层过程所面临的许多挑战，指出当前模式大多数次网格物理过程参数方案还存在差距，此外还缺少相应的观测数据来定义所需的参数变量。最后，如何将高分辨率模式的预报能力转换为QPF预报技巧也需要评估总结，模式分辨率的提升并不总意味着QPF技巧评分的提高，如Colle等[30]发现，模式格距从12 km下降到4 km，其降水量预报能力提升并不明显；而对于一些地形复杂地区，模式格距的降低也不能很好地表现出QPF技巧的提高[31]。

近年来，针对上述条件限制，发展出多种技术方法来改进高分辨率模式的预报能力，特别是对流尺度模式(Convection-Permitting Models, CPMs)的发展和业务应用，将推动短期QPF的精细化预报更上一个台阶[32]。CPMs定义为不使用对流参数化方案而直接模拟大气对流动力过程的模式系统。第一个用于实时的CPM由美国俄克拉何马大学的风暴分析与预报中心在1989年建立[33]，后来发展成为ARPS(Advanced Regional Prediction)[34]。随之更多的能够进行对流尺度模拟的模式发展起来，如美国天气研究与预报模式WRF(Weather Research and Forecasting)[35]、英国业务化的UKV(United Kingdom Model)对流模式[36]、欧洲联合小尺度模式COSMO(Consortium for Small-scale Modeling)[37]、法国研究的业务应用中尺度模式AROME(Applications of Research to Operations at Mesoscale)[38]和日本非静力中尺度模式[39]。尽管高分辨率与低分辨率模式相比 QPF 评分没有表现出较明显的提高幅度(这与如何评价也有关系),但能够帮助预报员建立各种中小尺度天气的预报概念模型,并且提供接近真实的 QPF 精细化特征。Weisman 等[40]对比实时 4 km WRF-ARW 模式与业务 12 km ETA 模式的 2003—2005 年春夏季预报结果发现,尽管 CPM 对对流系统的触发时间和位置没有表现更好的预报能力,但能够提供对流系统模态的重要信息(即为飑线、弓形回波还是中尺度对流涡旋等),以及表现对流的日变化特征。Clark 等[32]也总结出英国业务化的 UKV 模式用于 QPF 的优势:①提供接近真实的降水分布特征。如全球模式只能描述"未来可能会出现区域性的阵雨",而 CPM 则会描述"未来会出现区域性的对流性阵雨"。②提供降水接近真实的演变特征。若将图进行动画显示可发现,由于全球模式的对流参数化方案只对大气不稳定性有所反映而不能分辨次网格内的对流演变,因此会造成降水忽然出现或消失;而 CPM 能模拟对流性阵雨的平流运动,以及对流系统的生消和发展(如子对流的生成及对流的后向传播)。此外,基于空间分析的检验方法表明,CPM 的 QPF 在升尺度分辨率上要优于低分辨率模式[41]。

1.3 发展集合模式提供降水的预报不确定性

除高分辨率模式之外,发展集合模式预报是提高 QPF 的另一条重要途径。已有的研究表明,即使是低分辨率的集合模式也可显示出比单个高分辨率模式更高的预报技巧。例如 Grimit 等[42]对比 5 个成员 12 km 的 MM5 集合模式与单个 4 km 的 MM5 模式在西北太平洋区域的预报,结果表明集合模式具有更好的预报技巧。此外,相对于确定性模式,集合预报的主要优势在于能够提供预报的不确定性信息,并可直接计算概率预报[43-45],这将有利于用户基于概率预报以及自身的花费与损失比来进行更科学的决策。另外 Buizza[46]的研究表明,基于不同起报时刻的集合概率预报较确定性模式更加连续且稳定。赵琳娜等[44]指出,基于水文集合预报的洪水预报增加了预报附加值,并能够延长预警提前时间。

集合预报的构建方式多种多样。早期典型的有时间滞后法[47]和"穷人"集合法[48]。随着计算机能力的不断提高,目前的主流方法是基于同一模式,而采用不同的初值条件、边界条件或物理参数来构建的集合预报[49]。杜钧等[50]在其综述中详细总结了暴雨集合预报系统的建立。此外,Du[51]提出了"双分辨率混合集合预报法",前提是做一个高分辨率的单一预报和一个低分辨率的集合预报,把两者结合起来——基础预报由高分辨率提供(比较精确)而预报不确定性信息由低分辨率集合预报提供(计算机资源允许)——产生一个新的高分辨率集合预报。Fang 等[52]采用相似的思路,进一步设计了双分辨率的集合预报系统来预报台风暴雨,即用 36 km 低分辨率的 32 个成员的集合预报来估计最有可能的台风路径以及用同集合平均路径(位置)相近的成员来估计暴雨分布的空间结构,再用高分辨率集合成员来调整暴雨雨量预报。

目前,许多国家或地区的预报中心都建立了业务应用的集合预报系统。最早从 1992 年开始,ECMWF[53]和 NCEP[49]都相继建立了全球集合预报系统,随后加拿大气象中心 CMC(Canadian Meteorological Centre)[54]和澳大利亚气象局 BoM(Bureau of Meteorology)分别在 1995 年和 1998 年建立了本国的集合预报业务系统。中国数值预报中心在 2005 年底建立了在全球 T213 模式基础上的集合预报系统,投入实时运行,并在 2014 年进一步升级为 T639 集合模式系统。此外,为了加速提高中短期 1~14 d 高影响天气的预报能力,世界气象组织在 2003 年决定建立为期 10 年的观测系统研究与可预报性试验(The Observing-System Research and Predictability Experiment,THORPEX)计划,其中,交互式全球大集合(The THORPEX Interactive Grand Global Ensemble,TIGGE)项目[55]是 THORPEX 主要组成部分,建立的三个归档中心(欧洲中期天气预报中心、美国和中国)收集来自全球十多个主要预报中心的集合预报产品。Su 等[56]对比评估 TIGGE 各中心集合模式对北半球的降水预报能力,结果显示 ECMWF 的全球集合模式预报通常情况下最优,而 CMC 在短期 QPF 和小雨量级 PQPF 上表现较好。Hamill[57]的研究表明,基于多模式超级集合的 PQPF 较单个模式的 PQPF 有更好的可靠性和预报技巧。

1.4 高分辨率与集合预报相结合

高分辨模式与集合预报具有优势互补性[7]。以1999年9月16日登陆美国的热带风暴Floyd为例，Colle[31]采用高分辨率MM5模式显示，对湿对称不稳定、高空倾斜中性层结以及强烈锋生等中尺度过程的正确表现能够模拟出接近实况的狭长降水带。同时，一个较低分辨率的集合预报（9个成员32 km格距）的所有成员都在沿海地区报出了较大降水（尽管其强度不如高分辨率模式），因此，可以为预报员指示该区域强降水预报的潜力（即预报可信度的信息）；此外，集合预报还显示了降水极值中心位置的不确定性信息，因此，预报员可以得到强降水中心概率的信息。从这个例子可以看到，从低分辨率的集合预报获得关于降水区域和位置的不确定性信息，从高分辨率模式预报获得降水的精细中小尺度特征，预报员可以做出更为准确的QPF。

21世纪以来，随着计算能力的快速提高以及模式技术、集合技术的不断发展，直接基于高分辨率模式进行集合预报已经成为趋势。与主要基于初值误差的全球集合预报不同，针对短期预报的高分辨率模式集合除初值误差之外，主要考虑模式本身误差，如采用不同物理参数过程的组合[54,58-60]或不同模式的组合[61,62]。此外，在高分辨率集合预报系统中，尽管不同的对流参数化能够有效生成降水预报的离散度[63]，但会在降水预报中引入系统误差[64]，因此发展能显式模拟对流的CPM集合预报系统可进一步促进概率降水预报的能力[65]，如Clark等[66]对比了5个成员4 km的CPM集合预报和15个成员20 km的区域尺度集合预报，结果显示CPM集合预报对对流性降水的时空统计特征有更好的表现，对小尺度系统的移动预报有更大的离散度，且能够提供更准确、更可靠的概率预报。

近年来，已经有多个气象中心开始运行业务高分辨率集合预报[67-73]。如何构建高分辨率集合预报仍然是一个非常新的研究领域，主要方法包括将高分辨率模式嵌套一个相对低分辨率的集合[67,73]；使用集合卡尔曼变换滤波方法[74]、随机性扰动[71]、系统性扰动[67,75,76]或多种物理方案参数[77]扰动高分辨率模式初始条件。研究表明，在模式的6～12 h时效之前，物理过程或高分辨率模式格点上的初始扰动起主要影响，之后则是提供背景或边界条件的低分辨率模式（或集合模式）占主导作用。由于高分辨率模式集合预报对中小尺度系统的描述不确定性较大，若采用常规等权重概率计算方法会使得PQPF误差较大，因此可使用领域法[68,72,77]或升尺度法[78]来获得更有预报技巧的PQPF。

2 定量降水预报的统计后处理技术

初始场误差、数值计算近似和物理化学过程不完善的存在，使数值模式输出结果存在一定随机性和系统性误差。数值模式释用就是利用统计方法对数值预报的结果进行后处理和订正，消除系统性误差，从而给出更为精确的预报结果。Scheuerer等[79]指出，所有预报要素中降水的统计后处理相比温度和风速等要素更加复杂：首先，由于降水的不连续偏态分布特征，很难找到一个分布函数来拟合整个降水分布过程；其次，预报的不确定性随着降水量级的增加而增加，因此，需要在建模时予以考虑；最后，强降水例如暴雨量级降水的后处理理论上需要大量的训练样本。在实际业务中，通常采用三种技术进行降水预报后处理，包括定量降水订正与集成技术、概率预报处理技术和统计降尺度技术。

2.1 定量降水订正与集成技术

定量降水订正与集成技术，是针对单个或多个模式预报，进行系统偏差订正和权重集成，获得准确度更高和最有可能出现的单一或确定预报结果。表1给出了目前在业务中被广泛应用的几种订正与集成技术，包括：通过对数值模式输出的各种产品（压、温、湿及诊断量）利用统计方法建立预报模型，得到客观预报的降水量,如模式输出统计方法和基于"配料法"的强降水等级预报;在数值模式直接输出的降水预报结果上进行订正或集成,得到更为精确的降水预报产品,如概率匹配集合平均方法和频率订正技术。近年来,中国国家气象中心也参考国内外的工作,研发了多种业务应用的QPF客观预报技术,如基于"配料"逻辑回归方法、多模式动态权重集成方法、最优百分位方法等。其中,基于"配料"逻辑回归方法重点考虑强降水的物理条件;多模式动态权重集成方法考虑了不同模式预报之间雨带形状和降水强度的相似;最优百分位方法依据对集合QPF的检验评估建立集成规则,用于提取不同降水等级的优势集合QPF统计量,具有很好的应用前景。

表1 定量降水订正与集成技术概要

名称	主要工作	技术特点
模式输出统计方法(MOS)	2000年,Antolik等[80]回顾了美国国家天气局的统计定量降水预报	• 利用预报因子和降水建立回归方程; • 可以针对站点或是格点的降水进行建模,而精细化的格点MOS方法依赖于高分辨率的定量降水估测产品; • 对有降水量的简单的偏差订正不适用于无降水的情况
基于"配料法"的强降水等级预报	1996年,Doswell[81]引入气象研究,并用于暴洪预报; 2010年,张小玲等[82]应用于中国的暴雨预报	• 依据预报量和指示量之间的物理联系来建立相应的预报方程,而不是简单地依赖回归分析; • 具有清晰的物理意义
概率匹配集合平均方法	2001年,Ebert[48]首先提出并应用于多模式"Poor Man"的集成; 2013年,Fang等[52]针对台湾登陆台风的强降水提出了改进的频率匹配方法	• 相对于单一模式能够提高降水预报技巧,消减系统性的误差; • 既能保留降水集合平均的空间分布,也能保留集合成员的概率分布,好于集合平均方法; • 过滤掉了不可预报的小尺度信息[83]
频率订正技术	Zhu等[84]提出并应用于NCEP的GFS和GEFS模式的QPF预报	• 利用观测降水量的频率分布来校正预报降水量的频率分布; • 有效减小了降水预报误差,降水的分布也更加准确

2.2 概率预报处理技术

由于降水预报的不确定性较大,确定性模式QPF的应用仍受限。集合预报在近几十年得到了快速发展,PQPF是未来的重要发展方向。PQPF可以描述降水预报的不确定性,因而是更加科学的一种预报形式。但是,集合预报由于分辨率较低以及初始场等问题,仍然有必要进行后处理来提高可靠性和概率预报的技巧[85]。Stensrud等[86]将PQPF的模式释用方法归纳为两种:一种是回归模型方法,例如逻辑回归模型;另一种是参数估计方法,即假设降水服从一定的分布函数,然后估计分布的参数。

针对回归模型方法,Gahrs等[87]比较了逻辑回归、分箱方法和线性回归,结果表明前两者的Brier评分明显高于线性回归方法。类似方法还有分位数回归[88]。这些回归方法都是通过训练样本来估计降水的条件概率或分位数,缺点是不能对一些极端降水事件进行外推;都需要对不同的概率阈值和分位数进行拟合,因而需要估计很多的参数。延伸逻辑回归方法[89,90]将降水阈值作为预报因子,相对于传统逻辑方法来说可以减少参数,但对回归系数有一定的限制。

相对于回归模型方法,参数估计方法可以进行外推,但是预报效果很大程度上取决于对降水分布的假设。贝叶斯模型平均(Bayesian Model Averaging,BMA)[91]是一种重要的参数估计方法,相对而言,BMA方法更为灵活,但实现起来较为复杂。Liu等[92]利用TIGGE数据将BMA方法应用到淮河流域的PQPF中,结果发现BMA方法的效果优于原始的集合预报和逻辑回归方法。Zhu等[93]利用集合离散度提出了分级采样BMA方法,可以有效解决传统BMA方法低估大量级PQPF的问题。其他参数估计方法还有EMOS(Ensemble Model Output Statistics)方法[94]、Censored Shifted Gamma函数方法[79]等。

此外,Peel等[95]采用核密度估计来计算降水概率,对极端降水有很好的订正作用。Yuan等[96]利用人工神经网络方法来开展PQPF。基于相似法的PQPF技术也取得了较好的效果[97,98]。针对对流尺度集合预报的释用方法包括机器学习方法[99]、QPF-POF(Generate Probability of Precipitation)方法[100]等。WPC利用NCEP的短期集合预报系统SREFs(Short Range Ensemble Forecast),基于预报员的主观QPF预报,制作PQPF预报[101]。

利用实时历史回报(再预报)数据进行统计后处理是目前模式释用的一个重要发展趋势。Hamill等[102]首先提出了基于回报数据进行概率预报的思想。历史回报数据一方面可以极大地扩充训练样本,另一方面有利于诊断模式的误差,因此,被认为应该作为数值预报体系的一部分[103]。Hamill等[104]的研究表明:利用历史回报数据可以有效提高各个预报时效降水预报的技巧,特别是中雨以上量级降水的预报技巧,因为历史回报可以为较强量级的降水提供更多的样本进行训练。此外,Fundel等[105]利用30年回报数据对降水的重现期进行了订正。

2.3 统计降尺度技术

由于目前模式输出时空分辨率无法满足用户对精细化QPF的要求[106],因此,需要发展关键的统计降尺度技术。该技术主要是为了实现粗网格降水格点场到细网格降水场的转化。传统的提高分辨率的思路与方法通过双线性数学插值实现,忽略了低分辨率降水与高分辨率降水之间的特殊的气象关系,结果的准确性有限。当前主要使用降尺度技术提高降水的空间分辨率。由于动力降尺度需要成熟可靠的中尺度模式的支持,耗费大量的计算机资源和计算时间[107],一般预报业务中利用统计降尺度技术来提高降水的空间分辨率。

目前,有多种统计降尺度技术在业务中得到了应用,如Clark等[108]提出利用随机函数排序重构细网格降水;美国海洋和大气局下属的水文预报中心使用地形地图投影技术进行降水降尺度[109]。中国国家气象中心使用尺度矢量降尺度技术建立大尺度降水到小尺度降水的气象统计关系,并确定降尺度比例矢量,能体现出局地由于地形或者气候特点导致的降水精细尺度的变化,体现不同空间尺度降水的气象意义关系,此外计算也简便。

3 定量降水预报的检验技术

检验是监控预报性能、理解模式误差以及改进模式预报的主要途径。在所有检验当中,降水检验尤其重要。可以说QPF越精细,对应的检验技术越复杂。这是因为随着降水尺度的减小,其随机和分形特征将更加突出[110]。传统检验方法主要基于低分辨率的全球数值模式发展而来[111,112]。然而随着模式分辨率的提高,QPF检验问题变得更加复杂,如相对于以中长期预报为主的全球模式,以短期预报为主的中尺度高分辨率模式应该能提供更多的预报信息(如锋面降水的结构、位置和时间等中小尺度特征),但传统的评分方法不足以表现出这种预报优势,较小的位置将造成所谓的"双惩罚"现象,即对模式提供的中小尺度信息趋于惩罚而不是奖励。此外,随着20世纪90年代初集合模式预报的推广应用,如何检验评估集合预报带来的不确定性信息也变得越来越重要。

3.1 检验评分技术

有很多相关的书籍[112,113]和文章[114-116]详细地介绍了传统的QPF检验评分方法,Rossa等[117]对不同的检验评分方法进行系统的回顾,分为两种:一种是对降水事件(降水量大于某一阈值)检验,基于击中、空报、漏报以及正确拒绝四个要素的列联表,可以定义多种评分,如频率偏差指数(FBI)、正确比例(PC)、识别概率(POD)和错误预警率(FAR)等,从不同角度反映QPF性能;另外,威胁评分(TS)也是常用的方法,但其缺点是对本地气候很敏感,为此,可利用相当威胁评分(ETS)来消除气候背景的部分影响。另一种是将降水作为连续变量进行检验,方法包括平均误差(ME)、平均绝对误差(MAE)、平均误差均方根(RMSE)和相关系数等,需要注意,降水非高斯分布的特性使得这些连续变量评分方法对大误差的预报非常敏感。

近年来,新的科学问题不断涌现,促进一些新的检验评分技术发展,主要解决三个方面的问题[118]:如何发展一体化的评分方法来检验具有不同气候背景的降水,这样便于在同一个模式系统中监测不同区域和季节的降水预报性能;如何评估小概率事件或者极端事件的预报性能;如何对不同类型的预报产品(如确定性预报和概率预报、等级预报和具体数值的预报)给出一致的评估信息。表2给出针对不同检验问题的新方法概要说明。

表2 新的定量降水检验评分技术概要

检验目的	方法概要
检验不同气候背景的降水预报	• Rodwell等[119,120]发展了概率空间稳定相当误差(SEEPS)评分方法。该方法通过降水长期气候特征分布来确定降水等级阈值,便于比较不同区域和不同季节的模式QPF性能; • 目前成为ECMWF、芬兰等气象部门的主要检验评分手段,陈法敬等[121]也将该方法用于中国的降水预报检验
检验极端降水的预报能力	• Stephenson等[122]提出,用极端依赖评分(EDS)来检验小概率事件; • EDS不断得到改进,发展了稳定EDS(SEDS)评分方法[123],以及基于极端依赖指数和稳定极端依赖指数多变量的评分方法[124],可较好地分辨不同模式对极端事件的预报能力[114,125]
对比不同类型的预报评估	• Mason[126]和Weigel[127]提出通用辨别评分(GDS),用于度量两种不同实况能被相应的预报正确分辨的可能性大小,适用于不同预报类型的对比

3.2 空间检验技术

为避免传统检验评分方法的"双惩罚"现象,并反映检验结果的物理意义,一些新的基于降水空间结构分析的检验方法发展起来。Gilleland等[128]对现有空间检验技术的性能进行了对比分析,并将已有的17种方法分为4类:场变形技术、特征检验技术、尺度分离技术和邻域检验技术。同时,Gilleland等[128]指出,不同类型的检验技术各有优缺点,其性能与预报特性和检验关注的问题有关。表3给出了4类空间检验技术的概要说明。

表3 定量降水预报的4类空间检验技术概要

方法名称	方法概要
场变形技术	• 将误差分解为位置、量级和残余误差,给出较为清晰的物理意义; • Hoffman等[129]在检验方法中首先引入,标志着空间检验技术的开端
特征检验技术	• 识别降水目标,然后进行配对和对比,评估目标个体之间的属性差异; • 采用不同方法来识别降水目标,如简单阈值[130]、圆柱卷积滤波[131]等
尺度分离技术	• 采用空间滤波分解尺度,评估模式对不同尺度上降水结构的预报能力; • 因采用的滤波方法不同而有所差异,如Casati等[132]基于二维小波发展了强度-尺度检验技术,已经应用于QPF临近预报检验[133]
邻域检验技术	• 对比检验考虑了邻近空间或时间范围内的预报和观测降水事件[134]; • 包含降水固有的时空分布不确定性,尤其适用于高分辨率模式的检验

3.3 集合预报检验技术

集合预报检验技术来自于早期的概率预报检验方法,随着集合预报的兴起和广泛应用,概率预报检验技术得到了新的发展[135]。与确定预报检验不同,评价集合预报系统的性能应从可靠性(Reliability)、锐度(Sharpness)、解析度(Resolution)三个方面进行。目前主要的集合预报检验方法可归纳为三种类型。

(1)用于检验集合分布是否来自于某个概率分布函数的方法,包括排序直方图[136]、连续排序概率评分(CRPS)及其相对技巧评分[137]、最小生成树(MST)[138]和边界框方法[139]。这里,排序直方图常用于集合预报的离散能力评分,而 CRPS 因将检验信息总结成为单个数而受到业务应用的欢迎。

(2)用于评估概率预报的概率密度函数的方法,如无知评分[140]、Wilson 概率评分[141]。该方法同时考虑了概率预报的锐度和准确性,即概率分布越平滑(增大预报的不确定性)会导致不可能得到较高的评分,而相对高锐度(预报不确定性较小)但同时准确度较低的预报会受到更严重的惩罚。

(3)事件概率预报检验方法,包括 Brier 评分[142]、可靠性曲线[143]、受试者工作特征曲线 ROC[144]、排列概率评分[145]等。其中,Brier 评分作为最常用的二元事件概率预报检验方法,得到不断发展。Murphy[146]将其分解为可靠性、分辨率和不确定性三项;Stephenson 等[147]指出,需要额外的项来使分解更加准确;Ferro 等[148]的研究表明,应用 Brier 评分结果时必须考虑集合预报成员的数量。

4 定量降水预报中预报员的作用

早期 QPF 以主观为主,预报员发挥重要作用。Funk[149]在 1991 年总结了预报员在模式指导基础上进行 QPF 的主要预报分析技巧,包括天气型识别、水汽条件分析、低空急流和辐合分析、高空急流结构分析、低层相当位温和大气层结厚度分析等。应用这些主观预报经验,预报员对强降水的预报能力得到了明显的提高,较模式指导预报也呈现显著的优势。

至 21 世纪初,Reynolds[150]在 2003 年计算了 1993—2001 年美国 NCEP 的 HPC(Hydrometeorological Prediction Center,WPC 的前身)预报员较 NCEP 全球 AVN 模式和区域 Eta 模式的 QPF 预报 TS 评分提高百分比,结果表明:20 世纪 90 年代早期至中期,预报员预报提高率为 30%,至 90 年代中后期,下降至 20%左右;结合预报评分趋势分析,预报员将数值模式的预报水平提升了 14 年左右。由此可见,预报员在数值模式基础上所体现的作用仍是很重要的[151]。但 Reynolds[150]同时指出,随着模式更新频率的加快,预报员越来越难以掌握模式预报的误差特征,因此只依靠模式使用经验来获得附加值变得越加困难,需要预报员进一步理解中小尺度暴雨的物理过程以及模式预报的优缺点。

至 2014 年,Novak 等[152]再次对 WPC 的 QPF 预报进行了评估,结果表明:在过去 20 年间,预报员对 25 mm/d(大雨)的 1 d 时效预报较北美中尺度模式和 GFS 全球模式提高了 20%~40%的精度,而较 ECMWF 确定性模式提高幅度相对较小(5%~10%)。此外,预报员对极端降水事件(76 mm/d)的预报在短期时段体现出更大的提高幅度。但需要关注的一个问题是,预报员相对于经过偏差订正和降尺度方法处理后的客观 QPF 预报没有体现出优势,因此需要进一步通过测试研究来考虑预报员角色的转变。同样的问题在中国国家气象中心的 QPF 业务中出现,2015 年夏季,无论对 08:00 时还是 20:00 起报的 1 d 时效 QPF,表现最好的集合最优百分位方法在 TS 评分上已经接近或略微超过预报员,且其 BIAS 评分较预报员更接近于 1,表明预报员与客观 QPF 方法相比,并没有表现出评分优势。

未来预报员在 QPF 中如何发挥作用,是一个长期以来有争议的问题,预报员未来所扮演的角色也具有很大的不确定性[153]。Mass[154]认为,"人类无法在大脑中直接积分大气运动原始方程,因此,很难改进全分辨率的数值模式预报(包括消除偏差)。此外,预报员在中尺度集合预报的基础上订正概率预报也不现实"。但同时 Mass[154]也考虑到人类在物理图像处理上的优势,认为预报员可以在 0~12 h 时效发挥一定的作用。Roebber等[7]认为,尽管人类处理海量数据比较困难,但熟练的预报员适应了快捷地评估和解释预报信息,而目前的自动化系统还很难实现相同的功能;此外模式自身的预报缺陷、集合预报离散度偏小还会存在相当长一段时间,客观后处理技术在业务中心应用还不够深入,因此在可见的将来,预报员在1~3 d预报时效还能够发挥作用。Homar等[155]设计实验让预报员可以直接参与短期集合预报的实时构建,结果表明,在强降水的概率预报上较业务模式有明显改善。Novak等[152]的调查研究表明,预报员未来工作的一个重要部分是如何将预报的不确定性信息传递给用户。Sills[156]指出,现代预报员的主要作用在于维持对当前及未来大气演变的综合理解和掌握,而这需要通过基于广泛检验评估对模式预报深入理解的能力和周密设计的人工智能对海量信息挖掘的辅助;预报员的工作应主要关注高影响天气事件(无论短期或长期)。此外,Stuart等[157]提出,若要维持预报员的关键作用,需要高质量的预报培训,帮助预报员理解新的物理概念模型,掌握最新的预报工具以及深刻理解数值模式的运行机制。可以确定的是,尽管目前预报员在整个QPF流程中还处于核心位置,但随着预报技术的发展,预报员能够为QPF添加附加值的空间将会越来越小,其角色也将发生转变,逐渐从一个预报产品制作者向基于预报信息的传递和解释以帮助用户进行科学决策的角色转变[153]。

5 结论与展望

数值模式预报技术、统计后处理技术、检验评估技术和预报员主观订正构成QPF技术体系的主要组成部分。本文对这四个部分的研究工作进行了归纳,主要进展总结如下。

(1)数值模式预报是QPF技术的基础和核心。尽管降水是模式最难预报的变量之一,但通过持续不断的模式升级(包括提高分辨率、改进数据同化能力、完善数值技术和物理过程等),其对降水的预报技巧不断得到提高。如ECMWF全球确定性模式在过去10年间降水预报的可用时效相当于提高了一天左右。此外,随着计算机能力的提高和科学研究的深入,限制高分辨率模式发展的数据同化、物理过程描述以及应用评估等问题逐步得到解决,特别是对流尺度模式的发展将推动QPF的精细化水平更上一个台阶。发展集合预报是提高QPF水平的另一条重要途径。它能够提供预报的不确定性信息和最有可能的预报结论,并且是制作PQPF的基础。将高分辨率与集合预报结合起来优势互补,能够帮助预报员对中小尺度系统造成的降水进行精细预报,是未来短期QPF的主要发展方向。

(2)统计后处理技术是改善模式QPF的重要途径。QPF订正与集成技术已经不再是传统的MOS预报方法,而是通过数据挖掘技术,在检验评估的基础上消除模式系统误差,并将不同预报来源的结果进行优势集成,获得最优的预报结果。概率预报处理技术通过回归模型构建或者概率分布参数估计等方法,获得更加可靠及准确的PQPF预报。此外,再预报技术的发展提高了强降水事件的样本量,有利于促进中雨以上量级降水的预报技巧提升,在未来统计后处理中将扮演重要角色。统计降尺度技术通过加入降水的气候、地形分布等信息,制作更高分辨率的产品以满足用户的需求。

(3)检验评估技术是监控预报性能、理解模式误差以及改进模式预报的基础。除传统检验评分之外,新的评分方法发展起来以解决不同气候背景下降水预报评估、极端事件的预报评估、不同产品类型之间的评估对比等科学问题。为了反映精细化QPF的预报能力,场变形、特征识别、尺度分离和邻域分析四类基于降水空间结构分析的检验方法得到发展应用,避免了传统评分的"双惩罚"现象,反映了检验结果的物理意义。此外,随着集合预报的兴起和广泛应用,概率预报检验技术得到了新的发展,从可靠性、锐度和解析度等方面评价集合预报性能。

(4)随着数值模式的发展和统计处理方法的深入应用,预报员在最优预报的基础上添加附加值的空间将越来越小。但在可预见的将来,预报员在整个QPF业务流程中仍将处于核心的位置。基于对数值模式预报的深入理解和利用人工智能对海量信息挖掘,预报员能够在模式预报的薄弱环节发挥作用(如高影响天气事件的预报)。此外,通过预报信息的传递和解释以帮助用户进行科学决策的工作比重将会加大。

需要指出的是,QPF 技术的发展还需要对一些科学难点问题的深入认识和解决,包括如何利用现代化的观测手段改善数值模式对大气水汽的同化和模拟[158,159],如何提高暖季(或暖区)降水的预报能力[65],如何改进复杂地形地区降水的预报能力[160],这对山洪、泥石流等灾害预警尤其重要。

参考文献

[1] ARNAUD P,BOUVIER C,CISNEROS L,et al. Influence of rainfall spatial variability on flood prediction[J]. J Hydrol,2002,260:216-230.

[2] NOVAK D R,BAILEY C,BRILL K F,et al. Precipitation and temperature forecast performance at the Weather Prediction Center[J]. Wea Forecasting,2014,29:489-504.

[3] GOLDING B W. Quantitative precipitation forecasting in the UK[J]. J Hydrol,2000,239:286-305.

[4] EBERT E E,DAMRATH U,WERGEN W,et al. The WGNE assessment of short-term quantitative precipitation forecasts[J]. Bull Amer Meteor Soc,2003,84:481-492.

[5] SANDERS F. Trends in skill of Boston forecasts made at MIT,1966-84[J]. Bull Amer Meteor Soc,1986,67:170-176.

[6] APPLEQUIST S,GAHRS G E,PFEFFER R L,et al. Comparison of methodologies for probabilistic quantitative precipitation forecasting[J]. Wea Forecasting,2002,17:783-799.

[7] ROEBBER P,SHULTZ D M,COLLE B A,et al. Toward improved prediction:high-resolution and ensemble modeling systems in operations[J]. Wea Forecasting,2004,19:936-949.

[8] ROADS J O,MAISEL T N. Evaluation of the National Meteorological Center's medium range forecast model precipitation forecasts[J]. Wea Forecasting,1991,6:123-132.

[9] OLSON D A,JUNKER N W,KORTY B. Evaluation of 33 years of quantitative precipitation forecasting at the NMC [J]. Wea Forecasting,1995,10:498-511.

[10] MCBRIDE J L,EBERT E E. Verification of quantitative precipitation forecasts from operational numerical weather prediction models over Australia[J]. Wea Forecasting,2000,15:103-121.

[11] DAMRATH U,DOMS G,FRUEHWALD D,et al. Operational quantitative precipitation forecasting at the German Weather Service[J]. J Hydrol,2000,239:260-285.

[12] 王雨. 2004 年主汛期各数值预报模式定量降水预报评估[J]. 应用气象学报,2006, 17(3):316-324.

[13] FORBES R,HAIDEN T,MAGNUSSON L. Improvements in IFS forecasts of heavy precipitation[J]. ECMWF Newsletter, 2015, 144:21-26.

[14] FORBES R,TOMPKINS A. An improved representation of cloud and precipitation[J]. ECMWF Newsletter, 2011, 129:13-18.

[15] MALARDEL S,WEDI N,DECONINCK W,et al. A new grid for the IFS[J]. ECMWF Newsletter, 2016, 146:23-28.

[16] 刘艳,薛纪善,张林,等. GRAPES 全球三维变分同化系统的检验与诊断[J]. 应用气象学报,2016 (1):1-15.

[17] KAUFMANN P,SCHUBIGER F,BINDER P. Precipitation forecasting by a mesoscale numerical weather prediction (NWP) model:eight years of experience[J]. Hydrol Earth Syst Sci,2003,7:812-832.

[18] RICHARD E,COSMA S,BENOIT R,et al. Intercomparison of mesoscale meteorological models for precipitation forecasting[J]. Hydrol Earth Syst Sci,2003,7:799-811.

[19] UCCELLINI L W,KOCIN P J,SIENKIEWICZ J M. Advances in forecasting extratropical cyclogenesis at the National Meteorological Center[R]//The Life Cycles of Extratropical Cyclones. Amer Meteor Soc,1999.

[20] MASS C F,OVENS D,WESTRICK K,et al. Does increasing horizontal resolution produce more skillful forecasts? The results of two years of real-time numerical weather prediction over the Pacific Northwest[J]. Bull Amer Meteor Soc,2002,83:407-430.

[21] ROEBBER P J, GEHRING M G. Real-time prediction of the lake breeze on the western shore of Lake Michigan[J]. Wea Forecasting, 2000, 15: 298-312.

[22] NIELSEN-GAMMON J W, STRACK J. Model resolution dependence of simulations of extreme rainfall rates[R]. Preprints, 10th PSU/NCAR Mesoscale Model Users Workshop, Boulder, CO, PSU/NCAR, 2000, 110-111.

[23] ZHANG F, SNYDER C, ROTUNNO R. Mesoscale predictability of the "surprise" snowstorm of 24-25 January 2000 [J]. Mon Wea Rev, 2002, 130: 1617-1632.

[24] WEYGANDT S S, SHAPIRO A, DROEGEMEIER K K. Retrieval of model initial fields from single-Doppler observations of a supercell thunderstorm. Part Ⅱ: thermodynamic retrieval and numerical prediction[J]. Mon Wea Rev, 2002, 130: 454-476.

[25] GALLUS W A, SEGAL M. Impact of improved initialization of mesoscale features on convective system rainfall in 10-km Eta simulations[J]. Wea Forecasting, 2001, 16: 680-696.

[26] LARSON V E, WOOD R, FIELD P R, et al. Systematic biases in the microphysics and thermodynamics of numerical models that ignore subgrid scale variability[J]. J Atmos Sci, 2001, 58: 1117-1128.

[27] LYNN B H, KHAIN A P, DUDHIA J, et al. Spectral (Bin) microphysics coupled with a Mesoscale Model (MM5). Part Ⅰ: model description and first results[J]. Mon Wea Rev, 2005, 133: 44-58.

[28] LYNN B H, KHAIN A P, DUDHIA J, et al. Spectral (bin) microphysics coupled with a mesoscale model (MM5). Part Ⅱ: simulation of a CaPe rain event with squall line[J]. Mon Wea Rev, 2005, 133: 59-71.

[29] MARSHALL C H, CRAWFORD K C, MITCHELL E, et al. The impact of land surface physics in the operational NCEP Eta Model on simulating the diurnal cycle: evaluation and testing using Oklahoma Mesonet data[J]. Wea Forecasting, 2003, 18: 748-768.

[30] COLLE B A, MASS C F. The 5-9 February 1996 flooding event over the Pacific Northwest: sensitivity studies and evaluation of the MM5 precipitation forecasts[J]. Mon Wea Rev, 2000, 128: 593-617.

[31] COLLE B A. Numerical simulations of the extratropical transition of Floyd (1999): Structural evolution and responsible mechanisms for the heavy rainfall over the northeast United States[J]. Mon Wea Rev, 2003, 131: 2905-2926.

[32] CLARK P, ROBERTS N, LEAN H, et al. Convection-permitting models: a step-change in rainfall forecasting[J]. Meteorol Appl, 2016, 23(2): 165-181.

[33] LILLY D K. Numerical prediction of thunderstorms-has its time come[J]? Q J R Meteorol Soc, 1990, 116: 779-798.

[34] XUE M, WANG D H, GAO J D, et al. The advanced regional prediction system (ARPS), storm-scale numerical weather prediction and data assimilation[J]. Meteorol Atmos Phys, 2003, 82: 139-170.

[35] MICHALAKES J, CHEN S, DUDHIA J, et al. "Development of a next generation regional weather research and forecast model" in developments in teracomputing[C]//Proceedings of the Ninth ECMWF Workshop on the Use of High Performance Computing in Meteorology. Singapore: World Scientific, 2001.

[36] TANG Y, LEAN H, BORNEMANN J. The benefits of the met office variable resolution NWP model for forecasting convection[J]. Meteorol Appl, 2013, 20: 417-426.

[37] BALDAUF M, SEIFERT A, FÖRSTNER J, et al. Operational convective-scale numerical weather prediction with the COSMO model: description and sensitivities[J]. Mon Wea Rev, 2011, 139: 3887-3905.

[38] SEITY Y, BROSSEAU P, MALARDEL S, et al. The AROME-France convective-scale operational model[J]. Mon Wea Rev, 2011, 139: 976-991.

[39] SAITO K, FUJITA T, YAMADA Y, et al. The operational JMA nonhydrostatic mesoscale model[J]. Mon Wea Rev, 2006, 134: 1266-1298.

[40] WEISMAN M L, DAVIS C, WANG W, et al. Experiences with 0-36-h explicit convective forecasts with the WRF-ARW Model[J]. Wea Forecasting, 2008, 23: 407-437.

[41] ROBERTS N M. Assessing the spatial and temporal variation in the skill of precipitation forecasts from an NWP model [J]. Meteorol Appl, 2008, 15: 163-169.

[42] GRIMIT E P, MASS C F. Initial results of a mesoscale short-range ensemble forecasting system over the Pacific North-

west［J］. Wea Forecasting,2002,17:192-205.

［43］ TRACTON M S,KALNAY E. Operational ensemble prediction at the National Meteorological Center:Practical aspects ［J］. Wea Forecasting,1993,8:379-400.

［44］ 赵琳娜,刘莹,党皓飞,等. 集合数值预报在洪水预报中的应用进展［J］. 应用气象学报,2014,25(6):641-653.

［45］ PALMER T N. The economic value of ensemble forecasts as a tool for risk assessment:from days to decades［J］. Q J R Meteorol Soc,2002,128:747-774.

［46］ BUIZZA R. The value of probabilistic prediction［J］. Atmos Sci Lett,2008,9:36-42.

［47］ HOFFMAN R N,KALNEY E. Lagged average forecasting,an alternative to Monte Carlo forecasting［J］. Tellus A,1983,35A(2):100-118.

［48］ EBERT E E. Ability of a poor man's ensemble to predict the probability and distribution of precipitation［J］. Mon Wea Rev,2001,129:2461-2479.

［49］ TOTH Z,KALNAY E. Ensemble forecasting at NMC:the generation of perturbations［J］. Bull Amer Meteor Soc,1993,74:2317-2330.

［50］ 杜钧,李俊. 集合预报方法在暴雨研究和预报中的应用［J］. 气象科技进展,2014,4(5):6-20.

［51］ DU J. Hybrid ensemble prediction system:a new ensembling approach［C］//Symposium on the 50th Anniversary of Operational Numerical Weather Prediction,University of Maryland,College Park,2004.

［52］ FANG X Q,KUO Y H. Improving ensemble-based quantitative precipitation forecast for topography-enhanced typhoon heavy rainfall over Taiwan with a modified probability-matching technique［J］. Mon Wea Rev,2013,141:3908-3932.

［53］ PALMER T N,MOLTENI F,MUREAU R,et al. Ensemble Prediction［R］//Proc. of the ECMWF Seminar on Validation of Models over Europe:Vol. 1,pp. 21-66,7-11 September 1992,Reading,UK.

［54］ HOUTEKAMER P L,LEFAIVRE L,DEROME J,et al. A system simulation approach to ensemble prediction［J］. Mon Wea Rev,1996,124:1225-1242.

［55］ BOUGEAULT P,and Coauthors. The THORPEX Interactive Grand Global Ensemble (TIGGE) ［J］. Bull Amer Meteor Soc,2010,91:1059-1072.

［56］ SU X,YUAN H,ZHU Y,et al. Evaluation of TIGGE ensemble predictions of Northern Hemisphere summer precipitation during 2008-2012［J］. J Geophys Res Atmos,2014,119:7292-7310.

［57］ HAMILL T M. Verification of TIGGE multimodel and ECMWF reforecast-calibrated probabilistic precipitation forecasts over the contiguous United States［J］. Mon Wea Rev,2012,140:2232-2252.

［58］ STENSRUD D J,BAO J W,WARNER T T. Using initial condition and model physics perturbations in short-range ensemble simulations of mesoscale convective systems［J］. Mon Wea Rev,2000, 128:2077-2107.

［59］ DU J,and Coauthors. The NOAA/NWS/NCEP Short Range Ensemble Forecast (SREF) system:evaluation of an initial condition vs multiple model physics ensemble approach［C］//1138 Wea Forecasting Volume 24 Preprints,20th Conf. on Weather Analysis and Forecasting/ 16th Conf. on Numerical Weather Prediction,Seattle,WA,Amer Meteor Soc,21. 3,2004.

［60］ JONES M S,COLLE B A,TONGUE J S. Evaluation of a mesoscale short-range ensemble forecast system over the northeast United States［J］. Wea Forecasting,2007,22:36-55.

［61］ WANDISHIN M S,MULLEN S L,STENSRUD D J,et al. Evaluation of a short-range multimodel ensemble system［J］. Mon Wea Rev,2001,129:729-747.

［62］ ECKEL F A,MASS C F. Aspects of effective mesoscale,short-range ensemble forecasting［J］. Wea Forecasting,2005,20:328-350.

［63］ JANKOV I,GALLUS W A,SEGAL M,et al. The impact of different WRF model physical parameterizations and their interactions on warm season MCS rainfall［J］. Wea Forecasting, 2005,20:1048-1060.

［64］ CLARK A J,GALLUS W A,CHEN T C. Comparison of the diurnal precipitation cycle in convection-resolving and non-convection-resolving mesoscale models［J］. Mon Wea Rev,2007, 135:3456-3473.

［65］ FRITSCH J M,Carbone R E. Improving quantitative precipitation forecasts in the warm season:a USWRP research and development strategy[J]. Bull Amer Meteor Soc,2004,85:955-965.

[66] CLARK A J,GALLUS W A,XUE M,et al. A comparison of precipitation forecast skill between small convection-allowing and large convection-parameterizing ensembles[J]. Wea Forecasting,2009,24:1121-1140.

[67] GEBHARDT C,THEIS S,KRAHE P,et al. Experimental ensemble forecasts of precipitation based on a convection-resolving model[J]. Atmos Sci Lett,2008,9:67-72.

[68] CLARK A J,KAIN J S,STENSTRUD D J,et al. Probabilistic precipitation forecast skill as a function of ensemble size and spatial scale in a convection-allowing ensemble[J]. Mon Wea Rev,2011, 139:1052-1081.

[69] 王晨稀,姚建群,梁旭东. 上海区域降水集合预报系统的建立与运行结果的检验[J]. 应用气象学报,2007,18(2):173-180.

[70] 邓国,龚建东,邓莲堂,等. 国家级区域集合预报系统研发和性能检验[J]. 应用气象学报,2010,21(5):513-523.

[71] BOUTTIER F,VIE B,NUISSIER O,et al. Impact of stochastic physics in a convection-permitting ensemble[J]. Mon Wea Rev,2012,140:3706-3721.

[72] DUC L,SAITO K,SEKO H. Spatial-temporal fractions verification for high-resolution ensemble forecasts[J]. Tellus A,2013,65:18171-18193.

[73] GOLDING B W,BALLARD S P,MYLNE K,et al. Forecasting capabilities for the London 2012 Olympics[J]. Bull Amer Meteor Soc,2014,95:883-896.

[74] CARON J F. Mismatching perturbations at the lateral boundaries in limited-area ensemble forecasting:a case study[J]. Mon Wea Rev,2013,141:356-374.

[75] GEBHARDT C,THEIS S E,PAULAT M E,et al. Uncertainties in COSMO-DE precipitation forecasts introduced by model perturbations and variation of lateral boundaries[J]. Atmos Res,2011, 100:168-177.

[76] LEONCINI G,PLANT R S,GRAY S L,et al. Ensemble forecasts of a flood-producing storm:comparison of the influence of model-state perturbations and parameter modifications[J]. Q J R Meteorol Soc,2013,139:198-211.

[77] SCHWARTZ C S,KAIN J S,WEISS S J,et al. Toward improved convection-allowing ensembles:model physics sensitivities and optimizing probabilistic guidance with small ensemble membership[J]. Wea Forecasting,2010,25:263-280.

[78] BEN BOUALLÈGUE Z,THEIS S E. Spatial techniques applied to precipitation ensemble forecasts:from verification results to probabilistic products[J]. Meteorol Appl,2014,21:922-929.

[79] SCHEUERER M,HAMILL T M. Statistical postprocessing of ensemble precipitation forecasts by fitting Censored, shifted gamma distributions[J]. Mon Wea Rev,2015,143(11):4578-4596.

[80] ANTOLIK M S. An overview of the National Weather Service's centralized statistical quantitative precipitation forecast [J]. J Hydrol,2000,239:306-337.

[81] DOSWELL C A. Flash flood forecasting:an ingredient-based methodology[J]. Wea Forecasting,1996,11:560-581.

[82] 张小玲,陶诗言,孙建华. 基于"配料"的暴雨预报[J]. 大气科学,2010,34(4):754-764.

[83] SURCEL M,ZAWADZKI I,YAU M K. On the filtering properties of ensemble averaging for storm-scale precipitation forecasts[J]. Mon Wea Rev,2014,142:1093-1105.

[84] ZHU Y,LUO Y. Precipitation calibration based on the frequency-matching method[J]. Wea Forecasting,2015,30(5):1109-1124.

[85] BENTZIEN S,FRIEDERICHS P. Generating and calibrating probabilistic quantitative precipitation forecasts from the high-resolution NWP model COSMO-DE[J]. Wea Forecasting,2012, 27(4):988-1002.

[86] STENSRUD D J,YUSSOUF N. Reliable probabilistic quantitative precipitation forecasts from a short-range ensemble forecasting system[J]. Wea Forecasting,2007,22(1):3-17.

[87] GAHRS G E,APPLEQUIST S,PFEFFER R L,et al. Improved results for probabilistic quantitative precipitation forecasting[J]. Wea Forecasting,2003,18(5):879-890.

[88] FRIEDERICHS P,HENSE A. A probabilistic forecast approach for daily precipitation totals[J]. Wea Forecasting, 2008,23(4):659-673.

[89] ROULIN E,VANNITSEM S. Postprocessing of ensemble precipitation predictions with extended logistic regression

based on hindcasts[J]. Mon Wea Rev,2012,140(3):874-888.

[90] BEN BOUALLÈGUE Z. Calibrated short-range ensemble precipitation forecasts using extended logistic regression with interaction terms[J]. Wea Forecasting,2013,28(2):515-524.

[91] SLOUGHTER J M,RAFTERY A E,GNEITING T,et al. Probabilistic quantitative precipitation forecasting using Bayesian model averaging[J]. Mon Wea Rev,2007,135:3209-3220.

[92] LIU J,XIE Z. BMA probabilistic quantitative precipitation forecasting over the Huaihe basin using TIGGE multimodel ensemble forecasts[J]. Mon Wea Rev,2014,142(4):1542-1555.

[93] ZHU J,KONG F,RAN L,et al. Bayesian model averaging with stratified sampling for probabilistic quantitative precipitation forecasting in Northern China during summer 2010[J]. Mon Wea Rev,2015,143(9):3628-3641.

[94] GNEITING T,RAFTERY A E. Weather forecasting with ensemble methods[J]. Science,2005,310:248-249.

[95] PEEL S,WILSON L J. Modeling the distribution of precipitation forecasts from the Canadian ensemble prediction system using kernel density estimation[J]. Wea Forecasting,2008,23(4):575-595.

[96] YUAN H,GAO X,MULLEN S L,et al. Calibration of probabilistic quantitative precipitation forecasts with an artificial neural network[J]. Wea Forecasting,2007,22(6):1287-1303.

[97] VOISIN N,SCHAAKE J C,LETTENMAIER D P. Calibration and downscaling methods for quantitative ensemble precipitation forecasts[J]. Wea Forecasting,2010,25(6):1603-1627.

[98] FERNÁNDEZ-FERRERO A,SÁENZ J,IBARRA-BERASTEGI G. Comparison of the performance of different analog-based Bayesian probabilistic precipitation forecasts over Bilbao,Spain[J]. Mon Wea Rev,2010,138(8):3107-3119.

[99] GAGNE D J,MCGOVERN A,XUE M. Machine learning enhancement of storm-scale ensemble probabilistic quantitative precipitation forecasts[J]. Wea Forecasting,2014,29(4):1024-1043.

[100] SCHAFFER C J,GALLUS W A,SEGAL M. Improving probabilistic ensemble forecasts of convection through the application of QPF-POP relationships[J]. Wea Forecasting, 2011, 26:319-336.

[101] IM J S,BRILL K,DANAHER E. Confidence interval estimation for Quantitative Precipitation Forecasts (QPF) using Short-Range Ensemble Forecasts (SREF) [J]. Wea Forecasting,2006, 21(1):24-41.

[102] HAMILL T M,WHITAKER J S. Probabilistic quantitative precipitation forecasts based on reforecast analogs:theory and application[J]. Mon Wea Rev,2006,134(11):3209-3229.

[103] HAMILL T M, WHITAKER J S, MULLEN S L. Reforecasts:an important dataset for improving weather predictions[J]. Bull Amer Meteor Soc,2006,87(1):33.

[104] HAMILL,T M,HAGEDORN R,WHITAKER J S. Probabilistic forecast calibration using ECMWF and GFS ensemble reforecasts. Part Ⅱ:precipitation[J]. Mon Wea Rev,2008,136:2620-2632.

[105] FUNDEL F,WALSER A,LINIGER M A,et al. Calibrated precipitation forecasts for a limited-area ensemble forecast system using reforecasts[J]. Mon Wea Rev,2010,138(1):176-189.

[106] 宗志平,代刊,蒋星. 定量降水预报技术研究进展[J]. 气象科技进展,2012,2(5):29-35.

[107] PAVLIK D, SÖHL D, PLUNTKE T,et al. Dynamic downscaling of global climate projections for Eastern Europe with a horizontal resolution of 7 km[J]. Environmental Earth Sciences,2012,65(5):1475-1482.

[108] CLARK M,GANGOPADHYAY S,HAY L,et al. The Schaake shuffle:a method for reconstructing space-time variability in forecasted precipitation and temperature fields[J]. J Hydrol,2004,5(1):243-262.

[109] SCHAAKE J,HENKEL A,CONG S. Application of PRISM climatologies for hydrologic modeling and forecasting in the western US[C]//Preprints, 18th Conf on Hydrology,2004.

[110] ZAWADZKI I,Statistical properties of precipitation patterns[J]. J Appl Meteorol,1973, 12:459-472.

[111] STANSKI H R,WILSON L J,BURROWS W R. Survey of common verification methods in meteorology[R]. World Weather Watch Tech. Rept. No. 8,WMO/TD No. 358,World Meteorological Organization,Geneva,Switzerland,1989.

[112] WILKS D S. Statistical methods in the atmospheric sciences. An introduction[M]. San Diego:Academic Press, 2006.

[113] JOLLIFFE I T,STEPHENSON D B. Forecast verification[M]//A practitioner's guide in atmospheric science. Hoboken:Wiley and Sons Ltd,2003.

[114] NURMI P. Recommendations on the verification of local weather forecasts[R]. ECMWF Tech Memo,2003.

[115] BOUGEAULT P. WGNE survey of verification methods for numerical prediction of weather elements and severe weather events[R]. CAS/JSC WGNE Report No. 18,2002,Appendix C.

[116] WILSON C. Review of current methods and tools for verification of numerical forecasts of precipitation[R]. COST717 Working Group Report on Approaches to verification. 2001.

[117] ROSSA A,NURMI P,EBERT E. Overview of methods for the verification of quantitative precipitation forecasts [M]//MICHAELIDES S. Precipitation:advances in measurement, estimation and prediction. Dordrecht:Springer,2008.

[118] EBERT E E,GALLUS W A. Toward better understanding of the contiguous rain area (CRA) method for spatial forecast verification[J]. Wea Forecasting,2009,24(5):1401-1415.

[119] RODWELL M J,HAIDEN T,RICHARDSON D S. Developments in precipitation verification[J]. ECMWF News l,2011,128:12-16.

[120] RODWELL M J,RICHARDSON D S,HEWSON T D,et al. A new equitable score suitable for verifying precipitation in numerical weather prediction[J]. Q J R Meteorol Soc,2010,136:1344-1363.

[121] 陈法敬,陈静."SEEPS"降水预报检验评分方法在中国降水预报中的应用试验[J]. 气象科技进展,2015,5(5):6-13.

[122] STEPHENSON D B,CASATI B, FERRO C A T,et al. The extreme dependency score:a non-vanishing measure for forecasts of rare events[J]. Meteorol Appl,2008,15:41-50.

[123] HOGAN R,O'CONNOR E J,ILLINGWORTH A J. Verification of cloud-fraction forecasts[J]. Q J R Meteorol Soc,2009,135:1494-1511.

[124] FERRO C A T,STEPHENSON D B. Extremal dependence indices:improved verification measures for deterministic forecasts of rare binary events[J]. Wea Forecasting,2011,26:699-713.

[125] NORTH R,TRUEMAN M,MITTERMAIER M,et al. An assessment of the SEEPS and SEDI metrics for the verification of 6h forecast precipitation accumulations[J]. Meteorol Appl, 2013,20:164-175.

[126] MASON S J,WEIGEL A P. A generic forecast verification framework for administrative purposes[J]. Mon Weather Rev,2009,137:331-349.

[127] WEIGEL A P,MASON S J. The generalized discrimination score for ensemble forecasts[J]. Mon Weather Rev,2011,139:3069-3074.

[128] GILLELAND E,AHIJEVYCH D,BROWN B G, et al. Intercomparison of spatial forecast verification methods[J]. Wea Forecasting,2009,24:1416-1430.

[129] HOFFMAN R N,LIU Z,LOUIS J F,et al. Distortion representation of forecast errors[J]. Mon Wea Rev,1995,123:2758-2770.

[130] EBERT E E,MCBRIDE J L. Verification of precipitation in weather systems:determination of systematic errors[J]. J Hydrol,2000,239:179-202.

[131] DAVIS C,BROWN B,BULLOCK R. Object-based verification of precipitation forecasts. Part I:methodology and application to Mesoscale Rain Areas[J]. Mon Wea Rev,2006,134:1772-1784.

[132] CASATI B,ROSS G,STEPHENSON D B. A new intensity-scale approach for the verification of spatial precipitation forecasts[J]. Meteorol Appl,2004,11:141-154.

[133] 孔荣,王建捷,梁丰,等. 尺度分解技术在定量降水临近预报检验中的应用[J]. 应用气象学报,2010,21(5):535-544.

[134] EBERT E E. Fuzzy verification of high resolution gridded forecasts:a review and proposed framework[J]. Meteorol Appl,2008,15:53-66.

[135] ZHU Y,TOTH Z. Ensemble based probabilistic forecast verification[R]. 19th Conference on Probability and Statistics,2008.

[136] HAMILL T M. Interpretation of rank histograms for verifying ensemble forecasts[J]. Mon Wea Rev,2001,129:550-560.

[137] HERSBACH H. Decomposition of the continuous rank probability score for ensemble prediction systems[J]. Wea Forecasting,2000,15:559-570.

[138] WILKS D S. The Minimum Spanning Tree (MST) histogram as a verification tool for multidimensional ensemble forecasts[J]. Mon Wea Rev,2004,132:1329-1340.

[139] WEISHEIMER A,SMITH L A,JUDD K. A new view of forecast skill:bounding boxes from the DEMETER ensemble seasonal forecasts[J]. Tellus,2004,57(3):265-279.

[140] ROULSTON M S,SMITH L A. Evaluating probabilistic forecasts using information theory[J]. Mon Wea Rev,2002,130:1653-1660.

[141] WILSON L J,BURROWS W R,LANZINGER A. A strategy for verification of weather element forecasts from an ensemble prediction system[J]. Mon Wea Rev,1999,127:956-970.

[142] BRIER G W. Verification of forecasts expressed in terms of probability[J]. Mon Wea Rev, 1950,78:1-3.

[143] BROCKER J,SMITH L A. Increasing the reliability of reliability diagrams[J]. Wea Forecasting, 2007,22(3):651-661.

[144] SWETS J A,PICKETT R M. Evaluation of Diagnostic Systems:Methods from Signal Detection Theory[M]. New York:Academic Press,1982.

[145] EPSTEIN E S. A scoring system for probability forecasts of ranked categories[J]. J Applied Meteorology,1969,8:985-987.

[146] MURPHY A H. A new vector partition of the probability score[J]. J Applied Meteorology,1973,12:595-600.

[147] STEPHENSON D B. Definition,diagnosis,and origin of extreme weather and climate events[M]//Climate Extremes and Society. DIAZ H F,MURNANE R J. New York:Cambridge University Press,2008.

[148] FERRO C A T,RICHARDSON D S,Weigel A P. On the effect of ensemble size on the discrete and continuous ranked probability scores[J]. Meteorol Appl,2008,15:19-24.

[149] FUNK T W. Forecasting techniques utilized by the forecast branch of the national meteorological center during a major convective rainfall event[J]. Wea Forecasting,1991,6(4):548-564.

[150] REYNOLDS D. Value-added quantitative precipitation forecasts:how valuable is the forecaster[J]? Bull Amer Meteor Soc,2003,84:876-878.

[151] DOSWELL C A Ⅲ,BROOKS H E. Budget cutting and the value of weather services[J]. Wea Forecasting,1998,13,206-212.

[152] NOVAK D R,BRIGHT D R,Brennan M J. Operational forecaster uncertainty needs and future roles[J]. Wea Forecasting,2008,23:1069-1084.

[153] STUART N A,et al. The future role of the human in an increasingly automated forecast process[J]. Bull Amer Meteor Soc,2006,87:1497-1502.

[154] MASS C F. IFPS and the future of the National Weather Service[J]. Wea Forecasting,2003, 18:75-79.

[155] HOMAR V,STENSRUD D J,LEVIT J J,et al. Value of human-generated perturbations in short-range ensemble forecasts of severe weather[J]. Wea Forecasting,2006,21:347-363.

[156] SILLS D M L. On the MSC forecasters forums and the future role of the human forecaster[J]. Bull Amer Meteor Soc,2009,90:619-627.

[157] STUART N A,SCHULTZ D M,KLEIN G. Maintaining the role of humans in the forecast process[J]. Bull Amer Meteor Soc,2007,88:1893-1898.

[158] FUJITA T,STENSRUD D J,DOWELL D C. Using precipitation observations in a mesoscale short-range ensemble analysis and forecasting system[J]. Wea Forecasting,2008,23:357-372.

[159] MARCUS S,KIM J,CHIN T,et al. Influence of GPS precipitable water vapor retrievals on quantitative precipitation forecasting in Southern California[J]. J Appl Meteor Climatol,2007, 46:1828-1839.

[160] RICHARD E,BUZZI A,ZÄNGL G. Quantitative precipitation forecasting in the Alps:the advances achieved by the mesoscale alpine programme[J]. Q J R Meteorol Soc,2007,133:831-846.

中国无缝隙精细化网格天气预报技术进展与挑战

金荣花 代刊 赵瑞霞 曹勇 薛峰 刘凑华 赵声蓉 李勇 韦青

(国家气象中心,北京,100081)

摘 要:本文总结了2014年以来中国无缝隙精细化网格天气预报业务的技术进展,讨论了未来发展所面临的关键技术难点。无缝隙精细化网格预报技术的发展,得益于综合气息观测数据和多源资料融合分析网格实况产品的支撑,更依赖于多尺度数值预报模式和实时快速更新同化预报系统的快速发展。经过近5年的探索和努力,中国已经初步建立了针对不同预报时效的无缝隙精细化网格预报技术体系。对0~4 h预报时效,主要基于全国雷达拼图和GRAPES-MESO模式预报,发展临近分钟级滚动外推预报技术;对4 h到30 d预报时效,主要通过对区域或全球不同时空分辨率模式预报进行偏差订正、客观解释应用以及降尺度分析,提高预报的准确度和精细度。与此同时,研发了自动化、智能化的交互式预报制作平台,满足客观高效制作与预报员对极端或高影响天气主观预报优势相结合的需求。发展了以格点实况分析场为参照的空间分析检验方法,初步实现了对高分辨率网格预报的质量跟踪和性能评估。未来的网格预报技术体系,需要吸纳前沿的技术研究成果,包括人工智能应用技术、高级多模式统计后处理技术和协调一致性关键技术等,并且建立统一完整的技术架构和开发标准等。

关键词:网格天气预报;技术进展;技术框架;订正平台;检验方法;技术难点

Progress and Challeng of Seamless Fine Gridded Weather Forecasting Technologies in China

JIN Ronghua DAI Kan ZHAO Ruixia CAO Yong XUE Feng
LIU Couhua ZHAO Shengrong LI Yong WEI Qing

(National Meteorological Center, Beijing, 100081)

Abstract:This paper reviews the development of the technology for seamless fine gridded forecasting in China since 2014. And the key technical difficulties in the future development are analyzed. It is pointed out that the high spatio-temporal resolution observations capturing the fine structure of weather systems, the analysis products by multi-source data fusion, the real-time rapid updating assimilation and prediction system, the high resolution regional model providing short-time and short-term weather prediction, the global numerical forecast model proving 10 day' weather forecasting, and the ocean-atmosphere coupled ensemble prediction system proving 46 day' weather prediction, have jointly established the premise and foundation of the seamless gridded weather forecasts. After nearly 5 years,

---

① 本文发表于《气象》2019年第4期。
资助项目:国家科技支撑计划课题(2015BAC03B04);气象预报业务关键技术发展专项(YBGJXM201804)。
第一作者:金荣花,主要从事天气分析与大尺度动力过程研究。E-mail:jinrh@cma.gov.cn。

exploration and constant efforts, the technology system of seamless fine gridded forecasting with different temporal resolution has been established. The high-frequency lagrangian extrapolation skills are used for 0-4 h forecasting based on GRAPES-MESO model forecast products and radar data over China. For the 4 h to 30 d lead time forecasting, it mainly depends on the downscaling, error correction, model output statistics and post-processing methodologies based on regional and global models of different spatio-temporal resolution to improve forecast skills and resolution. At the same time, automatic and intelligent interactive forecasting platform is developed to meet the demand of combining efficient objective forecasting with forecaster's subjective intelligence. In order to assess and track the performance of high resolution gridded forecasting, a spatial analysis verification method based on gridded observation data is developed. It is also stressed that the future gridded forecasting technology system should be able to reflect the latest technology developing including the artificial intelligence application, more advanced statistical post-processing skills, key technics for consistency forecasting and unified complete technical architecture and standards.

Key words: gridded weather forecasting; technology development; technical framework; gridded forecast editor; verification methodology; technical difficulties

引　言

天气预报在过去一二十年取得了巨大进展，其背后驱动力来源于气象科技和信息技术的快速进步，电子通讯、计算系统和观测系统的发展，以及用户对天气信息越来越精细和个性的专业需求。实现"以人为本、无微不至、无所不在"的精细化气象预报和个性化服务成为中国和世界各国气象部门共同的发展目标。然而，传统的固定站点和文字定性描述预报形式无法覆盖和表达精细的时间和空间信息，亟待发展时空和内涵无缝隙、质量和内容精细化的网格天气预报。无缝隙是对时空分辨率和多尺度的涵盖，也体现"要素"的完整性，但是必须兼顾到气象问题的科学属性；无缝隙程度依赖于预报时效，而网格精细程度也受可预报性和不确定性等条件制约(宇如聪[①]，2018)。

当前，无缝隙精细化网格气象预报已经成为国际主流趋势。首次世界天气开放科学大会(WWOSC-2014)的主题就是"地球系统无缝隙预报：从分钟到月"(Brunet et al.，2015)。2017年，地球系统科学家学会(YESS)联合世界气候研究计划(WCRP)、世界气象研究计划(WWRP)和全球大气观察计划(GAW)创建的地球系统科学前沿白皮书(Rauser et al.，2017)，更是将无缝隙预报作为未来几十年科学界指导方针的重要指标，提出从分钟级到世纪尺度、从米到全球空间尺度的预报发展趋势。美国最早从2003年开始发展国家数字预报数据库(NDFD)(Glahn et al.，2003)，提供逐1 h更新2.5/5 km分辨率从0时刻至45 d延伸期的无缝隙网格天气预报，包括常规气象要素、灾害性天气、台风海洋等预报产品。NDFD的制作首先基于多尺度天气预报模式和全球集合预报，在多源实况观测资料的基础上采用模式解释应用技术发布格点指导预报，之后各地天气预报办公室的预报员使用网格预报编辑系统进行主观订正，最后拼接生成统一的网格预报(Ruth，2002)。奥地利气象局发展了无缝隙概率预报系统(SAPHIR)(Kann et al.，2018)，逐10 min更新生成最小分辨率1 km的概率和确定性天气预报，预报时效由分钟级临近预报到72 h，同时也提供15 km分辨率的14 d中短期网格预报。该系统基于综合分析的集成临近预报技术(Haiden et al.，2011)和标准化距平MOS预报技术(Stauffer et al.，2017)方法，显著提高了临近和短期

---

① 2018年全国智能网格预报业务工作部署会议报告。

预报效果。德国基于强大的数值模式能力，利用多个模式解释应用和统计后处理预报的优化集成预报方法，提供逐5 min更新2.5 km分辨率的分钟级至延伸期30 d的订正网格预报。澳大利亚自2012年起也开展了基于业务集成预报技术（OCF）的5 km分辨率8 d内网格指导预报（Engel et al.，2012）。

中国无缝隙精细化网格预报业务技术的发展起步于2014年，主要得益于综合气象观测系统、多源实况融合分析技术和多尺度数值预报模式的快速发展和支撑。能够捕捉天气系统精细结构信息的高时空分辨率观测资料，以及多源资料融合分析的网格实况产品，是精细化网格预报发展的坚实基础；而快速更新同化预报系统、覆盖短时短期的高分辨率区域模式、预报时效达10 d的全球数值预报模式，以及超过30 d的海气耦合集合预报系统，则为精细化网格预报的发展提供了必要的前提条件。

中国的观测系统尤其是进入21世纪后取得长足发展，建立了地基、空基和天基相结合，门类比较齐全，布局基本合理的综合气象观测系统（行鸿彦 等，2017）。加上全国200多个站的新一代雷达数据（高玉春，2017），大幅增加了实况观测对气象要素的精细空间分布刻画能力。2012年起发展的多源资料融合实况分析业务系统（师春香[①]，2018；潘旸 等，2018；师春香 等，2018；韩帅 等，2018），融合了常规、雷达、卫星、闪电、GNSS/MET水汽、飞机、船舶等多源实况观测资料，实时提供包括地面常规要素、三维云量、天气现象、洋面风、海表温度等多种网格实况分析产品，为网格预报模型建立、实况信息更新和预报产品检验提供了的基础条件。

在数值预报模式方面，中国气象局自主研发了全球/区域同化预报系统GRAPES，在动力框架、物理过程和变分同化技术等方面均取得显著进展（刘艳 等，2016；沈学顺 等，2017；张进 等，2017；黄丽萍 等，2017；朱立娟 等，2017；万子为 等，2015），并且建立了体系完善的全球、区域和集合预报模式系统。北京、上海和广东区域中心还分别建立了华北和华东区域的WRF模式、华南区域的GRAPES_TMM模式等（代刊 等，2016），每天8/24次实时发布空间分辨率3 km、时间分辨率1 h的区域模式产品。另外，欧洲中期天气预报中心（ECMWF）全球模式和集合预报（Hólm et al.，2016）、美国环境预报中心NCEP全球分辨率预报模式（Lien et al.，2016）等，也实时接收并作为网格预报的原始资料。

然而，大气是典型的混沌系统（Lorenz，1963；Danforth，2013），基于大气动力学的数值预报模式，即使是高分辨率区域模式，仍然无法解决近地面的天气细节问题，具有系统和随机误差（丁一汇，2005；Boeing，2016）。因此，在模式预报基础上利用实况信息进行统计后处理，以求提高预报技巧，是十分必要也是有效的（Glahn et al.，1972；Carter et al.，1989；Vislocky et al.，1995；Krishnamurti et al.，1999）。此外，数值模式的空间和时间分辨率通常不能满足天气预报服务的精细需求，为了提供精细刻画不同地形以及下垫面属性条件下的近地面天气细节，实现下游气象服务的精细信息提取，必须进行空间和时间上的降尺度（Ben Alaya et al.，2015；曹勇 等，2016；Tang et al.，2018）。为此，需要发展无缝隙精细化网格预报技术，包括：①模式产品的偏差订正、客观解释应用以及降尺度分析技术，获取更加精准、精细的网格预报产品（Hamill et al.，2015）；②建立自动化和智能化交互式预报制作平台，满足客观高效制作和预报员主观智力相结合的需求；③发展质量检验技术，实时跟踪和评估网格预报产品性能；④构建0~30 d无缝衔接、精细化网格预报产品的技术框架和系统设计。

历经5年建设发展，中国初步建立了逐1 h滚动更新、实时共享的全国5 km分辨率0~30 d和每日两次滚动更新全球10 km分辨率0~10 d无缝隙精细化网格天气预报业务。本文主要针对2014年以来中国国家气象中心无缝隙精细化网格天气预报技术体系建设，归纳网格预报的技术进展，分析未来发展面临的关键技术难点，希望有助于了解中国无缝隙精细化网格天气预报技术体系，也为未来网格预报技术向纵深和智能化方向发展提供借鉴和参考。

---

① 2018年成都智能网格预报和实况数据分析业务研讨会议交流报告《全国智能网格实况分析产品研制与评估应用》。

1 主要技术进展

1.1 无缝隙精细化网格预报技术框架

综合利用多源观测、模式数据、新技术新方法,设计和发展了分不同预报时效及不同预报尺度的无缝隙、精细化网格预报技术框架(图1)。

在不同类型数据集的基础上,基于不同预报时效可预报性特征(Browning,1980),发展相应的客观后处理技术,包括:对0～4 h预报时效,主要基于全国雷达拼图和GRAPES-MESO模式预报,发展临近分钟级滚动外推预报技术;对4～24 h预报时效,主要基于GRAPES-MESO、GRAPES-3 km、华东区域等高分辨率中尺度或对流尺度模式系统,发展逐小时滚动订正的短时预报技术;对1～10 d预报时效,主要基于多中心全球确定性和集合中期数值模式系统,发展订正和集成的中短期预报技术,提取最有效的预报信息;对于10～30 d时效,主要基于月尺度集合数值预报模式,发展延伸期统计后处理技术,进行系统性偏差订正,并保留低频预报信号。需要说明的是,在中国天气预报业务规范中,0～2 h和0～12 h分别为临近预报和短时预报,但无缝隙网格预报是按照所采用的技术方案及其预报质量来确定预报时效边界的。

此外,对不同的气象要素采用的技术路线也不同。目前主要将不连续变量降水与连续变量温度、风等区分开来,降水主要采用频率匹配、最优百分位等面向非连续变量的偏差订正集成方法,而温度、风等变量则主要采用MOS建模等方法。为使网格预报产品不同时效之间的连接处平滑过渡,在实际业务中采用线性权重方法将不同时效产品融合起来,形成0～30 d的无缝隙网格化客观产品。

由于网格化产品的高频滚动、高时空分辨率、海量信息的订正集成等特征,传统人工制作天气预报产品的途径已经无法满足发展需求,需要设计和研发客观方法和预报主观价值相融合的网格预报交互式订正平台,通过海量数据挖掘、网格调整、降尺度等智能技术,帮助预报员在不同层面上实现价值。

为让网格预报流程顺利运行,预报员及时介入实现价值,以及发布的产品获得用户认可,需要建立从数值模式系统到客观预报输出和最终无缝隙网格化产品的全流程检验评估体系。其中如何评估高分辨率的网格化产品是最重要的研发内容。

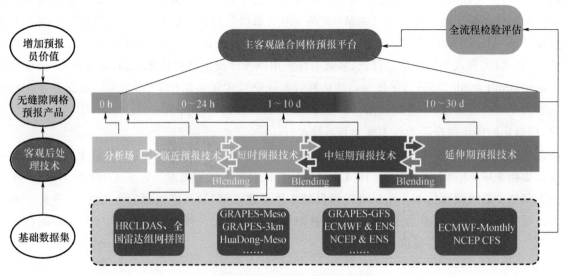

图1 无缝隙精细化网格预报技术框架

1.2 不同预报时效的客观后处理技术

1.2.1 0～24 h 网格客观预报技术

1.2.1.1 0～4 h 预报时效

对于0～4 h的临近时效,需要快速融入最新观测资料进行分钟级滚动更新。目前无论国际还是国内,都以基于实时观测或分析数据进行线性外推为主。临近时效重点关注降水演变外推技术,其中光流法因为较好的预报效果在业务中得到了广泛的应用(Bowler et al.,2004),但也存在一定缺陷,主要包括:对无降水区域如何给出最优的平流背景风场估测;如何避免半拉格朗日外推出现降水强度削弱的计算误差;传统光流法难以考虑降水系统的生消发展(Cheung et al.,2012)。为解决上述关键问题,从 2016 年发展了改进光流法的临近外推技术,包括三项关键技术:第一,利用金字塔架构的 LK 光流技术,给出无降水落区的最优平流背景风场,为较长时效外推提供可靠的背景风场;第二,利用强度守恒约束的半拉格朗日平流外推技术,解决传统半拉格朗日外推时出现的降水强度计算误差现象;第三,提出基于光流法回算残差以及基于 GRAPES-MESO 模式对流环境场预报,发展出降水强度变化预报技术,改进传统光流法无强度变化预报的缺陷。以上改进在提高外推位置预报准确率的同时,也可进一步提高临近时刻降水强度变化的准确率。图 2 对 2018 年 7 月 1 日 00 时至 9 月 30 日 23 时 0～4 h 时效的持续预报、传统光流及改进光流技术所输出的临近逐小时预报产品进行预报评分对比,可以看到,若以 TS=0.4 为基准,改进光流技术较传统光流技术将预报时效延长了 1 h 左右。

在人工智能、大数据和云计算迅速发展的技术背景下,中国国家气象中心已经开始探索将人工智能技术应用于网格预报业务,与清华大学合作,采用分布式深度学习框架、时空记忆深度循环网络算法,实现了外推时效延至 2 h 雷达回波不同尺度空间特征的有效提取,雷达外推预报准确率较交叉相关法(CO-TREC)平均提升 40%,有助于提高 0～2 h 降水要素的预报精准度(王建民[①],2018)。

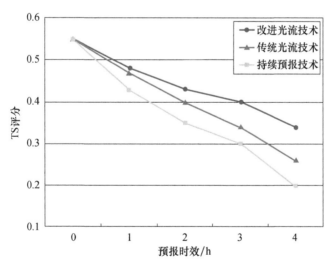

图 2　2018 年 7 月 1 日 00 时至 9 月 30 日 23 时 0～4 h 时效的外推雷达回波 TS 评分对比

1.2.1.2 4～24 h 预报时效

对于4～24 h的短时时效,主要基于快速滚动更新的高分辨率中尺度模式开展预报技术的研发。如降水要素,GRAPES-MESO 模式系统实现了逐 3 h 快速滚动更新运行。由于实时同化了最新观测资料,

---

① 2018 年 10 月 11 日全国气象台长会议交流报告《大数据与人工智能技术在天气预报中的应用》。

因此相比于全球模式,中小尺度对流系统在短时阶段对降水的预报能力更强。即便如此,GRAPES-MESO模式同样也存在预报误差,需要进行偏差订正。这里引入实时频率匹配订正技术。这一技术已在定量降水预报的模式后处理中得到了较广泛的应用(Zhu et al.,2015),其原理为:利用待订正量以及观测量样本资料,分别计算待订正量和观测量的经验累积概率分布函数,并利用两者之间的差异,对待订正量的数值修正,最终两者一致,计算公式如下:

$$x_c = F_o^{-1}(F_m(x_m)) \tag{1}$$

其中,x_m是待订正量;$F_m(x_m)$是待订正量的累积概率分布函数;F_o^{-1}观测累积概率分布函数的逆函数;x_c即为x_m对应的订正值。该方法的本质是实现待订正量和观测量对应分位数的映射。在实际业务中,根据中尺度模式快速滚动更新的特点,发展了实时动态频率匹配方案:即利用GRAPES_MESO模式前2 h的预报以及对应时段的实况观测数据,分别计算各自的累积概率分布函数,然后通过频率匹配技术,进行降水强度偏差订正。如图3所示,针对2016年4月13日08时出现在广东东部和南部的对流性降水过程,GRAPES-MESO模式较好地预报了系统的位置,但强度与实况相比明显偏弱。采用实时频率匹配订正技术之后,降水量级得到显著提高,更接近实况强降水值。

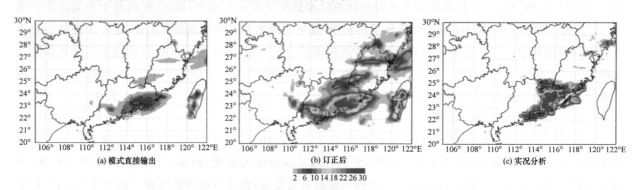

图3 2016年4月13日07—08时累积降水量的预报与观测实况对比
(a)GRAPES-MESO模式02时起报的降水预报;(b)05时滚动更新的动态频率订正后的降水预报,
其频率分布曲线利用03和04时的实况和预报构建;(c)为降水分析实况

1.2.2 1~10 d网格客观预报技术

在1~10 d的中短期预报时段,基于多中心全球确定性和集合预报模式,国家气象中心具有发展历史较长的多种类客观预报技术方法来支撑网格预报业务。

1.2.2.1 降水预报

目前支撑降水预报业务的客观预报方法包括:2009年开始研发、2012年业务运行的基于多中心确定性模式降水相似分析的多模式集成方法(陈力强 等,2005;林建 等,2013);2012年开始研发、2014年业务运行的基于确定性全球模式物理量统计建模的逻辑回归法(张芳华 等,2016)、2015年从南京大学引进的基于确定性全球模式的频率匹配订正算法(Zhu et al.,2015);2013年开始研发、2015年业务应用的基于集合模式的最优百分位方法(代刊 等,2018);以及2016年引入业务应用的基于确定性全球模式评分最优化的订正算法(简称OTS方法,吴启树 等,2017)。

根据每年的检验评估结果来看,每种方法都在进行不断的改进。例如对基于集合模式的最优百分位方法,起始于2013—2014年集合预报数据的业务应用研究,最初来自于预报员对集合预报产品的使用经验,即发现不同的集合预报统计量对不同等级降水量有优势,为此制定集成规则,将不同降水等级上的优势集合预报统计量集成起来,能够取得更好的评分技巧;2015年,引入动态百分位的概念,能够根据不同季节、不同时效的优势信息进行自动集成;2016—2017年,进一步采用频率匹配方法和局地概率匹配方法对最优百分位值分别进行量级和空间分布订正,实现多技术融合。图4给出了2018年1—9月四种定量降水客观预报方法与ECMWF模式直接输出的降水24 h预报时效24 h累计降水暴雨TS和BIAS评分对比,可以看到,所有客观方法结果较最先进模式的直接输出结果中,暴雨TS评分能够提升20%~28%,且能很好地修正模式的暴雨量级预报偏弱的问题。需要说明的是,不同客观方法优化目标不同,应用范围和适用的地区或天气类型也不相同;要找到最佳技术方法,首先要清楚如何定义"最佳"(多种评分的平衡)。此外,目前客观技术还存在如下问题:数值模式的有效预报信息集成度还不高,大多数客观方法都基于EC模式,没有考虑多中心或不同尺度模式;客观方法以降水量级订正为主,对空间分布的订正能力弱。

由于全球模式订正结果没有达到精细化网格预报要求的时空分辨率,在客观订正方法的基础上,还需要加入统计降尺度技术(详细技术说明见曹勇等(2016)的研究),包括:降水空间统计降尺度技术,实现粗分辨率降水预报具有气象意义的降尺度到细分辨率网格;降水时间拆分技术,实现降水预报由粗时间分辨率向细时间分辨率的转换。

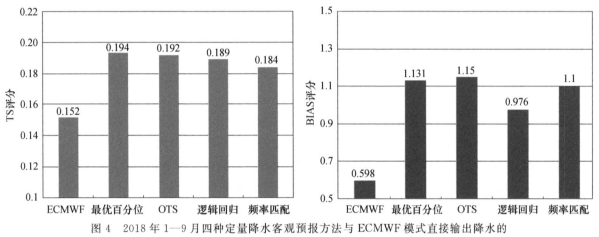

图4 2018年1—9月四种定量降水客观预报方法与ECMWF模式直接输出降水的
24 h预报时效暴雨(≥50 mm)TS和BIAS评分对比

1.2.2.2 其他气象要素客观预报技术

除降水要素外,对于温度、风、能见度等其他连续性变量,目前0~10 d预报时效均采用同一技术方案,尚没有在0~24 h内分不同时效进行区分处理。主要技术方案为:在模式背景场和城镇站点指导预报的基础上,通过考虑精细化地理信息订正的逐步插值分析方法,使网格点预报逐步向站点指导预报逼近,形成最终的网格预报结果。这里,城镇站点指导预报是基于传统MOS建模的预报结果(赵声蓉等,2012)。在此基础上选取对应的模式预报为插值分析的初始场,若对应模式预报要素和时效不存在,则选取替代的要素和时效;通过初始场插值分析得到与格点预报空间分辨率和区域一致的背景场;在插值分析过程中,考虑地理信息的订正作用,如对于温度要素,利用温度垂直递减率,把插值点温度订正到被插值点高度;采用反距离权重的逐步迭代方法,得到网格预报结果;为适应不同的站点疏密程度,在站点预报误差插值到格点的过程中采取变搜索半径、分等级的技术方案(Dallavalle et al.,2005)。对于能见度,采用区域模型的网格预报方案,即在适当的分区基础上,利用区域站点建立区域预报模型并应用于网格点,得到网格预报(赵翠光等,2012;Stauffer et al.,2017)。具体包括:利用历史能见度站点观测资料,应用REOF的分析方法进行客观分区。在站点分区结果的基础上,利用GIS系统确定区域内每个格点的区域归属;在站点观测和数值预报产品历史资料基础上,利用概率回归方法建立不同能见度等级的区域预报模型,将区域预报方程扩充到区域内任何格点上,得到格点的能见度等级预报结果。

1.2.3　10～30 d 延伸期网格客观预报技术

10～30 d 延伸期预报时效受到可预报理论、数值模式和历史资料等诸多因素的制约,该时段的降水、气温等要素网格预报,一直是无缝隙预报技术中的难点,也是国内外研究的前沿问题(杨秋明,2018;吴捷等,2018;Mariotti et al.,2018)。目前国际上均基于海气耦合的全球集合数值预报资料,采用偏差订正和降尺度技术获得 10～30 d 的降水和气温等基本气象要素的网格预报产品(Guan et al.,2015)。2017 年以来国家气象中心在延伸期网格预报技术方面进行初步探索,主要利用月尺度集合数值模式系统资料,通过多种统计后处理技术进行误差订正,并利用高分辨率实况资料,建立统计降尺度矢量关系,实现降水、气温等要素预报在延伸期时段的统计降尺度,提高空间分辨率。例如,对降水要素,为了消除由集合平均带来的延伸期小量级降水预报的湿偏差,采取逻辑回归法(Sloughter et al.,2007)和临界值法,设定降水临界值进行消空,一定程度上消除了集合平均带来的"小雨空报",同时结合统计降尺度技术提高预报产品的空间分辨率。对气温要素,则主要采用衰减平均偏差估计法(Cui et al.,2011;Guan et al.,2015),即对上一时刻相对分析场的偏差进行估计,使用具有权重系数的衰减平均值,结合预报前期偏差和最新的预报偏差计算当前时刻平均偏差,用以对预报场进行偏差订正。由此构成了每日两次滚动更新全国 10 km 分辨率 10～30 d 延伸期网格预报业务产品。

1.3　主客观融合的网格预报平台技术

Hoffman 等(2017)指出,利用最先进的探测手段也不可能对大气进行完整观测,同时基于最高级的计算机系统也不能完美预报天气。因此,天气预报是人和技术相互依赖、协同的结果。

针对时间、空间、频次及其精细气象要素格点预报,预报员开始借助可视化的软件系统来解决初猜场的编辑和修正。Ruth(2002)和 Mass(2003)在人机交互预报准备系统(Interactive Forecast Preparation System,IFPS)中应用基于可视化工具 AWIPS/GFE 来实现网格预报编辑功能,具体包括三方面:一是数据预处理,如将低分辨模式格点通过降尺度生成 5 km 分辨率预报;二是提供智能化编辑工具实现格点预报快速制作,如基于标准递减率对复杂地形下的气温预报插值、基于 MOS 开展区域调整等;三是 GFE 还实现网格预报产品处理加工、生成文字、图形制作等功能。丁建军等(2008)基于订正点与关联点之间相似区关联,通过文本、图形两种修改方式,完成对高时空密度、多预报要素值的修改订正。高嵩等(2014)基于 MICAPS 3.2 框架研发格点编辑平台,实现对原始的格点场进行基于单点、区域的编辑修改;也可以借助将等值线进行压缩、拉伸变形以及整体移动、删除等操作,将等值线的形变结果反演格点等方法。王海宾等(2016)提出基于 B/S 架构的大城市精细化格点预报系统,建立了依赖基准站的曲线订正反演模型和站点预报影响模型,将模型中基准站点反演到"面"预报,同时将面反演和时间序列反演结合,实现多时次预报快速订正。宁方志等(2017)采用 ArcGIS 地理信息系统技术开展网格预报编辑。张宏芳等(2017)把百度 WebGIS 应用于网格预报数据的展示。贺雅楠等(2018)基于 MICAPS 实现要素场协调性处理,集成降水、温度等客观预报算法,并提供丰富的主客观交互功能。

在网格预报所有要素中,预报员参与 QPF 产品订正的程度最高,也是发挥其关键作用的重要环节。毕宝贵等(2016)全面回顾了 QPF 预报技术的进展,指出数值模式不断发展和统计后处理技术的深入应用使得降水预报的精准度持续提升,预报员在其基础上能够提供的附加值越来越有限。为了发挥预报员在 QPF 网格化预报业务流程中的核心作用,2016 年起国家气象中心研发了新的主客观融合定量降水预报平台(唐健 等,2018)。该平台重新定义 QPF 预报员价值,通过客观手段、智能算法将预报员从繁重的手工操作中解放出来,重点发挥预报员对极端性、高影响或不确定性较高天气的预报经验和能力,并且控制整个 QPF 业务流程,实时监控无缝隙格点化产品预报质量,当客观预报方法失效时介入。主客观融合 QPF 系统的目标在于设计从全自动→半自动→手工的不同层级的预报平台,让预报员在数字化预报流程中,在关键环节或关键时间节点上能发挥其价值。主客观融合的定量降水预报平台有五种不同等级的工作模式(图5)。根据多个数值模式和客观方法预报的稳定性和一致性,降水的强弱等级和可预报性,以及模式等客观预报与预报员的天气分析概念模型的相近程度,预报员可以分别采取全自动客观网格化指导预报模式(L0)、选取优势预报数据源模式(L1)、指定权重或概率匹配平均合成预报模式(L2)、编辑格点模式(L3)及手工绘制降水落区等值线并反演主客观合成技术模式(L4),得到最终的网格预报产品。

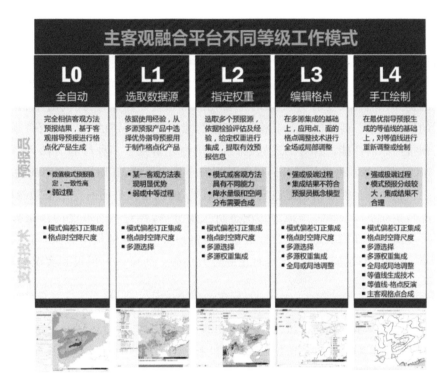

图5 主客观融合的定量降水预报平台的不同等级工作模式

1.4 高时空分辨网格预报产品的检验评估技术

随着精细网格预报业务的稳步推进,相应的质量检验和性能评估技术的研发工作也同时开展。

1.4.1 传统城镇预报检验技术的沿用

传统的城镇预报检验技术被继续应用,包括降水晴雨预报准确率、分级累加降水空报率、漏报率、TS评分和BIAS评分,最高气温、最低气温预报准确率,风向风速预报准确率,相对湿度预报准确率及能见度TS评分检验等。这些传统的检验指标一方面能够大致体现精细化网格预报的水平,同时也便于同历史上的城镇预报产品质量进行对比,从而表现精细网格预报技术进步带来的预报准确率的变化。然而传统的检验方法并不能充分反映预报结果精细化的质量和效益。这是因为网格预报带来的预报范围、时空精细度的巨大变化,对检验技术提出了新的要求。

1.4.2 精细网格预报检验参照实况方案的改进

传统的城镇预报重点关注站点附近的天气,而智能网格预报则覆盖预报区域的每个站点,如果仍然以国家站的观测数据作为实况参照,不能满足大部分区域的检验需求。为此,发展了两种针对精细化智能网格预报的参照实况方案。一种方案是将国家站和自动站作为检验参照,需要解决的技术问题是业务中许多自动站的观测存在明显的质量问题,为此通过对过去多年自动站观测资料进行梳理统计,挑选10461个站点组成一套用于检验的稠密站点序列。然后,针对上述站点实况,增加更为严格的质量控制流程,确保所用实况高度可信。此外,还根据站点分布密度和地形复杂性的差异对每个站点设置不同的权重,以便缓解中国东西部站点密度差异带来的检验公平性问题。另一种方案是将实况分析场作为检验参照,前提是准确地估算网格实况的误差范围,否则无法理解检验结果的含义。为此需要在业务产品中实时计算网格实况的误差方差,为基于网格实况的检验提供基础。

1.4.3 解决时空精细度变化的检验技术

目前精细化智能网格预报的空间分辨率达到5 km,时间分辨率达到逐小时,相邻时刻或相邻网格点之间的要素预报并非总是平滑过渡,而是包含高梯度细节,也就是说,它不是简单的分辨率的提升,而是空间精细度的提高。然而尺度越小,要素时空变化的可预报性越差,因此高时空精度对具体的单点而言意味着更大的误差方差。传统的点对点检验方法,经常会出现所谓的"双重惩罚"现象,其评分结果不能稳定地体现预报产品的性能。针对上述问题,近年国内外研发了多种检验方法,其中之一是点对面检验法(陈法敬 等,2019),基本思想是当某网格点预报有事件发生时,其周边一定距离内实况也有发生即考虑为预报正确。该检验方法能够给出更稳定的检验结果,但网格预报的时空细节被忽略。尺度分离检验法则采用傅里叶转换或小波分解将预报和实况进行尺度分离,再对不同尺度的预报分量进行检验,其中就包括对网格尺度的预报准确度的检验。针对精细网格预报内呈现的概率含义,可以采用邻域空间检验方法(Roberts et al.,2008;赵滨 等,2018),对比一定大小的空间窗口内预报和实况要素值达到某一阈值的比例。

然而,尺度分离法和邻域空间检验等方法并不能检验精细化网格预报的具体时空偏差,基于目标的检验方法则可满足该要求。该类方法是将预报和实况场中的要素值达到一定阈值的区域识别成多个目标,并相互匹配,进一步对预报目标的位置、形态和强度的误差进行检验。刘凑华等(2013)进一步改进了基于目标识别的检验评估方法,使得目标识别配对过程可以在网格预报和站点实况之间进行,避免实况插值带来误差的同时,拓展了此类方法的应用范围。但基于目标的检验方法并不能给出每个格点的位置误差,为此Keil等(2007)提出了变形场检验方法,其基本原理是通过由大到小各个尺度的平移变换,将预报场以整体最小的位移量变形至和实况场高度相似,而每个格点的位移量即代表了网格预报的位置误差。

Gilleland等(2009)将上述升尺度检验、尺度分离检验、邻域空间检验、基于目标的检验和变形场检验等方法统称为空间检验方法。上述方法侧重于空间上的对比,对天气过程在发生、持续和结束等时间特征上的检验非常薄弱,但对预报产品的用户而言,了解预报在时间上的误差幅度也是非常必要的。为此有必要开展针对精细化网格预报中天气过程的时间分布特征的检验研究。牛若芸等(2018)的研究提出了降水过程的识别方法,未来此类方法可以引入到精细化网格预报的时间特征检验中。表1归纳了目前网格预报中的常用检验方法。

表1 网格预报的常用检验方法一览表

类别	名称	方法概要
基于站点的常规检验方法	TS评分	在二分类预报中,用命中站次除以空报站次和实况发生站次之和
	BIAS	在二分类预报中,用预报发生站次除以实况发生站次
	准确率	记误差小于给定阈值的预报为准确预报,以准确预报站次除以总预报站次作为准确率
	均方根误差	在连续量预报中,用各站误差的平方求平均后取平方根

续表

类别	名称	方法概要
网格预报的空间检验方法	尺度分离检验方法	采用傅里叶转换或小波分解将预报实况场分解为不同尺度的场，用对关注尺度的网格预报信息进行检验
	邻域空间检验方法	对比一定大小的窗口内预报和实况要素值达到某一阈值的比例，以评估精细网格预报是否具有合理的概率意义
	基于目标的检验方法	将预报和实况场中的要素值达到一定阈值的区域识别成多个目标，并相互匹配，进一步对预报目标的位置、形态和强度的误差进行检验
	变形场检验方法	通过由大到小各个尺度的平移变换，将预报场以整体最小的位移量变形至和实况场高度相似，以每个格点的平移量作为格点预报的位置误差

2 未来发展面临的关键技术难点

人类社会正在步入人工智能、大数据、云计算和移动互联时代，这为网格预报技术新的发展以及气象与其他行业的深度融合、发展智慧气象带来机遇和挑战。当前，无缝隙精细化网格预报技术向纵深和智能化方向发展，面临诸多亟须破解的关键科学技术难点，也有亟待解决的系统设计升级和基础架构改造的迫切要求。

在科学技术和数据支撑方面：一是需要发展人工智能应用技术。近年来卷积神经元网络、深度学习和蒙特卡洛决策树等人工智能技术的发展，驱动以大数据为主的预报技术进步，也为智慧气象预报提供了新的技术思路。二是需要发展高级多模式统计后处理技术。网格预报技术需要对海量的观测信息、数值模式产品以及历史资料进行处理分析，需要统计学、大数据挖掘技术背景的专业人员加入，发展高级多模式统计后处理技术来支持偏差订正、海量有效信息的集成。三是需要解决协调一致性关键技术难题。多时空尺度的多源实况和预报数据的融合技术和网格化方法，以及不同气象要素、不同时间和空间尺度天气预报的一致性问题，例如时间上短时临近预报与短期预报的融合和一致性处理，空间上陆地与水体过渡带、不同粗糙度和热力属性过渡带网格气象预报的融合和一致性处理，都是中国无缝隙精细网格预报需要继续研究解决的技术挑战。四是需要长序列、高分辨、高质量的历史分析和模式资料支持。产品质量受到输入数据质量的限制，迫切需要实现长历史、统计特性一致的模式数据（如 Reforecasts）的定常化业务保障，以及整理和开发长序列、高质量、高分辨的观测和分析资料用于训练和检验。五是需要健全适应高频次、高分辨率的网格产品的检验技术体系。传统检验评估评分不能反映高时空分辨率网格产品精细化水平的提升，需要推动新技术（时间、空间分析）的研发、广泛理解和接受。

在系统设计和基础架构方面：一是发展统一、完整的技术架构和开发标准。采用现代编程语言、软件架构、版本控制、完整文档和协同开发无缝隙网格化的技术架构，发展统一、完整的技术架构和开发标准，形成开发合力，避免"筒仓现象"。二是重新设计和开发后处理业务系统。目前基于传统 MOS 预报技术主要面向站点建模，处理海量气象数据信息能力有限，需要扩展新的统计后处理技术、数据基础，重新设计和开发后处理业务系统。三是建立开放性、众创型的科研和业务结合平台。需要进一步推动研究成果到业务应用的转换，包括建立开放性、众创的后处理支持基础架构，如美国气象发展实验室（Meteorological Development Laboratory，MDL）正在开发的 WISPS（the Weather Information Statistical Post-processing System），建立跨行业、跨部门的团队来建设和维护后处理软件、测试数据和检验评估，提供资源用于训练相关人员的算法代码、软件构建水平。

与此同时，实现智慧气象服务的目标，提供更加精准精细的个性化专业气象服务产品，也给气象网格预报技术发展提出新的问题。预报空间方面，当前中国网格气象预报主要提供近地面天气预报，但是以航空为代表的空中作业气象保障,要求提供精细的三维空间气象观测和预报信息;以海上交通为代表的海上导航气象服务,要求将陆上天气预报向全球海洋延伸。预报服务方面,进一步将网格气象预报与不同行业的数据和服务需求相融合,提高基于网格气象预报面向农业、环境、水文、林业等多行业的智慧服务能力,也是中国无缝隙精细化网格预报发展有待加强的方向。

3　结论与展望

中国精细化无缝隙网格天气预报技术的发展任重而道远,需要在不断提高观测数据质量和数值模式预报水平的基础上,最大限度地借助和研发模式后处理与解释应用技术,尤其是补齐短临和延伸期客观预报技术短板,挖掘已有大数据信息,努力接近可预报性的极限边缘。

未来,较为成功的无缝隙精细化智能网格预报技术,将在高质量的实况客观分析和数值模式预报的基础上,附加最前沿的技术进展和科研成果,具体包括:业务产品能够反映最新的技术研究成果(分布式数据库、高级统计模型、人工智能 AI、数据挖掘、云计算等);预报员的理论分析和经验价值能够通过不同层级充分体现;高质量的训练数据集(气象观测与社会观测,模式预报与人工智能)用于高质量的加工处理流程;一流的气象、数学、信息及相关行业科学家一起研究该领域最重要的科学问题;具有先进、开放、标准化的技术基础架构,以便业务、研究、高校等部门协作,促进科研向业务转化(R2O)和预报技术研究领域的发展。

致谢:感谢国家气象中心王建捷、宗志平给予的技术指导和支持。

参考文献

毕宝贵,代刊,王毅,等,2016. 定量降水预报技术进展[J]. 应用气象学报,27(5):534-549.

曹勇,刘凑华,宗志平,等,2016. 国家级格点化定量降水预报系统[J]. 气象,42(12):1476-1482.

陈法敬,陈静,韦青,等,2019. 一种基于可预报性的暴雨预报评分新方法Ⅱ:暴雨检验评分模型及评估试验[J]. 气象学报,77(1):28-42.

陈力强,周小珊,杨森,2005. 短期集合预报中定量降水预报集合方法初探[J]. 南京气象学院学报,28(4):543-548.

代刊,曹勇,钱奇峰,等,2016. 中短期数字化天气预报技术现状及趋势[J]. 气象,42(12):1445-1455.

代刊,朱跃建,毕宝贵,2018. 集合模式定量降水预报的统计后处理技术研究综述[J]. 气象学报,76(4):493-510.

丁建军,罗兵,赵光平,等,2008. 精细化预报订正平台设计[J]. 气象,34(11):89-95.

丁一汇,2005. 高等天气学:第2版[M]. 北京:气象出版社.

高嵩,代刊,薛峰,2014. 基于 MICAPS 3.2 平台的格点编辑平台设计与开发[J]. 气象,40(9):1152-1158.

高玉春,2017. 气象业务发展对新一代天气雷达技术性能提升的要求[J]. 气象科技进展,7(3):16-21.

韩帅,师春香,姜志伟,等,2018. CMA 高分辨率陆面数据同化系统(HRCLDAS-V1.0)研发及进展[J]. 气象科技进展,8(1):102-108,116.

贺雅楠,高嵩,薛峰,等,2018. 基于 MICAPS4 的智能网格预报平台设计与实现[J]. 应用气象学报,29(1):13-24.

黄丽萍,陈德辉,邓莲堂,等,2017. GRAPES-MESO V4.0 主要技术改进和预报效果检验[J]. 应用气象学报,28(1):25-37.

林建,宗志平,蒋星,2013. 2010—2011 年多模式集成定量降水预报产品检验报告[J]. 天气预报,5(1):67-74.

刘凑华,牛若芸,2013. 基于目标的降水检验方法及应用[J]. 气象,39(6):681-690.

刘艳,薛纪善,张林,等,2016. GRAPES 全球三维变分同化系统的检验与诊断[J]. 应用气象学报,27(1):1-15.

宁方志,季民,陈许霞,2017. 基于 GIS 的精细化格点预报平台设计与实现——以青岛市为例[J]. 测绘与空间地理信息,40(5):33-35,40.

牛若芸,刘凑华,刘为一,等,2018. 1981—2015 年中国 95°E 以东区域性暴雨过程时、空分布特征[J]. 气象学报,76(2):182-195.

潘旸,谷军霞,徐宾,等,2018. 多源降水数据融合研究及应用进展[J]. 气象科技进展,8(1):143-152.

沈学顺,苏勇,胡江林,等,2017. GRAPES-GFS全球中期预报系统的研发和业务化[J]. 应用气象学报,28(1):1-10.

师春香,姜立鹏,朱智,等,2018. 基于CLDAS 2.0驱动数据的中国区域土壤湿度模拟与评估[J]. 江苏农业科学,46(4):231-236.

唐健,代刊,宗志平,等,2018. 主客观融合定量降水预报方法及平台实现[J]. 气象,44(8):1020-1032.

万子为,王建捷,黄丽萍,等,2015. GRAPES-MESO模式浅对流参数化的改进与试验[J]. 气象学报,3(6):1066-1079.

王海宾,杨引明,范旭亮,等,2016. 上海精细化格点预报业务进展与思考[J]. 气象科技进展,6(4):18-23.

吴捷,任宏利,许小峰,等,2018. MJO对中国降水影响的季节调制和动力-统计降尺度预测[J]. 气象,44(6):737-781.

吴启树,韩美,刘铭,等,2017. 基于评分最优化的模式降水预报订正算法对比[J]. 应用气象学报,28(3):306-317.

行鸿彦,张金玉,徐伟,2017. 地面自动气象观测的技术发展与展望[J]. 电子测量与仪器学报,31(10):1534-1542.

杨秋明,2018. 长江下游夏季低频温度和高温天气的延伸期预报研究[J]. 地球科学进展,33(4):385-395.

张芳华,曹勇,徐珺,等,2016. Logistic判别模型在强降水预报中的应用[J]. 气象,42(4):398-405.

张宏芳,李建科,陈小婷,等,2017. 基于百度地图的精细化格点预报显示[J]. 气象科技,45(2):261-268.

张进,麻素红,陈德辉,等,2017. GRAPES_TYM改进及其在2013年西北太平洋和南海台风预报的表现[J]. 热带气象学报,33(1):64-73.

赵滨,张博,2018. 邻域空间检验方法在降水评估中的应用[J]. 暴雨灾害,37(1):1-7.

赵翠光,李泽椿,2012. 华北夏季降水异常的客观分区及时间变化特征[J]. 应用气象学报,23(6):641-649.

赵声蓉,赵翠光,赵瑞霞,等,2012. 中国精细化客观气象要素预报进展[J]. 气象科技进展,2(5):12-21.

朱立娟,龚建东,黄丽萍,等,2017. GRAPES三维云初始场形成及在短临预报中的应用[J]. 应用气象学报,28(1):38-51.

BEN ALAYA M A,CHEBANA F,OUARDA T B M J,2015. Probabilistic multisite statistical downscaling for daily precipitation using a bernoulli-generalized pareto multivariate autoregressive model[J]. J Climate,28(6):2349-2364.

BOEING G,2016. Visual analysis of nonlinear dynamical systems:chaos,fractals,self-similarity and the limits of prediction[J]. System,4(4):37.

BOWLER N E H,PIERCE C E,SEED A,2004. Development of a precipitation nowcasting algorithm based upon optical flow techniques[J]. J Hydrol,288(1-2):74-91.

BROWNING K A,1980. Review lecture:local weather forecasting[J]. Proc Roy Soc A:Math Phys Sci,371(1745):179-211.

BRUNET G,JONES S,MILLS B,2015. Seamless prediction of the Earth system:from minutes to months[EB/OL]. http://library.wmo.int/pmb_ged/wmo_1156_en.pdf.

CARTER G M,DALLAVALLE J P,GLAHN H R,1989. Statistical forecasts based on the national meteorological Center's numerical weather prediction system[J]. Wea Forecasting,4(3):401-412.

CHEUNG P,YEUNG H Y,2012. Application of optical-flow technique to significant convection nowcast for terminal areas in Hong Kong[C]//The 3rd WMO International Symposium on Nowcasting and Very Short-Range Forecasting. Rio de Janeiro.

CUI B,TOTH Z,ZHU Y J,et al,2011. Bias correction for global ensemble forecast[J]. Wea Forecasting,27(2):396-410.

DALLAVALLE J P,GLAHN B,2005. Toward a gridded MOS system[C]//AMS 21st Conference on Weather Analysis and Forecasting. Washington,D. C:American Meteorological Society.

DANFORTH C M,2013. Chaos in an atmosphere hanging on a wall[EB/OL]. Mathematics of Planet Earth 2013. Retrieved 12 June 2018. http://mpe.dimacs.rutgers.edu/2013/03/17/chaos-in-an-atmosphere-hanging-on-a-wall/.

ENGEL C,EBERT E E,2012. Gridded operational consensus forecasts of 2-m temperature over Australia[J]. Wea Forecasting,27(2):301-322.

GILLELAND E,AHIJEVYCH D,BROWN B G,et al,2009. Intercomparison of spatial forecast verification methods[J]. Wea Forecasting,24(5):1416-1430.

GLAHN H R,LOWRY D A,1972. The use of model output statistics(MOS)in objective weather forecasting[J]. J Appl Meteor,11(8):1203-1211.

GLAHN H R,RUTH D P,2003. The new digital forecast database of the national weather service[J]. Bull Amer Meteor

Soc,84(2):195-201.

GUAN H,CUI B,ZHU Y J,2015. Improvement of statistical postprocessing using GEFS reforecast information[J]. Wea Forecasting,30(4):841-854.

HAIDEN T,KANN A,WITTMANN C,et al,2011. The integrated nowcasting through comprehensive analysis(INCA) system and its validation over the eastern alpine region[J]. Wea Forecasting,26(2):166-183.

HAMILL T M,SCHEUERER M,BATES G T,2015. Analog probabilistic precipitation forecasts using GEFS reforecasts and climatology-calibrated precipitation analyses[J]. Mon Wea Rev,143(8):3300-3309.

HOFFMAN R R,LADUE D S,MOGIL H M,et al,2017. Minding the weather:how expert forecasters think[M]. Cambridge,MA:MIT Press.

HÓLM E,FORBES R,LANG S,et al,2016. New model cycle brings higher resolution[J]. ECMWF Newsl,147:14-19.

KANN A,WANG Y,ATENCIA A,et al,2018. Seamless probabilistic analysis and forecasting:from minutes to days ahead[C]//EGU General Assembly 2018. Vienna:EGU.

KEIL C,CRAIG G C,2007. A displacement-based error measure applied in a regional ensemble forecasting system[J]. Mon Wea Rev,135(9):3248-3259.

KRISHNAMURTI T N,KISHTAWAL C M,LAROW T E,et al,1999. Improved weather and seasonal climate forecasts from multimodel superensemble[J]. Science,285(5433):1548-1550.

LIEN G Y,KALNAY E,MIYOSHI T,et al,2016. Statistical properties of global precipitation in the NCEP GFS model and TMPA observations for data assimilation[J]. Mon Wea Rev,144(2):663-679.

LORENZ E N,1963. Deterministic nonperiodic flow[J]. J Atmos Sci,20(2):130-141.

MARIOTTI A,RUTI P M,RIXEN M,2018. Progress in subseasonal to seasonal prediction through a joint weather and climate community effort[J]. npj Climate Atmos Sci,1:4.

MASS C F,2003. IFPS and the future of the national weather service[J]. Wea Forecasting,18(1):75-79.

RAUSER F,ALQADI M,AROWOLO S,et al,2017. Earth system science frontiers:an early career perspective[J]. Bull Amer Meteor Soc,98(6):1119-1127.

ROBERTS N M,LEAN H W,2008. Scale-selective verification of rainfall accumulations from high-resolution forecasts of convective events[J]. Mon Wea Rev,136(1):78-97.

RUTH D P,2002. Interactive forecast preparation-the future has come[C/OL]//Preprints,Interactive Symposium on the Advanced Weather Interactive Processing System. Orlando,Fl:American Meteorological Society:3.1. http://ams.confex.com/ams/pdfpapers/28371.pdf.

SLOUGHTER J M L,RAFTERY A E,GNEITING T,et al,2007. Probabilistic quantitative precipitation forecasting using Bayesian model averaging[J]. Mon Wea Rev,135(9):3209-3220.

STAUFFER R,UMLAUF N,MESSNER J W,et al,2017. Ensemble postprocessing of daily precipitation sums over complex terrain using censored high-resolution standardized anomalies[J]. Mon Wea Rev,145(3):955-969.

TANG B H,BASSILL N P,2018. Point downscaling of surface wind speed for forecast applications[J]. J Appl Meteor Climatol,57(3):659-674.

VISLOCKY R L,FRITSCH J M,1995. Improved model output and statistics through model consensus[J]. Bull Amer Meteor Soc,76(7):1157-1164.

ZHU Y J,LUO Y,2015. Precipitation calibration based on the frequency-matching method[J]. Wea Forecasting,30(5):1109-1124.

国家级格点化定量降水预报系统

曹勇　刘凑华　宗志平　谌芸　代刊　陈涛　杨寅

（国家气象中心，北京，10081）

摘　要：利用主客观融合降水反演、降水统计降尺度、降水时间拆分等技术构建了国家级格点化定量降水预报系统。该系统结构合理，模块功能明确，于2014年6月在国家气象中心投入业务使用，生成0～168 h时效，10 km分辨率，逐3 h的格点化定量降水预报产品。通过对2015年第13号热带气旋"苏迪罗"的格点化降水预报个例检验，结果显示，相比欧洲中期天气预报中心的确定性模式预报和预报员主观预报，该产品能更好地体现台风降水的时空精细化分布特点，对福建东北部和浙江东南部的特大暴雨中心位置表现更准确细致。通过对2015年4月-9月的格点化产品整体效果检验，结果显示，相比欧洲中期天气预报中心的确定性模式预报和由反距离客观分析后的预报员主观预报，该产品既能保持和预报员主观预报相同的准确率，同时也能较明显地提高降水预报的时空精细化程度。

关键词：定量降水预报；格点化；主客观融合；统计降尺度

State-Level Gridded Quantitative Precipitation Forecasting System

CAO Yong　LIU Couhua　ZONG Zhiping　CHEN Yun　DAI Kan　CHEN Tao　YANG Yin

(National Meteorological Centre, Beijing, 100081)

Abstract: A state-level gridded quantitative precipitation forecasting system has been built by using the fusion of subjective and objective precipitation inversion, precipitation statistical downscaling, time split method and other techniques. The technical structure of this system is reasonable, and its module features a clear division. It has been put into meteorological operation in the National Meteorological Center, China Meteorological Administration since June 2014, being able to generate gridded quantitative precipitation forecast products with forecast time length of 0-168 h, 10 km resolution and 3 h interval. The gridded precipitation forecast of the 13th tropical cyclone "Soudelor" in 2015 is tested and the results show that, compared to the deterministic model forecasts of the European Centre for Medium-Range Weather Forecast (ECMWF) and the subjective forecasts, this forecast can better reflect the spatio-temporal distribution of typhoon precipitation, and more accurately express the severe rainstorm center in northeast of Fujian and southeast of Zhejiang. The overall effect of the gridded forecasts from April to September 2015 is verified. The results indicate that, compared to the deterministic model forecasts of the ECMWF and the forecasters' subjective forecasts after inverse-distance objective analysis,

---

① 本文发表于《气象》2016年第12期。
资助项目：公益性行业（气象）科研专项（GYHY201306002）。
第一作者：曹勇，主要从事短期天气预报研究。E-mail: caoyong@cma.gov.cn。

the forecast products can maintain the same predictive accuracy as subjective forecasts, and meanwhile can obviously improve the degree of spatio-temporal precision of precipitation forecasts.

Key words:quantitative precipitation forecast; gridded;fusion of subjective and objective;statistical downscaling

引 言

当前中国发布的降水预报产品,在国家级和省级层面以落区等级预报为主,在市级和县级层面以站点预报为主,并配合文字性预报材料,对公众或政府部门提供预报决策服务(端义宏 等,2012)。由于这种形式的产品预报不够精细,内容相对刻板,已经难以适应当前高速发展的社会对降水预报的需求(宗志平 等,2012)。此外,以降水预报为前端的下游气象服务,如农业气象、水文气象、环境气象、交通气象等对降水预报的精细化程度需求也越来越高(曲晓波 等,2010;尹志聪 等,2015;尤凤春 等,2013),因此,当前亟须建立一套时空全覆盖、点线面预报相协调的格点化降水预报系统和产品,满足上述需求。美国国家海洋大气局(NOAA)下属的定量降水预报部门经过若干年的技术及平台的研究设计,已构建起较为成熟的格点化定量降水预报系统,能够提供 0～168 h 预报时效的格点化定量降水预报产品,最短时间分辨率在 6 h,最高空间分辨率在 5 km,并能提供如 grib2、shapefile、kml 等多种流行的气象存储格式供公众下载使用(Novak et al.,2014)。其格点化定量降水预报产品储存于国家数字预报服务器中,并可以网页互动的形式展示和访问。近些年,国内部分省份也开展对格点化定量降水预报相应的技术探索和研究,如,广东省建立以数据中心、网络释用、格点交互编辑、产品服务及检验反馈等部分构成的格点化预报业务系统,发布 5 km 分辨率、(部分产品)时间分辨率逐 1 h,预报时效 168 h 的格点化产品(吴乃庚,2015)。国家气象中心为适应当前社会对降水预报的需求发展,于 2012 年起开展国家级格点化定量降水预报系统的研究,通过近三年的技术摸索和储备,搭建起国家级格点化定量降水预报系统。

1 系统设计思路

由于数值模式不断地发展和成熟,其降水预报的准确率也在逐步提升(Bidlot et al.,2009),通过模式后处理技术,降水预报的准确率在模式基础上又进一步提高(赵声蓉,2012;Yuan,2007),但因大气固有的混沌效应、模式初值的误差及物理过程参数化等原因(Lorenz,1972),定量降水预报依旧需要预报员根据自身对大气物理过程的认识和经验,以及对模式误差特性的了解,进行主观订正。因此当前国内外降水预报业务的基本流程均是在模式降水预报基础上,预报员进行主观订正,增加预报员的订正信息,以提高降水预报的准确率。但是经过预报员主观订正的降水预报为落区等级预报,虽然提升了预报准确率(Novak et al.,2011;Brennan et al.,2008;Clark et al.,2004),但是精细化程度降低,降水分布的细节信息丢失。格点化定量降水预报系统的设计思路就是在不增加日常预报业务流程复杂程度的原则下,通过一系列统称为降水信息恢复技术的应用,在预报员主观落区等级预报的基础上,恢复降水分布的局部信息和特征,提高降水预报的精细化程度。

鉴于当前中国降水预报业务主要依托 Micaps 工作平台(李月安 等,2010),模式降水数据为 Micaps 第 4 类格式,而预报员降水落区等级预报为 Micaps 第 14 类数据,为此预期的目标主要是将定量降水格点化的过程融入现行的预报业务流程,即在预报员制作完成主观降水落区等级预报产品后,通过后期技术的处理,生成高时空分辨率的格点化降水预报产品。该产品既能体现预报员对模式降水预报的订正,完全遵从预报员的主观预报订正意图,又可以提高预报产品的精细化程度。这个过程完全由计算机自动化实现,无须预报员的干预。具体预报产品的指标如下:时间分辨率最短达到逐 3 h,空间分辨率达 10 km,

预报时效为0~168 h,每日随日常业务实现4次滚动更新。

2 系统结构及主要技术

2.1 系统结构

该系统主要实现降水信息恢复的功能,具体由三部分技术模块构成,如图1所示,分别为主客观融合降水反演技术模块、降水统计降尺度技术模块、降水时间拆分技术模块。技术模块之间都有相应的数据接口衔接,实现了技术的模块化分工,因此只要符合接口的数据标准,就可以实现单个技术模块的升级和完善,而不影响整个系统的架构和运行。

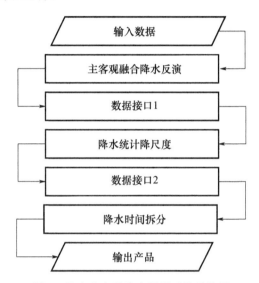

图1 格点化定量降水预报系统结构图

2.2 主客观融合降水反演技术

主客观融合降水反演技术主要实现与业务模式分辨率相当的粗网格格点化定量降水预报产品。图2为该技术流程图,主要由三步算法构成。

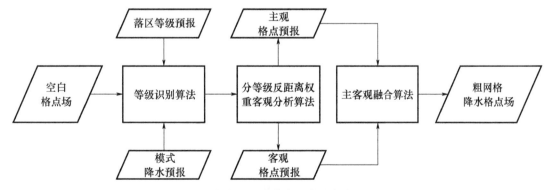

图2 主客观融合降水反演技术流程图

第一步为等级识别算法。利用预报员主观落区等级预报,识别空白格点场的每个格点降水所在等级并标记,根据中国现行的24 h降水等级划分标准,预报员主观预报等级分别为0.1、10、25、50、100以及250 mm,同时通过该步算法根据格点降水等级实现模式降水预报有效信息的筛选——若某格点上的模式降水符合预报员落区等级预报,则认为是有效降水信息,反之则认为是无效降水信息。如图3所示,A、B两格点被判别为 25～50 mm 等级,A 点模式降水为 32 mm,被筛选为有效信息,B 点降水为 18 mm,被筛选为无效信息。第二步为分等级反距离权重客观分析算法。该步算法分等级进行的目的是为了保证客观分析结果严格遵守预报员的主观落区预报,避免分析结果出现跨等级违背预报员预报意图的情形。使用反距离权重客观分析作为基本算法的原因主要是该算法设计相对简单,计算速度快,在有效信息较均匀分布时结果可靠,不足之处是算法主要以空间距离为权重系数,未考虑气象内在的空间分布特点,在有效信息分布不均匀时结果准确率会受影响。该步算法分别生成两类分析结果,第一类是将预报员的落区等值线离散化为网格密度的站点预报后,分等级反距离权重客观分析得到主观格点预报产品,如图3所示,使用 D、E、F 等点作为站点信息进行客观分析至网格点得到主观格点预报产品;第二类是将筛选的有效模式降水预报进行反距离权重客观分析得到基于模式有效信息的客观格点预报产品,如图3所示,使用 A 点等有效信息进行客观分析至网格点得到客观格点预报产品。第三步为主客观融合算法。该步算法假定,格点靠近预报员的落区等值线时应更多地参考主观格点预报,而远离预报员的落区等值线时应更多地参考客观格点预报,因此利用公式(1)通过权重系数动态加权生成最终的粗网格降水格点场。其中,$\mathrm{Var}_{i,j}^{s}$ 是主观格点场,$\mathrm{Var}_{i,j}^{o}$ 是客观格点场,$\mathrm{Var}_{i,j}^{f}$ 是最终格点场。而权重系数 ω 由公式(2)可知,是格点到最近等值线距离的 e 指数衰减函数,衰减半径 R 一般经验选取为业务模式的分辨率;$r_{i,j}$ 为格点到最近距离等值线的距离。

$$\mathrm{Var}_{i,j}^{f}=\omega\times\mathrm{Var}_{i,j}^{s}+(1-\omega)\times\mathrm{Var}_{i,j}^{o} \tag{1}$$

$$\omega=\mathrm{e}^{\dfrac{-r_{i,j}}{R}} \tag{2}$$

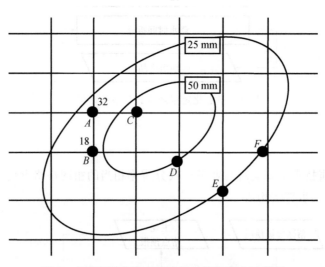

图3　主客观融合降水反演技术算法概念示意图

2.3　降水统计降尺度技术

降水统计降尺度技术主要是为了实现粗网格降水场到细网格降水场的转化。传统的思路与方法通过双线性数学插值实现,但这样的处理方式忽略了粗分辨率和细分辨率降水之间的气象关系,结果的准确性有限(Tobin et al.,2011)。当前主要使用降尺度技术提高降水的空间分辨率。由于动力降尺度需要成熟可靠的中尺度模式的支持,耗费大量的计算机资源和计算时间(Pavlik et al.,2012),实际预报业务中一般利用统计降尺度技术来提高降水的空间分辨率,如美国海洋和大气局下属的水文预报中心,使用地形地图投影技术进行降水降尺度(Schaake et al.,2004)。本系统使用降尺度比例矢量技术。该技术利用粗分辨率与细分辨率的定量降水估测产品,建立大尺度降水到小尺度降水的气象统计关系并确定降尺度比例矢量。由于比例矢量关系由实况降水所得,能体现出局地由于地形或者气候特点导致的降水的精细尺度的变化,体现不同空间尺度降水的气象关系。该技术建立降尺度比例矢量使用的数据为国家气象信息中心的定量降水估测产品。该产品利用地面观测和卫星反演降水各自的优势,在降水量值和空间分布上均更为合理,同时融合产品的平均偏差和均方根误差也均减小(沈艳 等,2013)。该数据的时间分辨率为1 h,空间分辨率为10 km,所用时间段为2008—2014年,共7年。图4为降水降尺度技术的流程图,主要通过三步算法实现功能。

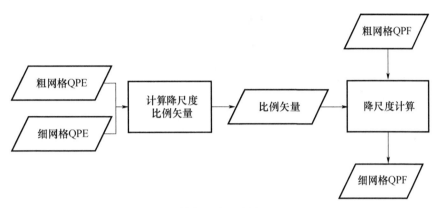

图4 比例矢量降尺度技术流程图

第一步使用定量降水估测产品,分别以每日为中心前后15 d的7年数据生成滑动平均的全国逐日的10 km分辨率细网格和30 km分辨率粗网格的降水气候分布。粗细网格场之间的关系如图5所示,任意一粗网格格点 $X_{i,j}$,对应9个细网格格点($A_{i,j}^1, A_{i,j}^2, A_{i,j}^3, \cdots, A_{i,j}^7, A_{i,j}^8, A_{i,j}^9$)。

第二步计算每个粗网格格点的降尺度比例矢量 $\boldsymbol{\beta}$,计算方法如公式(3)所示,主要是计算每个细网格格点与对应的粗网格格点之间的比例系数,构建降尺度矢量。

$$\boldsymbol{\beta}_{i,j} = (\beta_{i,j}^1, \beta_{i,j}^2, \beta_{i,j}^3, \cdots, \beta_{i,j}^7, \beta_{i,j}^8, \beta_{i,j}^9) \tag{3}$$
$$\beta_{i,j}^m = A_{i,j}^m / X_{i,j}, \quad m=1,2,3,\cdots,7,8,9$$

第三步使用待降尺度的粗网格降水,在每个粗网格格点乘以降尺度比例矢量,即获取降尺度后的细网格降水预报结果。

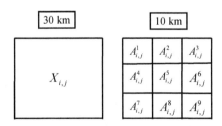

图5 粗网格和细网格降水空间关系示意图

2.4 降水时间拆分技术

降水时间拆分技术主要是提高降水预报的时间分辨率,能将预报员的订正信息合理地分配于各个精细化预报时段内,在提高降水预报时间分辨率的同时也提高精细化时段内的降水预报准确率。该技术假定,业务预报中使用的数值模式的降水资料为模式起报12 h后的预报资料,模式启动后的扰动影响已经明显减弱,因此能较好地反映出降水的时间变化过程,但是降水量的预报准确度不高。因此需要将预报员对降水预报的订正信息合理地分配到模式原始的降水预报中,已达到订正高时间分辨率模式降水的目的。该技术同样经过三个步骤实现功能,第一步利用模式降水信息计算每个格点逐3 h降水累计分布曲线;第二步利用模式逐 3 h 降水和总降水量获取逐 3 h 降水比例关系系数;第三利用上述逐 3 h 降水比例关系系数,将 24 h 格点化定量降水预报拆分成逐 3 h 格点化定量降水预报,以提高降水预报的时间分辨率和在该时段内的准确率。若某网格点的格点化降水预报不为零,而参考模式降水预报为零,则搜寻周边最近网格点的比例关系系数,作为该点的比例系数关系。

表 1 为降水时间拆分技术原理示意表,假定某格点模式 24 h 降水预报为 50 mm,而最终格点化 24 h 降水预报为 80 mm,经拆分技术处理后,逐 3 h 降水如上表所示。

表 1 降水时间拆分技术示意表

	3 h	6 h	9 h	12 h	15 h	18 h	21 h	24 h
模式分段降水/mm	5	10	20	0	0	5	10	0
模式累计降水/mm	5	15	35	35	35	40	50	50
拆分比例系数	0.1	0.2	0.4	0.0	0.0	0.1	0.2	0.0
格点化拆分降水/mm	8	16	32	0	0	8	16	0

3 应用及效果

2015 年第 13 号热带气旋"苏迪罗"登陆中国,对东部多个省(区、市)造成不同程度的影响,其中浙闽苏台等地受灾严重(胡余忠,2015)。本文选取包含登陆时间点在内的 8 月 08 日 08 时至 09 日 08 时研究。由于中国降水预报业务仅针对陆地区域,所以海洋降水未做讨论。格点化定量降水预报和预报员主观预报均是 07 日 08 时起报的 24 h 时效的预报。由于实际预报业务中模式资料获取的滞后性,一般都用业务预报时次的前 12 h 起报的模式资料作为参考预报,所以欧洲中期天气预报中心的确定性模式预报选取 07 日 20 时起报的 36 h 时效的预报,本文将上述三类预报结果进行对比分析。国家气象信息中心的定量降水估测产品(图 6a)显示,在福建东北部及浙江东南部普遍出现了 100 mm 以上的大暴雨,其中 A、B 两处为降水量级在 250 mm 以上的特大暴雨中心区域。经后期实况检验,国家气象中心对此时段台风的降水做出了较准确的预报,24 h 时效的大暴雨及以上量级预报的 TS 评分达到 0.47,但预报产品主要以落区等级预报为主(图 6b),在精细化预报上表现不足,尤其对于 A、B 两处特大暴雨中心区域没有做出更细致的预报。利用反距离空间插值分析至 10 km 分辨率,预报产品的均方根误差为 53 mm。欧洲中期天气预报中心的确定性模式预报结果偏弱(图 6c),且将降水中心区域考虑在福建沿海的中部区域,福建东北部及浙江东南部的降水量值预报偏弱。此外,该模式分辨率为 16 km 左右,也相对较粗。同样利用反距离空间插值分析至 10 km 分辨率,预报产品的均方根误差为 46 mm。通过国家级格点化定量降水预报系统处理出的格点化产品,相比于欧洲中期天气预报中心的确定性模式和预报员主观落区预报,能较明显地表现出浙江南部和福建东北部 A、B 两处的特大暴雨中心区域,体现出空间变化的局地性特点(图 6d)。通过计算,该预报产品的均方根误差为 41 mm,精细化程度较主观落区预报和确定性预报明显提高。由于格点化定量降水预报的核心思路是在预报员主观落区基础上恢复降水的空间细节分布,因此当预报员主观预报出现较明显空报时,该产品也不可避免地会出现空报现象,如在福建中部,格点化定量降水预报和预报员主观落区预报均较定量降水估测产品偏强。

用均方根误差及 TS 评分检验一段时间内的格点化定量降水预报产品的总体效果,检验时段为 2015 年 4 月 1 日—9 月 30 日(图 7)。为了体现检验结果的客观性,欧洲中期天气预报中心的确定性模式预报和预报员主观预报均利用反距离空间插值分析至 10 km 分辨率,权重系数为距离倒数。利用国家信息气象中心的定量降水估测产品对上述 3 种预报进行检验(国家信气象息中心的定量降水估测产品质量已在上文论述),结果发现:对于暴雨及以上量级 TS 评分,格点化定量降水预报和预报员主观预报一致为

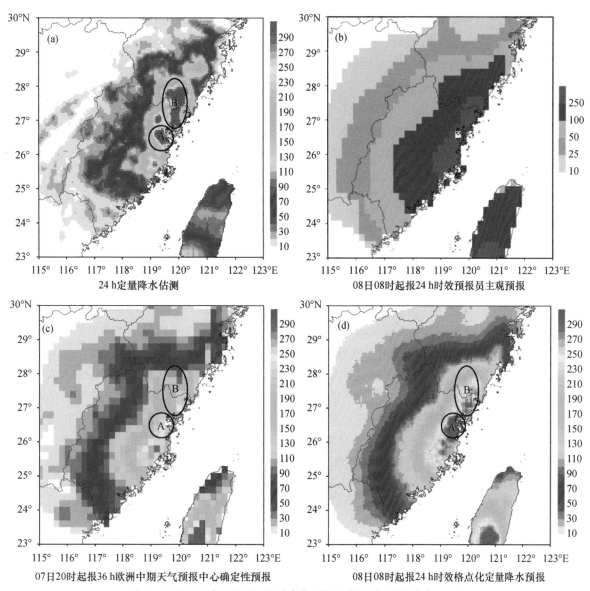

图6 08日08时至09日08时降水预报和实况对比图（单位：mm）

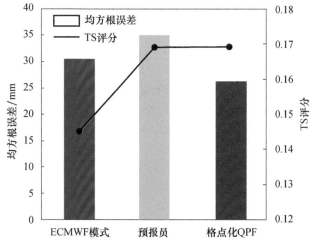

图7 2015年4—9月预报员、欧洲中期天气预报中心（ECMWF）确定性模式及格点化QPF的均方根误差及TS评分对比图

0.17,而欧洲数值预报中心确定性预报暴雨评分为0.14,预报员预报结果的均方根误差为35 mm,比欧洲中期天气预报中心的确定性模式的均方根误差30 mm高,但格点化定量降水预报能均方根误差能由预报员的35 mm下降至26 mm。对其他量级的降水检验(图略)发现,格点化定量降水预报和预报员主观预报TS评分较欧洲中期天气预报中心确定性模式都有一定提高,而中雨量级TS评分较欧洲中期天气预报中心确定性模式略偏低,因此总体而言,格点化定量降水预报产品的准确性较欧洲中期天气预报中心确定性模式提高。同时,通过均方根误差检验结果发现,格点化定量降水预报较预报员主观预报和欧洲中期天气预报中心确定性模式误差结果更小,因此在精准性上有提高。

4 结论和展望

本文详细介绍了国家气象中心格点化定量降水预报系统的构成以及实际应用效果。精细化的格点定量降水预报是现代天气预报服务需求中的一个重要问题,该预报系统较好地把模式定量化的优点和预报员主观预报在量级上的优势结合在一起,从主客观融合的角度来解决精细化格点定量降水预报问题。该系统于2014年6月在国家气象中心正式投入业务应用。通过该技术的应用,国家气象中心发布的产品由等级落区指导产品转变为高时空精度的格点化产品,可为省市级气象部门制作本地的精细化预报提供较好的指导产品,此外也为国家气象中心自身的精细化预报服务提供了有力的支撑。如2015年"9·3阅兵"预报保障服务、台风精细化降水预报服务等都以该技术和产品为基础进行制作,同时也为以降水预报为前端的下游气象服务,如农业气象、水文气象等对降水预报精度有较高需求的部门提供了较好的数据和技术支撑。该系统后期将进一步整合和完善定量降水预报格点化技术,使其在提高降水预报中的作用更合理和突出,另外还会进一步完善降水统计降尺度、主客观融合降水反演等技术,通过这些技术的升级来进一步提高格点化产品的时空分辨率和质量。

参考文献

端义宏,金荣花,2012. 中国现代天气业务现状及发展趋势[J]. 气象科技进展,2(5):6-11.
胡余忠,王祥,王立全,等,2015. 2015年13号台风"苏迪罗"洪水调查与评价[J]. 中国水利(19):19-22,9.
李月安,曹莉,高嵩,等,2010. MICAPS预报业务平台现状与发展[J]. 气象(7):50-55.
曲晓波,张涛,刘鑫华,等,2010. 舟曲"8·8"特大山洪泥石流灾害气象成因分析[J]. 气象,36(10):102-105.
沈艳,潘旸,宇婧婧,等,2013. 中国区域小时降水量融合产品的质量评估[J]. 大气科学学报,36(1):37-46.
吴乃庚,2015. 广东省精细化天气预报技术及业务工作进展情况[R]//2015年第十二届全国气象台长会议报告.
尹志聪,郭文利,李乃杰,等,2015. 北京城市内涝积水的数值模拟[J]. 气象,41(9):1111-1118.
尤凤春,郭丽霞,史印山,等,2013. 北京强降雨与道路积水统计分析及应用[J]. 气象,39(8):1050-1056.
赵声蓉,赵翠光,赵瑞霞,等,2012. 中国精细化客观气象要素预报进展[J]. 气象科技进展,2(5):12-21.
宗志平,代刊,蒋星,2012. 定量降水预报技术研究进展[J]. 气象科技进展,2(5):29-35.
BIDLOT J,FERRANTI L,GHELLI A,et al,2009. Verification statistics and evaluations of ECMWF forecasts in 2008-2009 [R]. ECMWF.
BRENNAN M J,CLARK J L,KLEIN M,2008. Verification of quantitative precipitation forecast guidance from NWP models and the Hydrometeorological Prediction Center for 2005-2007 tropical cyclones with continental US rainfall impacts [R]. 28th Conf on Hurricanes and Tropical Meteorology.
CLARK M,GANGOPADHYAY S,HAY L,et al,2004. The Schaake shuffle:a method for reconstructing space-time variability in forecasted precipitation and temperature fields[J]. Journal of Hydrometeorology,5(1):243-262.
LORENZ E. 1972. Predictability:does the flap of a butterfly's wing in Brazil set off a tornado in Texas[R]?
NOVAK D R,BAILEY C,BRILL K,et al,2011. Human improvement to numerical weather prediction at the Hydrometeo-

rological Prediction Center[C]//91th Annual AMS Meeting, AMS 24th Conference on Weather and Forecasting, 20th Conference on Numerical Weather Prediction.

NOVAK D R, CHRISTOPHER BAILEY, KEITH F. BRILL, et al, 2014. Precipitation and temperature forecast performance at the weather prediction center[J]. Weather Forecasting, 29: 489-504.

PAVLIK D, SÖHL D, PLUNTKE T, et al, 2012. Dynamic downscaling of global climate projections for Eastern Europe with a horizontal resolution of 7 km[J]. Environmental Earth Sciences, 65(5): 1475-1482.

SCHAAKE J, HENKEL A, CONG S, 2004. Application of PRISM climatologies for hydrologic modeling and forecasting in the western US Preprints[C]//18th Conf on Hydrology, Seattle, WA Amer Meteor Soc, CD-ROM, 5.

TOBIN C, NICOTINA L, PARLANGE M B, et al, 2011. Improved interpolation of meteorological forcings for hydrologic applications in a Swiss Alpine region[J]. Journal of Hydrology, 401(1): 77-89.

YUAN H, 2007. Analysis of precipitation forecasts from the NCEP global forecast system[C]//22nd Conference on Weather Analysis and Forecasting, 18th Conference on Numerical Weather Prediction.

OTS、MOS 和 OMOS 方法及其优化组合应用于 72 h 内逐 3 h 降水预报的试验分析研究

赵瑞霞[1] 代 刊[1] 金荣花[1] 韦 青[1] 张 宏[2] 郭云谦[1] 林 建[1] 王 玉[1] 唐 健[1]

(1. 国家气象中心,北京,100081；2. 中国气象局气象探测中心,北京,100081)

摘 要：开展了夏半年 72 h 内逐 3 h 降水预报试验,针对 ECMWF 模式预报、基于 ECMWF 的模式输出统计(MOS)预报、纳入超前空间实况信息的 OMOS 预报,以及三种预报的最优 TS 评分订正(OTS)预报,对比分析预报效果,探讨一种多方法结合能够提供良好预报性能的 3 h 定量降水预报技术方案。结果表明,在短期预报中,MOS 预报与 OTS 订正相结合的 MOS-OTS 综合预报方法的预报性能最好,而且 MOS-OTS 方法的 3 h 强降水预报与业务运行的城镇指导预报中融合主客观预报的降水预报相比,也具有一定优势；而在临近 3 h 预报中,OMOS 预报与 OTS 订正相结合的 OMOS-OTS 综合预报方法最优,3 h 内 0.1、3 和 10 mm 以上降水的 TS 评分最高比原始模式预报分别提高 73%、198% 和 483%,BIAS 评分接近于 1,在夏半年的逐日晴雨预报中,OMOS-OTS 方法在大部分日期都稳定优于 MOS-OTS 预报和城镇指导预报。

关键词：MOS 预报；OMOS 预报；OTS 订正预报；综合预报方法；精细降水预报

Comparison of OTS, MOS, OMOS Methods and Their Combinations Applied in 3 h Precipitation Forecasting out to 72 h

ZHAO Ruixia[1] DAI Kan[1] JIN Ronghua[1] WEI Qing[1] ZHANG Hong[2]
GUO Yunqian[1] LIN Jian[1] WANG Yu[1] TANG Jian[1]

(1. National Meteorological Center, Beijing, 100081;
2. CMA Meteorological Observation Center, Beijing, 100081)

Abstract: The performance of three statistical post-processing methods and their combinations for 3 h precipitation forecasts out to 72 h from May to September are compared in this paper. They are optimal threat score (OTS) correction, model output statistics (MOS) and MOS with prior-spatial observation predictors (OMOS). The 3 h precipitation forecasts of ECMWF model output (DMO), MOS, OMOS, and their OTS correction forecasts (DMO-OTS, MOS-OTS, OMOS-OTS) are evaluated. The results show that MOS-OTS method has the best performance in the short term forecast. At the same time, for the heavy precipitation forecast, MOS-OTS also obviously outperforms the operational guidance (GD) forecast which integrates subjective and objective predictions. In the first 3 h precipitation forecast, OMOS-OTS

---

① 本文发表于《气象》2020 年第 3 期。
基金资助：中国气象局预报员专项(CMAYBY2019-148)；国家科技支撑计划课题(2015BAC03B04)；国家自然科学基金项目(41575066)。
第一作者：赵瑞霞,主要从事模式统计后处理客观预报研究。E-mail：zhaorx@cma.gov.cn。

is the best method. For the first 3 h precipitation forecast, the TS scores of OMOS-OTS for thresholds of 0.1, 3 and 10 mm per 3 h are improved about 73%, 198% and 483% than DMO respectively. And the BIAS score of OMOS-OTS is close to 1. In the daily variation during summer time, the first 3 h precipitation forecast from OMOS-OTS outperforms both MOS-OTS and GD forecast in most days evaluated by TS and BIAS scores for the threshold of 0.1 mm per 3 h.

Key words: MOS forecast; OMOS forecast; OTS correction; integrated forecast method; fine precipitation forecast

引 言

精细化定量降水预报,是公众气象服务和防灾减灾决策服务中关注的重点内容。精准到逐3 h甚至逐1 h的降水量预报,是对天气可预报性和主客观预报能力的挑战。得益于不断丰富的观测资料和稳步发展的大气数值预报模式,天气预报水平不断提升(代刊 等,2016;金荣花 等,2019),但即使是国际上十分先进的ECMWF模式,其用户指南中也指出预报性能的种种局限性(https://confluence.ecmwf.int/display/FUG/9+Physical+Considerations+when+Interpreting+Model+Output;Buizza et al.,2018)。因此,结合多源实况信息对数值模式进行统计后处理以提高预报技巧,是十分重要且有效的。

自20世纪70年代开始,发展了数值模式输出统计(MOS)客观预报方法(Glahn et al.,1972;Carter et al.,1989;Vislocky et al.,1995)。多年来,世界各国客观预报业务中广泛使用基于确定性模式的MOS预报或者集合MOS预报方法(刘还珠 等,2004;Wilks et al.,2007;Charba et al.,2011;赵声蓉 等,2012;白永清 等,2013;Stauffer et al.,2017;代刊 等,2018),显著提高了临近和短期预报效果。在当前网格预报成为各国业务主流产品形式的大背景下(Glahn et al.,2003;Glahn et al.,2009;Engel et al.,2012),MOS预报、集合MOS预报方法仍然是客观指导预报业务中的有效手段(Ruth et al.,2009;金荣花 等,2019)。

传统意义上的MOS预报中,预报因子主要来源于数值模式预报产品。但是,鉴于零场实况的重要性,国际上有些研究,尤其是短时临近预报,将实况信息作为预报因子引入MOS预报(Schmeits et al.,2008;Chen et al.,2017;Trepte et al.,2018),主要纳入当前站点或者附近几个站点的实况信息。本文研发了一种提取时间上超前、空间上相关的实况信息作为预报因子的方法,利用空间上的距离实现时间上的滞后预报信息,延长对预报效果改进的时间。当然,滞后预报时效的长短与影响预报要素的天气系统的生命周期长短有关。由于纳入了起报时次之前的实况信息,本方法可以成为高频滚动预报的有效技术途径。

在降水客观预报业务中,目标评分阈值订正方法也十分常用(李俊 等,2014;毕宝贵 等,2016;唐文苑 等,2019)。这类方法消耗计算资源少,易于业务稳定运行。NCEP业务中使用的频率匹配拟合方法(FM)(Zhu et al.,2015),以降水预报偏差达到最优为算法核心,明显改善了NCEP全球预报模式和集合预报模式降水预报的系统性偏差,降水模态也更加真实。吴启树等(2017)分别设计了最优TS(OTS)和最优ETS评分订正法,并与FM进行对比,针对ECMWF等多个模式的24 h累积降水的订正预报试验结果表明,不论TS、ETS评分,还是概率空间的稳定公平误差评分,OTS方法在所有时效均能提高模式降水预报质量,是三者中最优的。

本文针对5—9月,开展72 h内逐3 h降水预报试验,对比ECMWF模式插值的原始预报DMO、基于ECMWF模式的MOS预报、进一步纳入超前空间实况信息后的OMOS预报,以及三种预报进行OTS订正的预报效果,分析几种方法的优劣,探讨一种将多类方法结合、能够提供良好预报性能的3 h定量降水预报方案。

1 资料

本文使用 2014—2018 年全国 2050 个自动站的逐 1 h 降水观测资料,合并生成逐 3 h 实况降水作为降水预报建模的预报真值。使用全国均匀分布的 930 个自动观测站的逐 1 h 降水观测,提取超前空间实况预报因子。另外,使用 2015—2018 年 ECMWF 模式的细网格预报产品,高空变量空间分辨率为 $0.25°×0.25°$,地面变量为 $0.125°×0.125°$,时间分辨率 72 h 内为 3 h 间隔,240 h 内为 6 h 间隔。需要说明的是,考虑存储空间限制等因素,地面变量在最初入库时,采取了跳格点方式存储,与高空要素一致采用 $0.25°×0.25°$ 的空间分辨率。

2 方法介绍

2.1 最优 TS 评分订正(OTS)预报方法

OTS 预报方法是一种以 TS 评分最优化为目标的偏差订正方案(吴启树 等,2017)。订正匹配公式如下:

$$y = \begin{cases} 0 & x < F_1 \\ O_k + (O_{k+1} - O_k) \times \dfrac{(x - F_k)}{(F_{k+1} - F_k)} & F_k < x < F_{k+1} \\ x \times \dfrac{O_M}{F_M} & x \geq F_M \end{cases} \tag{1}$$

式中:x 和 y 分别为模式降水的预报值和订正值;O_k 为第 k 量级降水阈值($k=1,2,\cdots,M$),共分为 M 级降水量级;F_k 为预报降水量订正到 O_k 时,该量级对应的模式降水量阈值。当预报值小于最小量级降水的阈值 F_1 时,订正预报 y 为 0,称作消空订正。训练过程中调整 F_k,使第 k 量级以上累积降水的 TS 评分达到最高时求得。

与吴启树等(2017)研究中最优 TS 阈值训练期的选取方法不同,本研究针对每个站点做了单独的最优 TS 阈值训练。考虑到训练样本量问题,选取 2015—2017 年的 5—9 月作为训练样本期,统计出最大 TS 评分对应的量级阈值,统一应用于该站点 5—9 月的降水订正。

2.2 模式输出统计 MOS 预报方法

MOS 预报的建模采用双重 F 检验因子筛选的最优逐步回归方法,生成如式(2)所示的多元线性回归预报方程。

$$y = b_0 + b_1 x_1 + b_2 x_2 + \cdots + b_n x_n \tag{2}$$

式中:y 为 MOS 预报值;$x_1,x_2,\cdots,x_n$ 为模式预报因子或者其他预报因子;b_0 为回归常量;$b_1,b_2,\cdots,b_n$ 为回归系数。

为提高建模和预报效率,对预报因子进行初选。选取 ECMWF 模式的低云量、总云量、3 h 累积降水量、总水汽含量、850、700 和 500 hPa 等压面层的相对湿度、垂直速度、散度、涡度、纬向风、经向风,以及三个等压面层的衍生物理变量温度平流和风垂直切变,作为备选预报因子。

需要指出的是,本研究中并没有如 Glahn 等(1972)研究中以及后来大部分 MOS 降水预报中一样(赵声蓉 等,2009),将各等级的降水预报处理为发生与否的 0、1 事件,而是直接对降水预报值进行建模。

2.3 纳入超前空间实况信息的 OMOS 预报方法

本文设计研发的 OMOS 预报方法与 MOS 预报方法的关键区别在于,增加了时间上超前、空间上相关的"超前空间实况因子"(以下简称 OPTOR 因子),主要用于提取和表征前期上游天气系统的移动影响效应。具体是指,针对某预报要素,检索起报时间之前的一段时间内与预报要素具有显著相关性的某些区域的实况信息,通过客观方法提取生成该信息的表征量,作为 OPTOR 因子。

OPTOR 因子的提取,首先需要根据历史实况数据进行超前空间相关信息统计。本研究利用 2014—2016 年的实况资料样本统计相关信息,以起报时间 08 时为准,针对每个预报站点,求取每个预报时效对应时间如 11 时、14 时、17 时等的 3 h 实况降水量,与全国 930 个实况因子备选站群(图略)"超前"1~12 h 的逐 1 h 实况降水的相关系数,对应于前一日 20 时至当日早上 07 时。实况因子站群的选取原则为到报率高且尽量均匀分布。其中,第 m 个预报站点、第 n 个预报时效的预报要素 $Y(m,n)$ 与第 i 个超前时刻、第 j 个因子站点的实况要素 $X(i,j)$ 的相关系数的求取如下:

$$r(Y(m,n),X(i,j)) = \frac{\mathrm{Cov}(X(i,j),Y(m,n))}{\sqrt{\mathrm{Var}[X(i,j)]\mathrm{Var}[Y(m,n)]}} \tag{3}$$

式中:$\mathrm{Cov}[X(i,j),Y(m,n)]$ 为 $X(i,j)$ 与 $Y(m,n)$ 的协方差;$\mathrm{Var}[X(i,j)]$ 为 $X(i,j)$ 的方差;$\mathrm{Var}[Y(m,n)]$ 为 $Y(m,n)$ 的方差。只保留通过显著性水平检验的相关信息。

然后,根据相关信息进行区域组合,检索同一超前时间点属于相同天气系统的多个超过显著性水平检验的相关站点群。针对每个超前时刻,分象限挑选最大相关系数站点,给定半径搜索邻近相关站,反复循环搜索选入站点群中所有站点的邻近站点群,最终将它们整合为空间上相近具备区域代表性的区域组合。多次循环上面过程直到将所有站点实现相近站点群的区域组合。

最后,基于区域组合信息设计特征因子方案,计算 OPTOR 因子。针对每个区域提取特征量,本文选取相关系数权重降水 V_r、最大相关系数站点降水和区域最大降水等 3 个指标,提取计算了 2015—2017 年的 OPTOR 因子纳入 MOS 建模过程。V_r 的计算如下:

$$V_r = \sum_{j=1}^{\mathrm{ns}} r_j V_j \tag{4}$$

式中:ns 为某组合区域内的站点总数;r_j 为区域内第 j 个站点观测值与预报要素 3 h 降水量的历史统计相关系数,V_j 为该站点的 1 h 降水量观测值。

2.4 试验方案

从国家级城镇指导预报站点中,挑选出空间上均匀分布的 2050 个站点(站点分布图略),针对"夏季类"样本期 5—9 月(车钦 等,2011),开展 08 时起报的 72 h 内逐 3 h 降水预报试验。考虑模式预报产品的实际生成时间滞后于起报时间,ECMWF 模式资料使用的是前一天 20 时起报的预报产品。以 2015—2017 年作为 OTS、MOS 及 OMOS 方法的阈值统计训练或建模期,2018 年作为独立预报检验样本期,对比开展 6 种方案的预报试验,分别为:ECMWF 模式双线性插值得到的 DMO 预报,基于 ECMWF 模式因子的 MOS 预报,在 ECMWF 模式因子基础上进一步纳入 OPTOR 因子的 OMOS 预报,以下分别简称为 ED、EM、EO 预报;以及分别基于以上三种预报的 OTS 订正预报,简称 ED_OTS、EM_OTS 和 EO_OTS 预报。文章采用 TS 评分、预报偏差(BIAS)、空报率和漏报率等参数,将 3 h 降水划分为 0.1、3 和 10 mm 以上累积降水量 3 个等级进行检验。

3 短期3h精细降水预报对比结果分析

3.1 逐3h有无降水预报效果对比

由图1a可以看到,对逐3h有无降水预报,ED和ED_OTS预报的TS评分比较接近,很多时效ED_OTS评分甚至略低于ED;但是ED_OTS预报明显减小了ED预报过分偏大的BIAS评分(图1b)。说明直接进行OTS订正对TS评分改进并不明显,但对BIAS评分的改进比较明显,只是仍然偏大,绝大部分时效大于2。这与吴启树等(2017)24 h降水OTS预报结果有所不同,可能和累积降水量时段不同有关。

EM和EO预报的TS评分大部分时效接近或者低于ED和ED-OTS预报,四种预报有无降水的TS评分都在0.17~0.21波动;同时EM和EO预报的BIAS评分整体而言比ED预报偏大更加严重。可见,仅仅进行MOS或者OMOS预报订正,对有无降水的预报效果较差。

进一步对EM和EO结果进行OTS订正,TS评分明显提高,最大比ED提高0.125左右,预报时效越短提高量越多,改进效果直至72 h都十分明显;BIAS评分的偏大程度也大幅减小,接近于1。EM_OTS与EO_OTS的预报效果比较接近,绝大部分时效EM_OTS预报略好,但临近预报时效EO_OTS的预报效果更好。

可见,对ECMWF有无3h累积降水预报,单独使用OTS或者MOS统计后处理方法,订正效果都不太理想,但是如果将两种方法结合,开展MOS-OTS及OMOS-OTS组合订正方法预报,则能够显著提高预报技巧。说明通过历史时期与实况的拟合建模,MOS和OMOS方法使得降水预报分布更加合理,但是由于建模中直接对原值进行拟合,会出现例如负值等不符合降水量特性的预报数据,利用OTS方法将EM和EO预报重新与实况进行量级拟合后,有效规范了降水阈值,明显提高TS评分,并且克服了小雨偏多特征。除临近预报时效外,EM_OTS的预报效果略好于EO_OTS预报。

3.2 逐3h累积3 mm以上降水预报效果对比

图1c和1d表明,与有无降水预报不同,逐3h累积3 mm以上降水预报的检验评分中,ED_OTS在绝大部分时效明显改进了TS评分,最大提高幅度约0.028,但是BIAS评分由原来小于2大幅增加为3.3~4.1;EM和EO的TS评分在所有时效都高于ED和ED_OTS,预报时效越短提高幅度越大,例如临近3 h的TS评分由ED的0.063分提高至EM的0.172和EO的0.188,而且BIAS评分也接近于1,整体预报性能显著优于ED和ED_OTS;EM和EO相比,除临近3 h外,其他时效EM的TS评分优于EO,BIAS评分两者几乎一样;在EM和EO基础上进一步进行OTS订正,TS评分提高不明显,但是BIAS评分更接近于1,改善了3 mm以上降水预报略多的问题,优化了无偏性能。

可以看到,在3 mm以上降水预报中,OTS订正方法在略微提高TS评分的同时会大幅增加BIAS评分,说明OTS方法主要通过调小阈值增加3 mm以上降水量来提高TS评分;MOS和OMOS方法十分有效,大幅提高TS评分的同时,还使得BIAS评分接近于1,可以有效优化ECMWF中3 mm以上降水预报分布;在MOS和OMOS预报基础上再进行OTS订正,可以进一步优化预报的无偏性能;EO和EO_OTS显著改进了临近降水预报性能,其他时效EM和EM_OTS预报效果更好。

3.3 逐3h累积10 mm以上降水预报效果对比

逐3h累积10 mm以上降水预报检验结果表明(图1e、1f),OTS方法在所有时效都显著提高了TS

评分,但同时预报偏差大幅增加。ED 的 BIAS 小于 1,ED_OTS 的 BIAS 评分则增加至 6~8,10 mm 以上降水预报严重偏多。

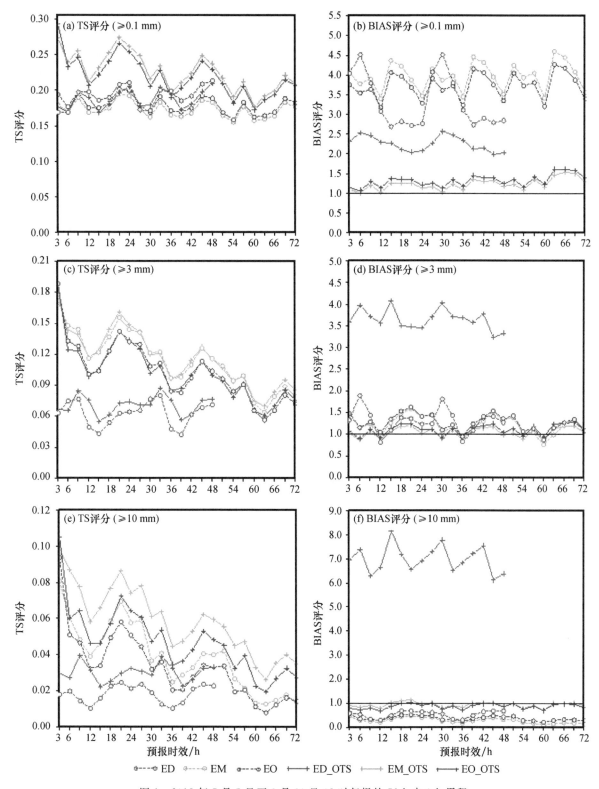

图 1　2018 年 5 月 5 日至 9 月 30 日 08 时起报的 72 h 内 3 h 累积
0.1 mm(a、b)、3 mm(c、d)、10 mm(e、f)以上降水的预报评分

(a、c、e)为 TS 评分;(b、d、f)为预报偏差(BIAS);ED、EM、EO、ED_OTS、EM_OTS、EO_OTS 分别代表 ECWMF
模式的插值预报、基于 ECMWF 预报的 MOS 预报、纳入超前空间实况因子的 OMOS 预报,以及相应的 OTS 订正预报

EM 和 EO 均显著提高了 10 mm 以上降水预报的 TS 评分,时效越短提高越明显,第一个预报时效由 ED 的 0.018,提高为 EM 的 0.081 和 EO 的 0.094,最大提高了 0.076,随着预报时效的增长,提高幅度逐渐减小;BIAS 评分与 ED 比较接近且整体略微减小;EM 与 EO 相比,除临近 3 h 预报外,EM 的 TS 评分更高,EO 的 BIAS 评分更接近于 1。

在 MOS 预报基础上进一步进行 OTS 订正,得到了更好的预报效果,一方面,TS 评分进一步显著提高,最大比 ED 提高达 0.087;另一方面,BIAS 评分由 ED 的明显小于 1 改进为接近 1,显著提高预报的无偏性。EM_OTS 与 EO_OTS 相比,前者的 TS 评分大部分时效明显高于后者,BIAS 评分两者接近,部分时效 EM_OTS 的 BIAS 评分更接近 1。

由上可知,对 ECMWF 模式 3 h 累积 10 mm 以上降水预报进行 OTS 订正,可以适当提高 TS 评分,但代价是严重增大 BIAS 评分,说明 OTS 方法主要通过大幅增加 10 mm 以上降水来提高 TS 评分;而 MOS 和 OMOS 方法,则通过合理优化 ECMWF 的 10 mm 以上降水预报分布显著提高 TS 评分,进一步通过 OTS 订正适度增加 10 mm 以上降水预报,再次提高了 TS 评分并且改善了预报偏小属性,可见当 OTS 订正方法基于降水分布更加合理的 MOS 或者 OMOS 预报时,比基于原始 ECMWF 模式预报更加有效。除临近预报时效外,EM_OTS 比 EO_OTS 订正效果更好。

3.4 MOS-OTS 组合方法与城镇指导预报对比分析

由 3.3 节分析可以看到,采取 MOS-OTS 组合方法,显著提高了 10 mm 以上降水的预报性能,本文进一步将 EM_OTS 预报结果与国家气象中心下发的城镇指导预报(以下简称 GD 预报)进行对比。GD 预报融合了客观预报和预报员的主观预报(赵声蓉 等,2012),48 h 内提供逐 3 h 降水量预报。由图 2a 可以看到,逐 3 h 累积 10 mm 以上降水预报中,16 个预报时效的大部分都是 EM_OTS 具有更高的 TS 评分,只有 5 个时效 GD 预报的 TS 评分大于等于 EM_OTS。BIAS 评分方面(图 2b),EM_OTS 远远优于 GD 预报,前者基本为 0.8~1.1,后者则在 0.5~1.8 波动。可见,MOS_OTS 在短期 3 h 强降水预报中,具有很好的订正性能,不仅超过其他几种客观预报方法,与 GD 预报相比也具有一定优势。

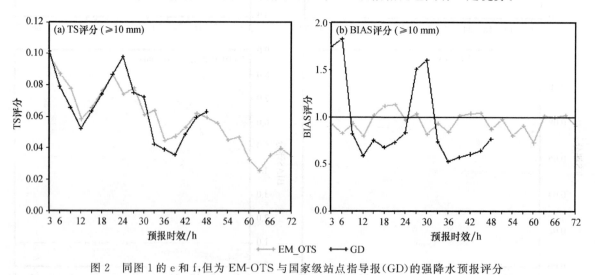

图 2 同图 1 的 e 和 f,但为 EM-OTS 与国家级站点指导报(GD)的强降水预报评分

4 临近 3 h 降水预报中超前空间实况因子的作用

4.1 临近 3 h 降水预报整体统计评分

降水属于不连续变量,夏季与之相关的天气系统生命周期较短,因此已经出现的实况降水对后期降

水的预报指示时效比较短。在MOS预报中引入OPTOR因子后,6 h以后的较长预报时效(图1)反而造成预报性能的下降,EO、EO_OTS的预报性能比相应的EM和EM_OTS预报略差。

但是,对于临近3 h预报时效内的精细降水预报,OPTOR因子的纳入显著提高了预报性能。考虑到临近3 h预报时效OTS订正方法可以明显提高ED、EM和EO的预报性能(图表略),选择ED_OTS、EM_OTS、EO_OTS与GD预报一起进行对比。图3表明,EO_OTS预报是几种预报中TS评分最高的,3 h累积0.1、3、10 mm以上降水量的TS评分分别为0.294、0.188和0.105,与原始ED预报的TS评分0.170、0.063和0.018相比(图表略),分别提高73%、198%和483%。EM_OTS的0.1和3 mm以上3 h降水量的TS评分仅次于EO_OTS,高于GD预报;10 mm以上降水的TS评分略低于GD预报。

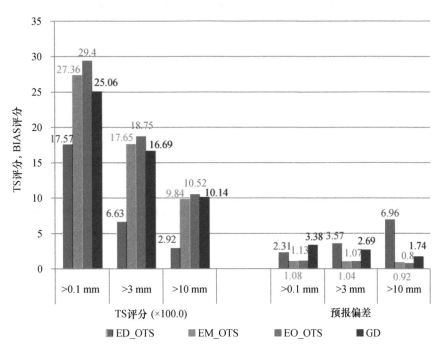

图3 2018年5月5日至9月30日08时起报的临近3 h降水量的不同量级以上累加降水预报的四种预报方法TS评分(TS评分扩大了100倍)和BIAS评分

EO_OTS临近3 h的0.1、3和10 mm以上降水的BIAS评分分别为1.13、1.07和0.8(图3),明显优于ED_OTS和GD预报。EM_OTS的BIAS评分与EO_OTS接近,更加靠近1。进一步对比EO_OTS和EM_OTS的空报率和漏报率(图4)发现,不论是0.1、3 mm还是10 mm以上降水,不论是空报率还是漏报率,EO_OTS的评分都更优于EM_OTS,具有更小的空报率和漏报率,只有10 mm以上降水的漏报率两者基本相同。

因此认为,EO_OTS临近3 h降水预报在不同量级的总体性能,均明显优于EM_OTS、GD和ED_OTS预报。

4.2 临近3 h降水预报逐日评分

为进一步了解OPTOR因子对逐日临近3 h降水预报的影响效果,将EO_OTS、EM_OTS和GD预报的逐日预报性能进行比较(图5)。针对2018年5月5日至9月30日EO_OTS有无降水预报的TS评分,分别计算与EM_OTS及GD预报的TS评分的逐日差值(图5a、5b)。可以看到,大部分日期EO_OTS的TS评分相对于EM_OTS及GD预报都是正技巧,明显优于其他两种预报。

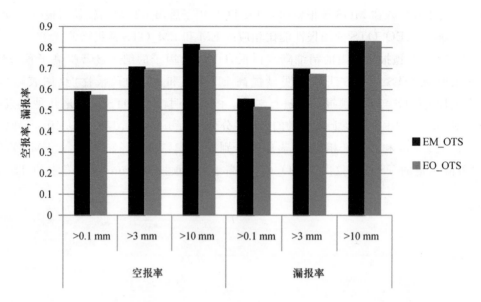

图 4　同图 3，但为 EM_OTS 和 EO_OTS 预报的空报率和漏报率对比

在临近 3 h 有无降水预报的预报偏差方面(图 5c)，GD 预报明显偏多，BIAS 评分远大于 1；EO_OTS 和 EM_OTS 的 BIAS 评分接近于 1，尤其是前者，几乎一直稳定在 1 附近，EM_OTS 在 5 月中旬至下旬 BIAS 评分出现偏大现象。

可见，在 2018 年 5 月 5 日至 9 月 30 日的逐日临近 3 h 预报中，TS 评分和 BIAS 评分都表明，EO_OTS 有无降水预报性能的整体表现持续优于 EM_OTS 和 GD 预报，OPTOR 因子的纳入使其具有稳定的预报优势。

5　结论和讨论

针对短期和临近 3 h 降水预报，对比分析了基于 ECMWF 模式的插值预报 ED、MOS 预报 EM、纳入超前空间实况预报因子 OPTOR 的 OMOS 预报 EO，以及在每种预报基础上进行最优 TS 阈值订正的 ED_OTS、EM_OTS 和 EO_OTS 预报六种方法的预报效果，其中 OMOS 预报是本文提出和改进的预报方法。得到以下结论：

(1) 单独采取 OTS 方法对原始模式预报进行订正，通过调整阈值大幅增加了 3 和 10 mm 以上降水预报，一定程度上提高了 TS 评分，但同时使 BIAS 评分明显增大；针对 ECMWF 模式有无降水预报偏多情况，OTS 方法通过调整阈值明显减小了 BIAS 评分，但没有改善 TS 评分效果，反而引起有无降水预报的 TS 评分降低。

(2) MOS 预报方法可以整体上合理优化 ECMWF 降水预报分布，明显改进 3 mm 以上降水预报的 TS 和 BIAS 评分，以及 10 mm 以上降水预报的 TS 评分，但对 ECMWF 模式有无降水预报 BIAS 评分偏大问题没有改进。

(3) 在 MOS 方法合理优化降水分布的基础上开展 OTS 订正，通过适当调整阈值，MOS-OTS 预报方法不论是 TS 评分还是 BIAS 评分，都显著提高了所有量级降水的预报性能；纳入 OPTOR 因子的 OMOS-OTS 方法显著提高了临近降水预报性能，0.1、3、10 mm 以上降水的 TS 评分最高比 ED 预报分别提高 73%、198% 和 483%，BIAS 评分接近于 1。

(4) 六种基于 ECMWF 模式的 3 h 降水预报方法中，对短期各量级预报来说 MOS-OTS 综合方法的预报性能最好，强降水预报效果与城镇指导预报 GD 中的主客观融合降水预报相比也略胜一筹。而对临

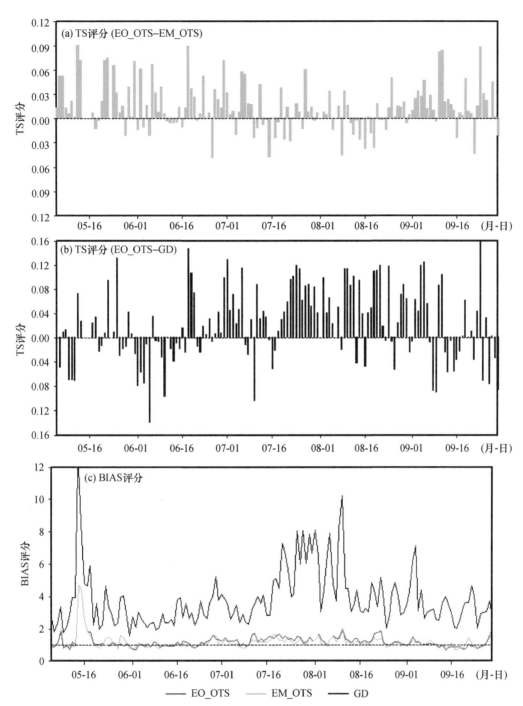

图 5 2018 年 5 月 5 日至 9 月 30 日 08 时起报的临近 3 h 降水量 EO_OTS、
EM_OTS 和 GD 预报的逐日评分对比

(a)EO_OTS 与 EM_OTS 有无降水预报的 TS 评分差值;(b)EO_OTS 与 GD 预报有
无降水预报逐日 TS 评分差值;(c)EO_OTS、EM_OTS、GD 有无降水预报的 BIAS 评分差值

近 3 h 各量级降水预报来说,OMOS-OTS 综合方法预报技巧最高,无论是 TS、BIAS 评分,还是空报率和漏报率,EO_OTS 预报都表现出最好的预报性能。而且,无论是整体统计评估,还是逐日有无降水预报对比,都表明 OMOS-OTS 方法的预报性能明显优于 GD 预报。

本次研究中 OPTOR 因子只选取了 1 h 降水,如果选取云量作为 OPTOR 因子,可能会为降水预报提供更好的超前相关预报信息。另外,由于 3 h 降水,尤其是 10 mm 以上的 3 h 降水的实际发生样本量少,导致部分站点 MOS 预报无法建模,OTS 预报也缺乏足够训练样本,今后可以采取区域建模的方法增加降水样本。OPTOR 因子的纳入,显著提高了基于 ECMWF 全球模式的 OMOS 临近 3 h 降水预报效果,应该是得益于滚动更新实况资料的临近"相关外推"效应,为滚动预报提供了一种有效方法。临近降水预报方面后续可以基于中尺度模式或者循环更新的对流尺度模式开展进一步试验。

参考文献

白永清,林春泽,陈正洪,等,2013. 基于 LAPS 分析的 WRF 模式逐时气温精细化预报释用[J]. 气象,39(4):460-465.
毕宝贵,代刊,王毅,等,2016. 定量降水预报技术进展[J]. 应用气象学报,27(5):534-549.
车钦,赵声蓉,范广洲,2011. 华北地区极端温度 MOS 预报的季节划分[J]. 应用气象学报,22(4):429-436.
代刊,曹勇,钱奇峰,等,2016. 中短期数字化天气预报技术现状及趋势[J]. 气象,42(12):1445-1455.
代刊,朱跃建,毕宝贵,2018. 集合模式降水预报的统计后处理技术研究综述[J]. 气象学报,76(4):493-510.
金荣花,代刊,赵瑞霞,等,2019. 中国无缝隙精细化网格天气预报技术进展与挑战[J]. 气象,45(4):445-457.
李俊,杜钧,陈超君,2014. 降水偏差订正的频率(或面积)匹配方法介绍和分析[J]. 气象,40(5):580-588.
刘还珠,赵声蓉,陆志善,等,2004. 国家气象中心气象要素的客观预报——MOS 系统[J]. 应用气象学报,15(2):181-191.
唐文苑,郑永光,2019. 基于快速更新同化数值预报的小时降水量时间滞后集合预报订正技术[J]. 气象,45(3):305-317.
吴启树,韩美,刘铭,等,2017. 基于评分最优化的模式降水预报订正算法对比[J]. 应用气象学报,28(3):306-317.
赵声蓉,赵翠光,邵明轩,2009. 事件概率回归估计与降水等级预报[J]. 应用气象学报,20(5):521-529.
赵声蓉,赵翠光,赵瑞霞,等,2012. 中国精细化客观气象要素预报进展[J]. 气象科技进展,2(5):12-21.
BUIZZA R,ALONSO-BALMASEDA M,BROWN A,et al,2018. The development and evaluation process followed at ECMWF to upgrade the Integrated Forecasting System(IFS)[C]//The 47th Scientific Advisory Meeting,Reading:ECMWF.
CARTER G M,DALLAVALLE J P,GLAHN H R,1989. Statistical forecasts based on the national meteorological center's numerical weather prediction system[J]. Wea Forecasting,4:401-412.
CHARBA J P,SAMPLATSKY F G,2011. High-resolution GFS-based MOS quantitative precipitation forecasts on a 4-km grid[J]. Mon Wea Rev,139(1):39-68.
CHEN M,BICA B,TÜCHLER L,et al,2017. Statistically extrapolated nowcasting of summertime precipitation over the eastern Alps[J]. Advances in Atmospheric Sciences,34:925-938.
ENGEL C,EBERT E E,2012. Gridded operational consensus forecasts of 2-m temperature over Australia[J]. Wea Forecasting,27(2):301-322.
GLAHN H R,LOWRY D A,1972. Use of model output statistics(MOS) in objective weather forecasting[J]. J Appl Meteor,11(8):1203-1211.
GLAHN H R,RUTH D P,2003. The new digital forecast database of the national weather service[J]. Bull Amer Meteor Soc,84(2):195-202.
GLAHN B,GILBERT K,COSGROVE R,et al,2009. The gridding of MOS[J]. Wea Forecasting,24(2):520-529.
RUTH D P,GLAHN B,DAGOSTARO V,et al,2009. The performance of MOS in the digital age[J]. Wea Forecasting,24(2):504-519.
SCHMEITS M J,KOK K J,VOGELEZANG D H P,et al,2008. Probabilistic forecasts of(Severe) thunderstorms for the purpose of issuing a weather alarm in the Netherlands[J]. Wea Forecasting,23(6):1253-1267.
STAUFFER R,UMLAUF N,MESSNER J W,et al,2017. Ensemble postprocessing of daily precipitation sums over complex terrain using censored high-resolution standardized anomalies[J]. Mon Wea Rev,145(3):955-969.
TREPTE S,ECKERT M,MAHRINGER G,2018. Using model output statistics for aerodrome weather forecasts[C]// Proceedings of 20th EGU General Assembly. Vienna2018. Vienna:EGU.
VISLOCKY R L,FRITSCH J M,1995. Improved model output statistics through model consensus[J]. Bull Amer Meteor

Soc,76(7):1157-1164.

WILKS D S, HAMILL T M, 2007. Comparison of ensemble-MOS methods using GFS reforecasts[J]. Mon Wea Rev, 135(6):2379-2390.

ZHU Y J, LUO Y, 2015. Precipitation calibration based on the frequency-matching method[J]. Wea Forecasting, 30(5):1109-1124.

基于集合预报系统的日最高和最低温度预报

熊敏诠

(国家气象中心,北京,100081)

摘 要:根据欧洲中期天气预报中心集合预报系统 2 m 温度预报的集合统计值,提出了 BP-SM 方法,针对中国 512 个台站 2016 年 3 月的日最高(低)温度作预报分析。将集合预报系统的模式直接输出、BP 和 BP-SM 方法得到的日最高(低)温度进行了比较,结果表明:当预报时效越长,BP-SM 方法较之 BP 方法的预报优势也更明显;在 1～5 d 的预报中,BP-SM 方法显著降低了预报绝对误差,误差在 2 ℃ 以内的准确率大部分在 60% 以上,部分站点达到了 90%;正技巧评分均值大多高于 30%,在青藏高原东部和南部地区超过了 60%。预报正技巧站点次数在绝对误差≤2 ℃(1 ℃)范围内有所提高,对日最高温度预报准确率的提高略好于日最低温度;BP-SM 方法有效地降低了预报系统偏差,较大预报误差出现次数显著减少。

关键词:集合预报系统;日最高温度;日最低温度;BP-SM 方法

Calibrating Daily 2 m Maximum and Minimum Air Temperature Forecasts in the Ensemble Prediction System

XIONG Minquan

(National Meteorological Center, Beijing, 100081)

Abstract: BP neural network-Self Memory method (BP-SM) is used to calibrate daily 2-m maximum (minimum) air temperature forecasts over 512 stations in China with the European Centre for Medium-Range Weather Forecasts (ECMWF) Ensemble Prediction System (EPS) in March 2016. Seven statistical characteristics used as predictors are calculated based on 2 m air temperature model output of EPS. Daily maximum (minimum) air temperature forecasts by BP-SM, BP and direct model output (DMO) are compared. The postprocessing with BP-SM is shown to improve the forecast accuracy. Compared with BP method, more advantages of BP-SM method are attained in longer predictable time. The accurate rate of daily maximum (minimum) air temperature forecasts with absolute errors less than 2 ℃ reaches above 60% and ever over 90% at some station. Compared with DMO, the forecasting skill score of BP-SM is 30% on average, and above 60% over the eastern Tibet Plateau. This program is obviously superior with forecast errors within 2 ℃(1 ℃). The calibrated daily 2 m maximum air temperature is slightly better than the daily 2 m minimum air temperature. By BP-SM method, the systematic deficiencies of daily 2 m maximum (minimum) air temperature forecasts are significantly reduced.

---

① 本文发表于《气象学报》2017 年第 2 期。
资助项目:发展格点预报融合和订正关键技术(YBGJXM201703)。
作者简介:熊敏诠,主要从事天气学研究。E-mail:minquanxiong@sina.com。

Key words: ensemble prediction system; daily maximum air temperature; daily minimum air temperature; BP neural network-Self Memory method

1 引言

不确定性是天气预报的特征,集合预报是对不确定性进行描述的概率预报系统。集合后处理技术的关注点是如何提高集合概率预报,有直方图法(Hamill et al.,1997,1998)、集合映射(Roulston et al.,2003)、贝叶斯模型平均法(Raftery et al.,2005)、逻辑回归(Hamill et al.,2004)、非奇次高斯回归(Gneiting et al.,2005)、相似法(Hamill et al.,2006)、"预报同化"法(Stephenson et al.,2005)等。近些年有较多的深入研究和发展,例如,Wilks 等(2007)对比三种不同的后处理技术在集合温度和降水概率预报中的优劣,得出训练样本长度和预报技巧有密切关系。Maurice 等(2010)根据 ECMWF 的 20 年总降水量集合再预报数据集,对比分析了 EPS 原始预报、修正的 BMA 和扩展逻辑方法在降水概率预报上的优劣。Emmanuel 等(2012)使用扩展的逻辑回归提高 EPS 降水概率预报。Mendoza 等(2015)使用不同的统计后处理技术,结合高分辨率的区域气候模式,实现降尺度的日降水概率预报。

在多模式集合、集成技术领域,中国学者有多方面进展,例如,陈超辉等(2010)开展了有限区域模式的多模式短期超级集合预报研究。王敏等(2012)基于集合 2 m 温度预报,对比了非均匀高斯回归方法与自适应卡尔曼滤波偏差订正方法的订正效果。刘琳等(2013)基于 T213 集合预报极端降水天气预报指数,对中国极端强降水天气进行预报试验和分析。刘建国等(2013)根据全球多模式集合预报(TIGGE)资料,进行地面日均气温贝叶斯模型平均概率预报及其检验与评估。刘永和等(2013)对 TIGGE 资料在沂沭河流域 6 h 降水集合预报能力作了分析。杜钧等(2014)揭示了"集合异常预报法"和集合预报在极端高影响天气预报中的优越性。马旭林等(2015)利用日本气象厅区域集合预报的 850 hPa 温度资料,分析了基于卡尔曼滤波递减平均法的集合预报综合偏差订正。

由于次网格的不确定性影响,集合预报系统对近地面气象要素预报的可靠性不如高层大气要素,如何提高近地面的温度和降水概率预报是统计后处理技术发展的主要方向。另外,确定性预报也有着巨大需求,特别是在短期预报中,人们总是根据确定性预报做出决策,集合预报提供了未来天气演变的多种可能性,通过适当的方法提炼出确定性信息也需要深入研究。本文针对近地面日最高(低)温度预报,根据集合预报系统的统计特征量,提出 BP-SM 方法,明显提高了预报准确率。

2 资料和方法

2.1 资料

根据 2015 年 12 月 1 日至 2016 年 3 月 31 日 08 时(北京时,下同)的欧洲中期天气预报中心集合预报产品,预报时效为 0~240 h,都是 6 h 间隔输出,共 51 个预报成员,可得到集合平均、最低、最高、中数、众数、离散度共 7 个统计量。得到集合预报系统 6 h 间隔的 2 m 温度的集合统计特征量,将每天 4 个时次的高(低)值作为日最高(低)值,作为预报模型输入量。在集合统计量得到的诸多日最高(低)值中,由集合平均得到的日最高(低)温度通常有较高的精度。当对集合预报系统进行评价时,就以此作为日最高(低)气温的 DMO(模式直接输出)。针对全国 512 个台站(图 1),在 2016 年 3 月 1 日至 3 月 31 日 08 时,逐日进行日最高(低)气温多个时效的预报。

2.2 方法

BP(Back-Propagaion)网络是人工神经网络的重要组成部分,广泛应用于诸多领域,其由 3 个环节构

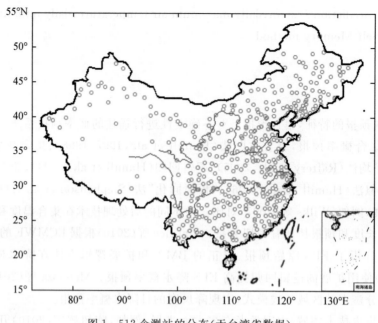

图1 512个测站的分布(无台湾省数据)

成,一是正向传播,信号从输入层经隐层到输出层;二为反向传播,预测误差从输出层,经原来的通道到达输入层;三是迭代过程,设立误差目标,当正向传播,未达到目标值时,通过反向传播,调整网络权值和阈值,反复迭代直至达到目标值。本文使用集合预报的7个统计量作为输入层的7个节点,依据Kolmogorov定理,隐层节点数设为15,输出层为1个节点,即预测和实际观测之差。构成3层网络进行日最高(低)气温预报,以预测误差(观测值减DMO)为目标值。出于实用考虑,迭代过程均为1000步。同时,网络学习过程采用经典的动量梯度法,步长为0.05。

泛化性是BP网络好坏的关键,一方面是BP算法本身固有缺陷造成的,如网络训练过程中较易陷入局部极小而"难以自拔",影响到网络性能;相关参数选取缺乏理论指导等。另一方面来自于问题的复杂性,首先是训练样本和独立(预报)样本差异,本文使用的滚动建模中70个训练样本距离预报日最近的也要提前1 d,当做120 h预报时,相差就达到了5 d。3月中国气温变化剧烈,在某些时段某个地区容易出现偏高(低)气温,那么,偏高(低)的输入量就易出现偏高(低)的输出,即异常值。同时,训练样本变化易致网络权值变动,对于动态建模而言,每天都有最新的样本增加,得到的训练结果会有差异,当新样本数值上出现较大差异时,网络的权值变动也大,表现为BP网络对历史样本"记忆"功能缺失及网络稳定性变化。由此,本文提出BP神经网络-自忆(简称BP-SM)建模方法。

自相关函数是描绘时间序列过去对未来的影响,EPS预报误差时序普遍有一阶或二阶记忆特征。如开展上海南汇(58369)2016年3月1日08时的24 h预报,以2015年12月21日至2016年2月29日08时的24 h集合平均日最高温度预报为训练样本,得到预报偏低(共有66 d)和偏高(4 d)两个序列。预报偏低的误差序列自相关函数(图2)表明,上一期会影响到下一期,当时滞大于3时,自相关函数位于两倍标准误差带中,为单边衰减、二阶截尾。可使用自回归动态建模,通过历史数据对未来进行预测:

$$y_t = \phi y_{t-1} + \varepsilon_t \tag{1}$$

其中,y_t 是预报误差;ϕ 为自回归系数,即 y_t 记忆能力的强弱;y_{t-1} 是一阶滞后的预报误差;ε_t 为现期;$\{\varepsilon_t\}_{t=-\infty}^{\infty}$ 满足 $E(\varepsilon_t)=0$、$E(\varepsilon_t^2)=\sigma^2$、$E(\varepsilon_t\varepsilon_{t'})=0(t\neq t')$ 的白噪声过程,序列 $y_0、y_1、y_2、\cdots、y_{t-1}、y_t$ 都以式(1)相类似的表达,使用递归替代法,y_t 可写成初始值 y_{-1} 和 ε 在第0期到第 t 期的历史值的函数:

$$y_t = \phi^{t+1} y_{-1} + \phi^t \varepsilon_0 + \phi^{t-1}\varepsilon_1 + \phi^{t-2}\varepsilon_2 + \cdots + \phi\varepsilon_{t-1} + \varepsilon_t \tag{2}$$

由 $\hat{y}_t = \phi y_{t-1}$,$\hat{y}_t$ 是 y_t 的估计值,可得

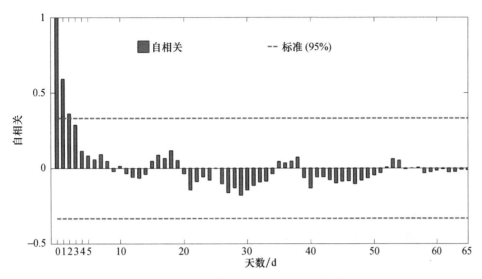

图2 2015年12月21日至2016年2月29日08时上海南汇(58369)DMO日最高气温24 h预报偏低的误差序列自相关函数

$$y_t - \varepsilon_t = \phi^{t+1} y_{-1} + \phi^t (y_0 - \hat{y}_0) + \phi^{t-1}(y_1 - \hat{y}_1) + \phi^{t-2}(y_2 - \hat{y}_2) + \cdots + \phi(y_{t-1} - \hat{y}_{t-1}) \quad (3)$$

$$\hat{y}_t = (\phi^{t+1} y_{-1} + \phi^t y_0 + \phi^{t-1} y_1 + \phi^{t-2} y_2 + \cdots + \phi y_{t-1}) - (\phi^t \hat{y}_0 + \phi^{t-1} \hat{y}_1 + \phi^{t-2} \hat{y}_2 + \cdots + \phi \hat{y}_{t-1}) \quad (4)$$

$$\hat{y}_t + (\phi^t \hat{y}_0 + \phi^{t-1} \hat{y}_1 + \phi^{t-2} \hat{y}_2 + \cdots + \phi \hat{y}_{t-1}) = (\phi^{t+1} y_{-1} + \phi^t y_0 + \phi^{t-1} y_1 + \phi^{t-2} y_2 + \cdots + \phi y_{t-1}) \quad (5)$$

令
$$L_t = \beta_0 y_0 + \beta_1 y_1 + \beta_2 y_2 + \cdots + \beta_\tau y_\tau \quad (6)$$

其中，$\{\beta_i\}_{i=0}^{\tau} \approx \{\phi^{t-i}\}_{i=0}^{\tau}$，$\tau = t-1$，$\tau$是时间常数，即训练样本长度。式(5)左边第2项$\{\hat{y}_i\}_{i=0}^{t}$并非严格的平稳时间序列，在一些天气演变中，变化特点往往难以估计，通过自回归动态建模获得的$\{\hat{y}_i\}_{i=0}^{t}$总是达不到理想的精度。若$\{\hat{y}_i\}_{i=0}^{t}$为白噪声，式(5)左边第2项较小，可略去；将L_t作为式(5)右端项形式表达，即存在有限大的L_t，满足$\hat{y}_t < L_t$。式(6)表明，L_t和历史预报误差有密切关系。使用指数衰减方程描述构造函数随时间的变化：$\dfrac{\mathrm{d}F(t')}{\mathrm{d}t'} = -\dfrac{F(t')}{\tau}$。其中，$t' = \tau - t$；$F(t')$为$t'$时刻误差系数。方程的解析解：$F(t') = F(0)\exp(-t'/\tau)$。训练样本按时间顺序排列$i = 0,1,2,\cdots,\tau$，愈临近当前时间的样本序列号愈大。对于有限大$L_t$，设$\sum_{i=0}^{\tau}\beta_i = 1$，$\beta_i = \dfrac{\exp\left(-\dfrac{\tau-i}{\tau}\right)}{\sum_{i=1}^{\tau}\exp\left(-\dfrac{i}{\tau}\right)}$，当前时刻极大似然值$L_t = \sum_{i=0}^{\tau}\dfrac{y_i \times \exp\left(-\dfrac{\tau-i}{\tau}\right)}{\sum_{i=1}^{\tau}\exp\left(-\dfrac{i}{\tau}\right)}$。由于要对误差区间估计，将$y_i$按正负值分成两个序列，分别计算出上界$E_{\text{up}}$和下界$E_{\text{down}}$。将BP方法得到的预报值限定在$[E_{\text{down}}, E_{\text{up}}]$区域内，称为BP-SM方法。

采用滚动建模方法进行实时训练和预报，例如：当要做2016年3月1日08时的24 h预报时，就以2015年12月26日至2016年2月27日集合产品为训练样本建模，而2016年3月1日08时的120 h预报，其训练样本则是2015年12月21日至2016年2月22日的资料，依此类推。模型就可以反映出数值模式最近的误差特点，有助于提高预报精度。

预报评价方面，日常业务中使用预报绝对误差、准确率和技巧评分，具体如下：最高（低）气温预报平均绝对误差：$T_{\text{MAE}} = \dfrac{1}{N}\sum_{i=1}^{N}|F_i - O_i|$；预报准确率：$TT_K = \dfrac{N_{r_K}}{N_{f_K}} \times 100\%$。其中，$N$是参加检验预报总次数（例如计算3月1日全国24 h的气温预报平均绝对误差时，N即为512）；F_i为第i次的预报；O_i为第i次观测；$K=1$表示$|F_i - O_i| \leqslant 1\ ℃$，$K=2$表示$|F_i - O_i| \leqslant 2\ ℃$；$N_{r_K}$是预报正确的次数；$N_{f_K}$为总预

报次数（例如计算 3 月 1 日全国 24 h 的气温预报绝对误差≤1 ℃的准确率，N_{f_1} 为 512）。技巧评分 $SST = \dfrac{T_{BP} - T_{DMO}}{T_{DMO}} \times 100\%$。其中，$T_{BP}$ 是使用 BP-SM 方法得到的日最高（低）气温预报平均绝对误差；T_{DMO} 是通过集合平均得到的最高（低）气温预报平均绝对误差。

3　结果分析

3.1　BP-SM 和 BP 方法的比较

通过 BP-SM 方法得到的预报绝对误差小于相应的 BP 方法，称为"正效应"；反之，为"负效应"；若 BP 预测值位于期望区内，则两者一致（重合）。将"正效应"中的各单站预报绝对误差降低值求和，再取平均，即得"正效应"产生的预报误差减少值；同理，有"负效应"站点的预报绝对误差增加值的均值，为"负效应"增加的误差值，由此，可以直观地分析两种方法的预报优劣。全国 512 个台站 1～10 d 的日最高气温预测有如下特点（图 3a）：首先，对于中长时效（3～10 d），BP-SM 法有一定优势，例如第 10 天的预报，"正效应"绝对误差降低值 2.41 ℃，而"负效应"引起的增加值是 1.71 ℃，两者数值相差较大，对应预报准确率的明显提高；其次，随着时效的延长，"负效应"预报次数逐渐下降，而"正效应"次数总体均衡或略有增长，如第 1 天"正（负）效应"次数是 2688（3091），第 10 天是 3814（1064）；最后，BP 算法并不是十分完善的神经网络，"正效应"和"重合"次数大致是 3000：10000，占有相当比例，即便是在较高的"重合"次数中还有异常值，表明 BP-SM 方法的积极意义和深入分析之价值。

上海南汇日最高气温 24 h 预报误差逐日变化曲线（图 3b 主图）：DMO 误差起伏大，最大值 8.97 ℃（第 5 天），最小值 −0.470 ℃（第 6 天），预报值大多低于实况；BP 方法只在前几天有较高的预报精度，但中下旬误差曲线走势完全不同于 DMO，而且预报值明显比实况高；BP-SM 的误差曲线和 DMO 的线型走势相近，特别是在中下旬，有力地修正了原始 BP 误差曲线，较大幅度地降低了误差，主要原因是 BP-SM 考虑了 EPS 预报误差的历史信息。5 个不同时效的误差月均值 DMO（BP-SM、BP）分别是 2.833（0.637、−1.648）、2.965（0.666、−1.464）、2.838（0.483、−1.656）、2.580（0.293、−2.005）、2.155（−0.360、−2.198）℃。平均绝对误差（简称"MAE"，下同）月均值为 2.863（1.589、1.935）、2.987（1.656、2.06）、2.909（1.632、2.057）、2.670（1.627、2.287）、2.554（1.796、2.624）℃。BP-SM 的误差在所有 5 个时效中是最小的，显著降低了误差（图 3b 子图）；在较长时效（第 4 天和第 5 天）预报中，BP-SM 方法的有效性增强。

在日常天气预报业务中，BP-SM 方法分两步进行。以上海南汇站为例，在 2016 年 3 月 30 日 08 时，进行日最高气温的 24 h 预报，下文较详细地介绍预报过程。第 1 步（历史资料的 BP 建模和 SM 方法计算）：选择 2016 年 1 月 19 日至 2016 年 3 月 29 日 08 时的日最高温度 24 h 预报，共计 70 d 资料作为训练样本集，每天的日最高气温 7 个预报统计量作为 BP 方法的输入值，而 BP 网络的目标值是 DMO 误差，完成 BP 网络的构建；根据 DMO 误差正负值划分成预报"偏低"和"偏高"两个序列，"偏低"序列有 67 d，只有 3 d 属于预报"偏高"，使用式（6）可分别得到两个边界值 −0.32 ℃ 和 2.27 ℃。第 2 步（实时预报）：将当天 7 个日最高气温 24 h 预报的统计量输入已建好的模型中，2016 年 3 月 30 日 08 时 EPS 在上海南汇的 7 个预报统计量依次是 16.27、16.41、18.15、15.1、16.12、15.8、0.67 ℃，BP 网络预报值是 23.11 ℃；根据上一步 SM 方法求得的边界值加上 DMO 的 24 h 预报（16.27 ℃），即有 [15.95 ℃，18.54 ℃]；BP 法预报已经超出 SM 方法确定的边界，可判定 BP 网络输出属于异常值，就以邻近边界的 18.54 ℃ 作为 BP-SM 方法最终预报。3 月 30 日 08 时—31 日 08 时，上海南汇日最高气温观测值是 18.7 ℃，那么，DMO（BP-SM、BP）预报误差是 2.43 ℃（0.16 ℃，−4.41 ℃），BP-SM 法预报精度高。

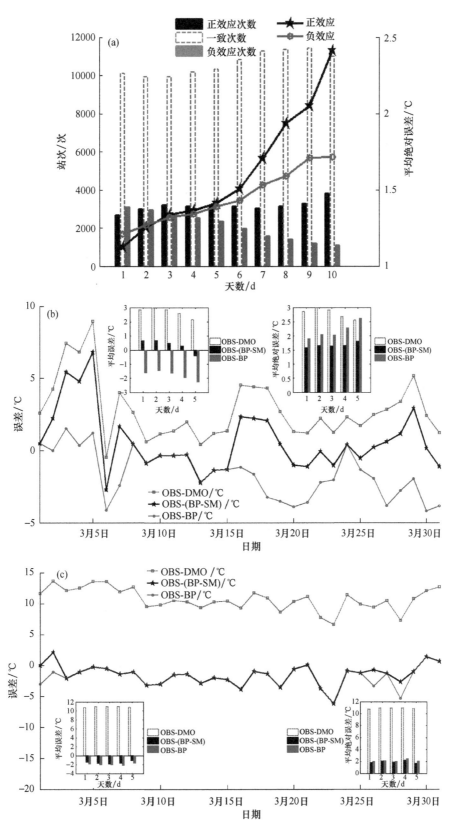

图3 2016年3月DMO、BP方法和BP-SM方法的日最高温度预报比较

(a)1~10 d的BP-SM方法"正(负)效应"预报次数和误差增减值；
(b)上海南汇24 h预报误差的逐日比较(主图)，1~5 d的预报误差和绝对误差月均值比较(子图)；
(c)拉萨市24 h预报误差的逐日比较(主图)，1~5 d的预报误差和绝对误差月均值比较(子图)

分析拉萨(55591)日最高气温24 h预报误差逐日变化曲线(图3b主图)发现,DMO误差曲线总体平稳,BP-SM和BP方法预报基本重合,只有3 d的预报有较大的差别,1日、26日、28日的BP-SM(BP)方法预报误差值对应是0.053(−2.974)、−0.714(−3.299)、−2.655(−5.292)℃,BP-SM法的预报值更接近于实况。5个不同时效的误差月均值DMO(BP-SM、BP)分别是：10.725(−1.476、−1.844)、10.928(−1.672、−2.003)、10.893(−1.790、−2.008)、10.904(−1.661、−2.096)、10.812(−1.047、−1.605)℃。MAE月均值为10.725(1.759、1.987)、10.928(2.034、2.083)、10.893(1.881、2.024)、10.904(2.204、2.455)、10.812(1.647、2.0312)℃。在1～5 d的预报中(图3 c子图),BP-SM方法都有优势。

BP网络预测精度和数值模式预报能力密切相关。有较大误差的训练样本反馈在网络权值上,网络不稳定性增大,导致较大的输出偏差,表现为网络学习和记忆是不稳定的,有别于人类大脑记忆的稳定性,也是泛化性问题的反映;中长期(2 d以上)预报中,数值模式预报误差逐渐增大,BP-SM方法有较好的效果;对于短时(1 d)预报,BP方法预报有较高精度,需要更细致的分析,BP-SM方法才能达到预期。

3.2 绝对误差、准确率和技巧评分空间分布

日最高气温未来5 d的预报中,DMO的MAE≤2 ℃的站点位于东北地区和华北平原大部分地区、长江中下游平原北部,但是,在青藏高原东部和南部,DMO的MAE超过了4 ℃,局部地区在16 ℃以上。在BP-SM方法得到的误差分布上,MAE≤2 ℃的站点数在DMO相对应的区域都有所扩大,并且新增了青藏高原中部、内蒙古高原、黄土高原、四川盆地和云南的部分地区,全国大部分地区的误差都小于4 ℃,较之DMO,有明显的提高。从技巧评分来看(图4 a～e),对青藏高原上站点气温预报提高显著,一般为超过60%的正技巧,局部达到了80%以上;其他大部地区也有较高的正技巧;只有东北北部、内蒙古中部、华北平原西南部和华南西部的部分地区呈现负技巧,数值在0～−20%,随着预报时效的延长,负技巧区的面积向东略有扩大。负技巧区出现在DMO预报准确率比较高的地区:负技巧站点的DMO绝对误差≤2 ℃准确率基本在70%以上,而在华北平原及其以北地区相应的站点DMO准确率大多超过80%,部分达到90%,反映了当模式预报有较高准确率、预报偏差微弱时,通过BP-SM方法达到更高的准确率还需要深入的研究。

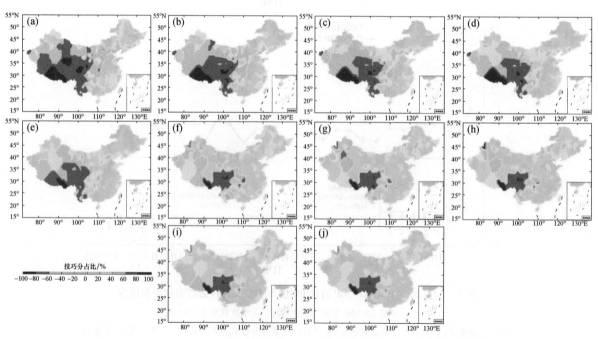

图4 2016年3月BP-SM方法在1～5 d气温预报的月平均技巧评分图
(a)～(e)为最高气温,(f)～(j)为最低气温

最低气温预报 MAE 低于最高气温，在江南、华南地区 DMO 的 5 d 预报中，MAE 大都在 2 ℃ 以内，准确率也在 70% 以上，在中国南方的部分地区超过了 80%（日最高气温准确率高分值区出现在长江以北的平原地区），青藏高原东部和南部的误差值也有所下降。BP-SM 方法在中国大部分地区都小于 2 ℃，即使在青藏高原也低于 4 ℃，明显降低了 DMO 预报误差；不同时效都有较高的准确率，在中国中部和东部沿海部分地区的预报准确率超过了 80%。在技巧评分方面（图 4 f~g），青藏高原东部超过了 60%，负技巧评分也出现在原始 DMO 准确率较高的站点，主要位于中国南方、华北和西北的部分地区，但是负技巧分都较小（0~−10%）。

从日最高（低）气温预报技巧空间分布来看，在青藏高原东部和南部都有较高的正技巧，BP-SM 方法对日最高气温预报准确率提高能力好于日最低气温；负技巧评分站点有相对固定的"区域性"特点，随着预报时效的延长（1~5 d 的预报），该区域逐渐向附近扩散、面积也增大；而且负技巧区通常出现在 DMO 准确率较高的站点。

3.3 不同误差范围准确率的变化特点

预报技巧评分根据所有预报绝对误差的总和计算，没有考虑预报误差的离散度，结果是某个较大的预报误差可能会导致整个预报评分的大幅变化；基于神经网络方法预报的过拟合现象是普遍存在的，个别时次的预报异常容易对评分产生负面影响，技巧评分就难以全面反映该方法预报的优劣。同时，日常实践中，预报异常值往往可以通过主、客观方式得到有效抑制，人们更为关心的是某个具体误差段的准确率，因此只是笼统地给出所有预报误差段的技巧评分难以全面反映方法的优劣。本文以日常使用的准确率为基准，绘出了绝对误差≤2 ℃（1 ℃）准确率不同变化方向的站点数比率。

图 5a 显示了各个预报时效中 512 个台站预报评分正负技巧站点数和分值。日最高气温中，BP-SM 在 1~5 d 的预报正技巧比例（站点数）是 86%（440）、82%（420）、77%（394）、74%（379）、68%（348），而相应的预报技巧分占比是 38%、35%、33%、32%、30%。

通过 BP-SM 方法得到的预报绝对误差≤1 ℃ 的准确率呈现上升趋势的比例（站点数）有 87%（445）、84%（430）、82%（420）、81%（415）、75%（384）；绝对误差≤2 ℃ 的准确率也有相似的变化特点，图形上处于"绞合"状态。即在较小预报误差范围内，BP-SM 方法优势突出。对负技巧评分而言（图 5b），1~5 d 站点比例依次是 14%、18%、23%、26%、32%，平均分值是 −11%、−12%、−12%、−13%、−14%；而绝对误差≤1 ℃ 准确率呈下降趋势的站点比例为 14%、16%、18%、19%、26%。所以，在较小预报误差范围内，BP-SM 方法优势突出。

日最低气温的预报误差比较也有上述类似特点。BP-SM 方法预报正技巧（图 5c）1~5 d 的站点比例为 71%、67%、61%、64%、59%；绝对误差≤1 ℃ 的准确率有提高的站点比例依次是 73%、70%、68%、66%、63%；平均正技巧分值是 27%、27%、28%、26%、26%。负技巧站点数则为 29%~41%；准确率下降（绝对误差≤1 ℃）的站点数介于 26% 和 37%；而负技巧得分大约为 −11% 左右。

综上所述，BP-SM 方法在 5 d 的预报中正技巧站点数有较高的比例，分值也大多位于 30% 附近，日最高气温预报的提高幅度要略好于最低气温。关于绝对误差≤2 ℃（1 ℃），正技巧站点数占比和分值都要高于总体评价，同时，负技巧站点数也比较低，说明在人们较关注的预报误差范围内，BP-SM 方法表现更佳。结合实际气象工作者通行的预报准确率分析，BP-SM 方法表现出了更优越的预报特性，说明 BP 神经网络的泛化问题中，BP-SM 方法虽然能大幅减少由此引发的异常值，但是还不能完全排除。另外，由于 BP 网络的高度非线性难以得到精确的数值解，所以表现为 DMO 预报准确率比较高的部分站点出现负技巧现象。

3.4 系统偏差比较

根据日最高气温的 24 h 预报资料，对预报系统偏差进行分析，加深对 BP-SM 方法的认识，有助于完

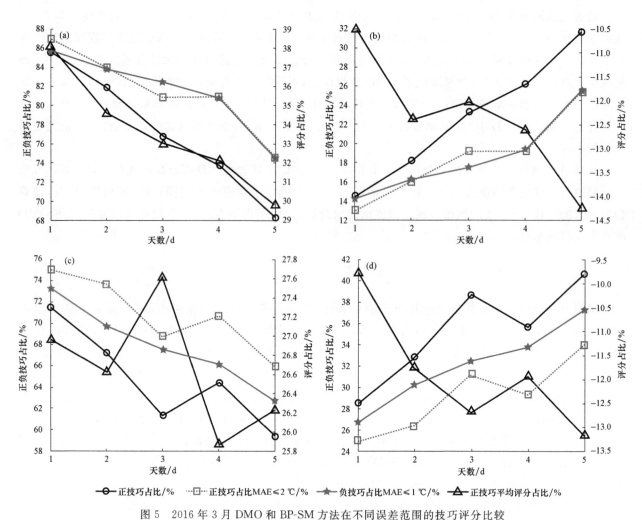

图 5 2016 年 3 月 DMO 和 BP-SM 方法在不同误差范围的技巧评分比较
(a)日最高气温的正技巧比较;(b)日最高气温的负技巧比较;(c)日最低气温的正技巧比较;(d)日最低气温的负技巧比较

善方法和逐步提高预报精度。

散点图(图6)直观地展现了预报的系统偏差,在对称轴($y=x$)附近(图6a),离散点呈发散、非对称性,图中使用不同颜色表示误差范围。由于 DMO 预报误差普遍较大,远距离的数值点显得较多;另外,数值点大多位于对称轴线的下方,即 DMO 预报"偏冷"。BP-SM 方法(图6b)减小了系统性偏差,数值点在对称轴附近呈现收敛、对称性,远离对称轴的离散点较少,大多数集中于 0~3 ℃误差范围之内,点的密度变得更紧凑、密集;数据点以对称轴为轴心呈"准对称"分布,DMO 预报"偏冷"现象得到有效校正。DMO 在 6 个误差范围站次数(图6c)依次是 4070、3630、2590、1830、1290、2440,而 BP-SM 为 6810、4660、2410、1080、488、397,误差在 0~1 ℃、1~2 ℃范围的预报站次数增加明显,同时又大幅降低了>5 ℃误差站次。

在全国的预报误差空间分布中,中国中西部地区 DMO 预报能力较弱。以 100°E 为界,分析中国西部、东部地区预报误差变化特点。西部地区,尤其是青藏高原东部地区,大致在 0~20 ℃(图7a),实况和 DMO 差别很大,DMO 预报偏低,个别站次的误差超过了 20 ℃,经过 BP-SM 方法再预报后,得到极大的改进。图 7a 和 b 前后对比表明,在东部和西部,散点分布都变得更"收敛和对称",BP-SM 有效降低了数值模式预报系统性偏差。西部地区的 DMO 柱状图(图7c)在 6 个误差范围中站次数分别是 139、175、200、236、190、544,改进后为 562、392、255、145、76、54;DMO 预报误差迥异于其他地区,其中误差大于 5 ℃站次较多,改进后,预报精度提高十分明显,特别是在较大误差预报发生时。东部地区(图7d)有相似的特点。

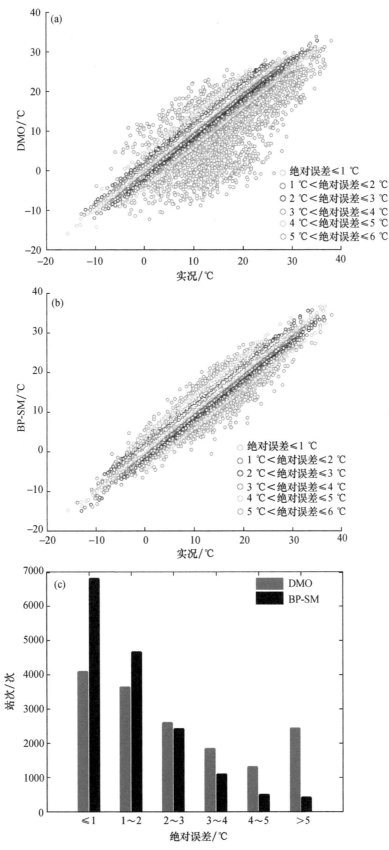

图 6 2016 年 3 月日最高气温 24 h 预报-实况散点图和不同预报误差范围站次数对比
(a)DMO 和实况；(b)BP-SM 和实况；(c)DMO 和 BP-SM 比较

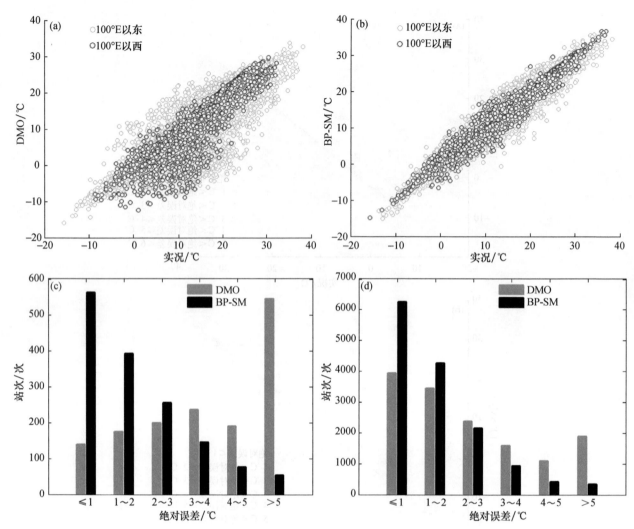

图 7 2016 年 3 月 100°E 以西和以东地区日最高气温 24 h 预报-实况散点图
(a)DMO 和实况；(b)BP-SM 和实况；(c)100°E 以西地区和
(d)100°E 以东地区 DMO 和 BP-SM 方法的日最高气温 24 h 预报在不同误差范围站次数对比图

不同区域的 DMO 系统性偏差表现各异，可能与次网格不确定性、中国天气及地理复杂性密切相关，散点图呈发散、非对称性分布。将预报误差由小到大划分成 6 个区段，相应的预报站次数并非均匀地落在每个区段中，DMO 预报次数随着误差增大而逐渐衰减，只是到了＞5 ℃段有"翘尾"现象，因为文中将误差＞5 ℃都统一划归到此段。在西部地区 DMO 预报误差次数在 6 个区段的分布形态比较奇特，可能是误差分段和集合预报在此区域的低准确率两方面原因引起的。BP-SM 方法则将 DMO 误差各异分布都转换成一致的衰减形态，而且衰减率也明显增大。

4 结论和讨论

（1）从 BP-SM 和 BP 方法预报精度对比分析可以发现，随着预报时效的延长，BP-SM 方法优势也逐渐增大，"负效应"预报次数明显下降，表明 BP-SM 法能有效地减少 BP 网络预报异常值。从上海南汇多个预报时效比较中发现，BP 网络的不稳定性导致较大的预报误差；而 BP-SM 法避免了上述现象，在较长时效预报中，BP-SM 法有更大的正向作用。分析上海南汇和拉萨两单站的 DMO 预报误差变化特点可初步获悉，EPS 误差时空特点和 BP 网络预报准确度有密切关系。以上海南汇为例，较细致地介绍了 BP-SM 方法日常预报过程。

（2）BP-SM方法十分显著地提高了日最高（低）气温预报准确率，尤其是在青藏高原东部和南部地区。和DMO相比，BP-SM方法在大部分地区都有较高的正技巧，部分地区超过了60%，负技巧站点通常出现在DMO预报准确率较高的部分地区，技巧分值较小（0～−20%）。

（3）具体比较绝对误差≤2 ℃（1 ℃）准确率变化的特点发现，在未来5 d预报中，BP-SM方法在此误差范围内的正技巧站点比例表现出更好的优势，同时，对日最高气温预报准确率的提高能力要略好于日最低气温。探讨了BP-SM方法出现负技巧预报的原因，说明BP-SM方法有较强的非线性映射能力，当面对函数逼近精度和模型容错性等问题时，还要和天气变化特点相联系，可能会达到更好的预报效果。

（4）在实况-预报散点图及在东、西部站点的比较中，DMO呈现发散、非对称性分布特点，而BP-SM方法减小了系统性偏差，数值点也收敛、对称地分布。BP-SM方法改变了DMO在6个误差区段分布形态的不一致性，使得误差越大的区段站点数越少，分布结构呈快速衰减趋势。

传统的动力统计释用方法通常根据模式输出的动力因子进行建模预报。实践中如何获得最优预报子集往往是难点，也是预报质量优劣的关键。当面临集合预报海量的因子输出时，预报因子集选择将显得尤为困难。BP-SM方法直接使用集合气温预报统计特征值，避免了上述问题。日最高（低）气温是日常气象研究和业务服务的重要预报量，受到广泛的关注，基于集合预报系统而提出的快速、准确建模和优化的BP-SM方法，可操作性强，实用性好，能在实践中发挥作用并将得到进一步完善。

参考文献

陈超辉,李崇银,谭言科,等,2010. 基于交叉验证的多模式超级集合预报方法研究[J]. 气象学报,68(4):464-476.

杜钧,GRUMM R H,邓国,2014. 预报异常极端高影响天气的"集合异常预报法"：以北京2012年7月21日特大暴雨为例[J]. 大气科学,38(4):685-699.

刘建国,谢正辉,赵琳娜,等,2013. 基于TIGGE多模式集合的24小时气温BMA概率预报[J]. 大气科学,37(1):43-53.

刘琳,陈静,程龙,等,2013. 基于集合预报的中国极端强降水预报方法研究[J]. 气象学报,71(5):853-866.

刘永和,严中伟,冯锦明,等,2013. 基于TIGGE资料的沂沭河流域6小时降水集合预报能力分析[J]. 大气科学,37(3):539-551.

马旭林,时洋,和杰,等,2015. 基于卡尔曼滤波递减平均算法的集合预报综合偏差订正[J]. 气象学报,73(5):952-964.

王敏,李晓莉,范广洲,等,2012. 区域集合预报系统2 m温度预报的校准技术[J]. 应用气象学报,23(4):395-401.

GNEITING T, RAFTERY A E, WESTVELD A H, et al, 2005. Calibrated probabilistic forecasting using ensemble model output statistics and minimum CRPS estimation[J]. Monthly Weather Review, 133:1098-1118.

HAMILL T M, COLUCCI S J, 1997. Verification of Eta-RSM short-range ensemble forecasts[J]. Monthly Weather Review, 125:1312-1327.

HAMILL T M, COLUCCI S J, 1998. Evaluation of Eta-RSM ensemble probabilistic precipitation forecasts[J]. Monthly Weather Review, 126:711-724.

HAMILL T M, WHITAKER J S, WEI X, 2004. Ensemble reforecasting: improving medium-range forecast skill using retrospective forecasts[J]. Monthly Weather Review, 132:1434-1447.

HAMILL T M, WHITAKER J S, MULLEN S L, 2006. Reforecasts: an important new dataset for improving weather predictions[J]. Bull Amer Meteor Soc, 87:33-46.

MAURICE J, SCHMEITS, KEES J, et al, 2010. A comparison between raw ensemble output, (modified)Bayesian model averaging, and extended logistic regression using ECMWF ensemble precipitation reforecasts[J]. Monthly Weather Review, 138(11):4199-4211.

MENDOZA P A, RAJAGOPALAN B, MARTYN P, et al, 2015. Statistical postprocessing of high-resolution regional climate model output[J]. Monthly Weather Review, 143(5):1533-1553.

RAFTERY A E, GNEITING T, BALABDAOUI F, et al, 2005. Using Bayesian model averaging to calibrate forecast ensembles[J]. Monthly Weather Review, 133: 1155-1174.

ROULSTON M S, SMITH L A, 2003. Combining dynamical and statistical ensembles[J]. Tellus, 55A: 16-30.

STEPHENSON D B, COELHO C A S, BALMASEDA M, et al, 2005. Forecast assimilation: a unified framework for the combination of multi-model weather and climate predictions[J]. Tellus, 57A: 253-264.

WILKS D S, HAMILL T M. 2007. Comparison of ensemble-MOS methods using GFS reforecasts[J]. Monthly Weather Review, 135: 2379-2390.

A Tropical Cyclone Similarity Search Algorithm Based on Deep Learning Method[①]

WANG Yu(王玉)[1]　　HAN Lei(韩雷)[2,*]　　LIN Yinjing(林隐静)[1]
SHEN Yue(沈悦)[3]　　ZHANG Wei(张巍)[4]

(1. National Meteorological Center of China Meteorological Administration,
Beijing, 100081; 2. College of Information Science and Engineering,
Ocean University of China, Qingdao, 266101; 3. China Meteorological
Administration Training Center, Beijing, 100081; 4. Department of
Computer Science and Technology, Ocean University of China, Qingdao, 266101)

Abstract: The tropical cyclone(TC) track forecast is still a challenging problem. For operational TC forecasts, it is useful for forecasters to find the similar TC in history and reference its data to improve TC forecasting. Considering the vast number of historical TC cases, it is necessary to design a suitable search algorithm to help forecasters find similar TC cases. A historical TC similarity search algorithm(named as SA_DBN) used deep learning approaches based on 500 hPa weather patterns was proposed in this study. Various weather features were automatically extracted by a deep belief network(DBN) without subjective influences. The Chebyshev distance was used to measure the similarity between two TCs. In order to show that similar-TCs retrieved by SA_DBN are helpful for forecasting, a modified WPCLPR method based on the standard WPCLPR and similar-TC track is designed. The modified WPCLPR improved the forecast result(at 85% confidence level) when the lead time was 54H, 60H or 66H. These results showed that the proposed algorithm could effectively retrieve similar TCs and be helpful to forecasters.

Key words: tropical cyclone track forecasting; deep learning; weather circulation; feature extraction

1 Introduction

　　The forecasting of tropical cyclone(TC) is still a challenging task as TCs are random and complex weather systems that have a serious impact on human society. An accurate track forecast is necessary, for if the track forecast is inaccurate, the intensity, storm surge, rainfall forecast will also be

---

① Atmospheric Research, 2018, 214(12): 386-398.

* Corresponding author address: HAN Lei, College of Information Science and Engineering, Ocean University of China, Qingdao, 266101, P. R. China. E-mail: hanlei@ouc.edu.cn.

inaccurate. The numerical weather prediction(NWP)models have been the main objective of TC forecasting methods. Using a consensus of NWP models, as well as ensemble members of the various models, can help reduce TC forecast errors(Elsberry, 2014). The TC track forecasting skill of operational NWP models has been steadily improved over recent decades(Goerss et al. ,2004;Elsberry,2014;Peng et al. ,2017). However, none of the prior techniques can achieve a satisfactory result. For example, models have not been able to routinely predict convectively-related "jumps" or reformations of the centers of TCs(Todd et al. ,2017).

The TC track forecasting methods, according to the main technique employed, can be grouped into the following categories: averaging across cyclones, statistical forecasting techniques, dynamic and numerical cyclone forecasting techniques, statistical-dynamic techniques and hybrid forecasting techniques (Roy et al. ,2012). In most of the time, historical TC track data is used to establish the regression equations(Neumann,1972;Xu et al. ,1985), the detailed track information provided by a similar TC is obscured. Feng et al. (2005)used retrieved similar-TC to forecast TC's intensity. Their study showed that the rational use of similar-TC data could improve the TC forecasts.

Although Carr et al. (2000)have shown that midlatitude circulation is an important factor that affects the forecast accuracy of TC track, the existing TC similarity search algorithms do not consider the similarity of circulation. For instance, the three criteria used in the National Meteorology Center of China Meteorological Administration(NMC of CMA)to identify similar TCs are geographical position similarity, seasonal similarity and moving speed similarity(Chen et al. ,1979;Xu et al. ,2013). This study made the first attempt to utilize the similarity of circulation in the TC similarity search algorithm. Our algorithm used a feature extraction(FE)algorithm, to be precise, a deep belief net(DBN), to extract 500 hPa circulation features, then retrieved the most similar historical TC case by comparing these features.

FE algorithms have been widely used in meteorology, including TC forecasting(Rios-Berrios et al. , 2016a;Rios-Berrios et al. ,2016b), weather pattern classification(Casado et al. ,2016;Demuzere et al. , 2011;Huth et al. ,2008;Philipp et al. ,2014), nowcasting of thunderstorms(Dixon et al. ,1993;Han et al. ,2009), satellite image processing(Lee et al. ,2004;Lazri et al. ,2018;Meyer et al. ,2016;Tao et al. , 2016;Xiao et al. ,2015;Zhuo et al. ,2014), and hydrographic information processing(Harish et al. ,2017). FE algorithms based on artificial neural network(ANN)are widely used in the scientific community. The advantage is that an unsupervised or semi-supervised FE algorithm is used instead of manual feature selection.

ANN is a computing system that was inspired by the biological NNs that constitute animal brains, and can be applied when dealing with numerical models. ANN is based on a collection of connected units or nodes called artificial neurons(analogous to biological neurons in an animal brain). This kind of network relies on the complexity of the system, through the adjustment of interconnections among internal nodes, to affect the degree of complicity of the network, thereby processing information. Based on ANN, Hinton proposed the concept of deep learning(Hinton et al. ,2006a,2006b;Lecun et al. ,2015). The goal of deep learning is to create a neural network(NN)that simulates the human brain, and mimics the human brain's mechanism of interpreting data. Based on DBN, Hinton(2009)proposed an unsupervised, greedy, layer-by-layer training procedure. In addition, a convolution neural network(CNN)is the first real multi-layer learning structure(Lecun et al. ,1990).

In terms of practical application, Kovordányi et al. (2009)used ANN to forecast the movement direction of TC. ANNs offered a more skillful rainfall forecasting in Australia(Abbot et al. ,2014,2017) and Greece(Nastos et al. ,2013,2014). Chen et al. (2015)used DBN in the classification of hyperspectral data and determined it to be an effective FE method, which reduced the dimensions of a feature and enabled a good reconstruction to be computed with extracted feathers. Le et al. (2017)used recurrent neural networks(RNNs)to investigate the complex interactions between the long-term trend in dryness and a projected, short but intense, period of wetness due to the 2015-2016 El Niño.

In this research, features extracted from 500 hPa circulation field were used to retrieve the similar TC case. Features were extracted by DBN from raw ERA-Interim 500 hPa geopotential height data(Dee et al. ,2011). The following section will introduce the methodology and DBN. The data and detailed methods involved in the design of the algorithm will be provided in the third section. The results of the algorithm will be presented in the fourth session.

2 Methodology and Theory

2.1 Methodology

This article aims to study a new TC similarity search algorithm which was based on the similarity of atmospheric environments. For traditional TC similarity search algorithms, more attention was paid on the similarity of characteristics of the TC itself(e. g. the geographical position similarity, seasonal similarity and moving speed similarity), the similarity of the TC's atmospheric environments was ignored.

By contrast, our algorithm started with the similarity of the atmospheric environment, by comparing the similarities of the 500 hPa circulation pattern to find the most similar circulation pattern, then determine the similar TC case(if there was a TC at that moment).

For ease of description, the TC similarity search algorithm is denoted as SA_DBN since DBN is used to extract circulation features; define target-TC(the algorithm's input in Fig. 1)as the TC that wants to search similar TC from historical TC dataset; define the most-similar-TC (the algorithm's output in Fig. 1)as the TC which is retrieved by SA_DBN.

The search algorithm SA_DBN can be divided into 3 main steps(Fig. 1):

Step 1: if the center pressure of target-TC $<$ 975 hPa, extract features from its 500 hPa geopotential height field.

Step 2: get the top 10 most similar 500 hPa geopotential height field from the archived historical dataset, the corresponding moments is named as $t_{min-1}, t_{min-2}, \cdots, t_{min-10}$, the similarity is ranked as $t_{min-1} > t_{min-2} > \cdots > t_{min-10}$.

Step 3: retrieve the historical TC dataset, checks whether there is a TC record at moment $t_{min-j}(j=0,1,\cdots,10)$, return the first TC record that satisfies criterion 2 and 3.

The reasons for using DBN as a feature extractor(in step 1, Fig. 1) were: Larsen et al. (2016)showed that feature-wise errors captured data distributions better than point-wise errors and they used an autoencoder to extract features. As a kind of autoencoder, DBN extract features from raw data, then try to restore the raw data from these features, which ensure these extracted features can represent the raw data as detailed as possible. In SA_DBN, only the FE process(the encoding) of DBN was used. For a detailed description of DBN, see section 2.3.

The reasons for using 3 criterion(in Fig. 1)were:

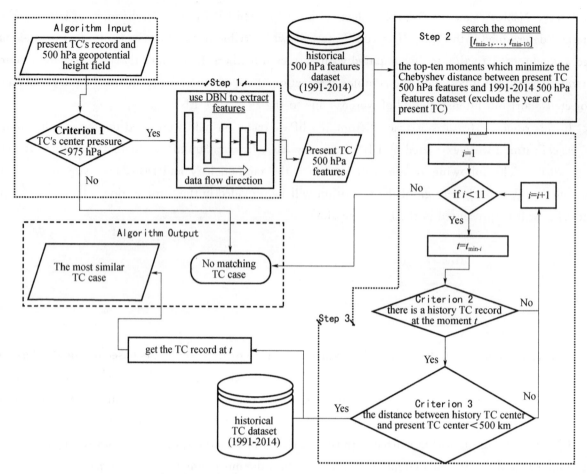

Fig. 1 Flowchart of the tropical cyclone(TC)similarity search algorithm

(1)Criterion 1: for a TC, there should be a corresponding closed low pressure center on the 500 hPa geopotential height field, then the FE algorithm was able to extract features with effective TC information. This required that TC should be strong enough, we found that using the TC center pressure as a threshold was more effective than the TC center wind speed. Thus the algorithm used the TC center pressure as a criterion for determining whether the TC was strong enough. 975 hPa was an empirical value.

(2)Criterion 2 and 3: make sure there was a history TC record at the moment t_{min} and excluded the existence of multiple TC records at the same time;

The main differences between SA_DBN and other common TC similarity search algorithm are:

(1)The similarity is determined by comparing the features of 500 hPa geopotential height field;

(2)The features of 500 hPa geopotential height field are extracted by an autoencoder(step 1 in Fig. 1), which exclude the influence of artificially specified thresholds;

As mentioned before, only the encoding of DBN was used in SA_DBN. But for a DBN, it should be fully trained before it was put into use(e. g. extract features). In the training process, the decoding needs to participate in training. The encoding and decoding will be described in section 2. 3.

Section 2. 2 will describe the general structure and the training method of RBMs, which is the major components of DBN, section 2. 3 will introduce the general structure and training method of DBNs.

2.2 RBMs

RBMs are some of the most common building blocks of deep probabilistic models(Goodfellow et al. ,2016). It is a generative stochastic ANN that can learn a probability distribution over its inputs.

As shown in Fig. 2, the RBM's nodes form two layers, referred to as visible and hidden layers, respectively. None of the nodes in the hidden layer has a connection with any other node, and the situation is identical for nodes in the visible layer. However, any two nodes from each of the two layers(one hidden layer node h_j and one visible layer node v_i) will have symmetric connections w_{ij} between them. Each node has a bias value, denoted as b_i and c_j for nodes v_i and h_j, respectively. A standard RBM has a binary-valued node state.

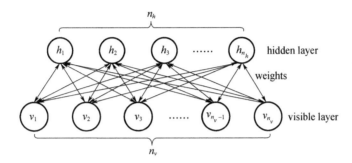

Fig. 2 Diagram of restricted Boltzmann machine(RBM) with n_v visible units and n_h hidden units

The bidirectional arrow indicates the symmetric connection between hidden and visible nodes.

n_v and n_h are the number of visible and hidden nodes, respectively.

Each node has two mutually exclusive states: activation(1) and deactivation(0)

Let $\theta = \{b_i, c_j, w_{ij}\}, i = 1, 2, \cdots, n_v, j = 1, 2, \cdots, n_h$ denote all parameters in a RBM, n_v and n_h are the number of visible and hidden nodes(Table 1), respectively. The vectors $v = [v_1, v_2, \cdots, v_{n_v}]^T$ and $h = [h_1, h_2, \cdots, h_{n_h}]^T$ indicate the states of the visible and hidden layer, respectively. The RBM's energy can be defined as:

$$E(v,h;\theta) = -\sum_{i=1}^{n_v} b_i v_i - \sum_{j=1}^{n_h} c_j h_j - \sum_{i=1}^{n_v}\sum_{j=1}^{n_h} w_{ij} v_i h_j \quad (1)$$

Table 1 The number of visible and hidden layer nodes for the four restricted Boltzmann machines(RBMs)

	RMB-1	RBM-2	RBM-3	RBM-4
the number of visible layer nodes	625	500	200	80
the number of hidden layer nodes	500	200	80	30

the probability distribution over state $\{v,h\}$ is:

$$P(v,h;\theta) = \exp[-E(v,h;\theta)]/Z \quad (2)$$

where $Z = \sum_{v,h} \exp[-E(v,h;\theta)]$ is a partition function defined as the sum of $\exp[-E(v,h;\theta)]$ over all possible states $\{v,h\}$. The individual activation probabilities of the visible node v_i and hidden node h_j are given by:

$$P(v_i = 1 \mid h;\theta) = \sigma(\sum_{j=1}^{n_h} w_{ij} h_j + c_i) \quad (3)$$

$$P(h_j = 1 \mid v;\theta) = \sigma(\sum_{i=1}^{n_v} w_{ij} v_i + b_j) \quad (4)$$

where $\sigma(x) = [1 + \exp(-x)]^{-1}$ denotes the logistic sigmoid function.

Because the RBM has no connections between nodes in the same layer, the hidden node activations are mutually independent given the visible node activations and vice versa (Carreira-perpinan et al., 2005). The probability of a visible layer state (not limited to the hidden layer state) v is:

$$P(v;\theta) = \frac{1}{Z}\sum_h \exp[-E(v,h;\theta)] \quad (5)$$

A fully trained RBM not only can extract features of the visible layer, but can also reconstruct the visible layer to be as similar as possible. Extracted features are stored in the hidden layer.

To train a RBM we need to find (Fischer et al., 2014; Sutskever et al., 2010):

$$\theta = \underset{\theta}{\mathrm{argmax}}\{\mathrm{Expect}[\log P(v;\theta)]\} \quad (6)$$

where $\mathrm{Expect}(x)$ is the expected value of variable x, while in $\underset{\theta}{\mathrm{argmax}}[f(\theta)]$ the value of θ is taken that maximizes the value of $f(\theta)$.

The FE algorithm adopted in this study used the contrastive divergence (CD) procedure to train RBMs. The CD (Hinton, 2012; Sutskever et al., 2010) procedure performed alternating Gibbs sampling for a very long time, and could provide an approximation of the maximum likelihood method that would ideally be applied to learn the weightings (Fischer et al., 2014).

2.3 DBN

DBN is a non-convolutional framework for effectively training deep architectures. It can be regarded as a cascade of several RBMs, where each RBM's hidden layer can be viewed as the visible layer for the next layer. Fig. 3 showed a DBN which could be regarded as a cascade of 4 RBMs: RBM-1 consisted of visible layer and hidden layer-1; RBM-2 consisted of hidden layer-1 and hidden layer-2 meanwhile the hidden layer of RBM-1 was the visible layer of RBM-2, and so on.

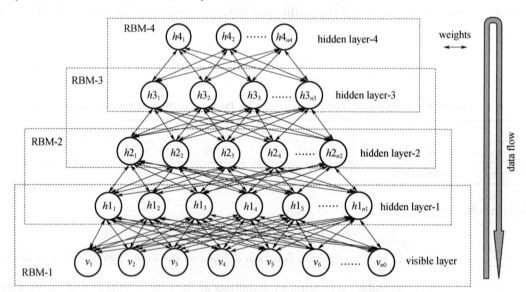

Fig. 3 Diagram of deep belief network (DBN)

The hidden layer of each RBM serves as the visible layer for the next

According to the direction of data flow (Fig. 3), the processes of DBN processing data can be divided into two process:

(1) Encoding: if a data is fed into DBN, the raw data is stored in the visible layer. Then the state of hidden layer-1 is updated by using Equation (4); after that, update the state of hidden layer-2 with the same method and so on until the state of hidden layer-4 is updated. During this process, features are

extracted from raw data, and saved in the hidden layer-4(this layer is named as bottleneck layer).

(2)Decoding: update hidden layer-3's state with Equation(3), RBM-3's parameters and hidden layer-4's state; use the same method to update hidden layer-2's state and so on, until the visible layer's state is updated. This process is the inverse of the encoding, also referred to as the decoding.

Fig. 4 showed the conceptual model of a 4-layer DBN which was expanded in the direction of dataflow. In this figure, the encoding, resp. decoding, corresponded with the description of encoding, resp. decoding. The state of hidden layer-4 is also regarded as features of raw data, stored in the bottleneck layer.

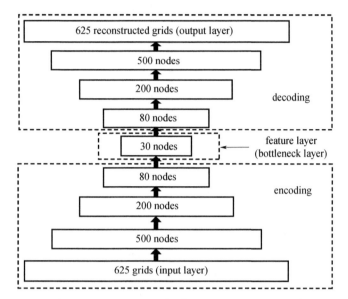

Fig. 4 The DBN structure in the feature extraction(FE)algorithm

This structure can be divided into three parts: the encoding, the feature layer(neck layer),
and the decoding. The encoding attempts to extract features and compress raw data.
Extracted features are stored in a feature layer. The decoding attempts to reconstruct raw data with these features

A DBN can be trained greedily(Hinton et al., 2006a). For example, if a DBN is a cascade of four RBMs(e. g., Fig. 3), RBM-1 should be trained first. After RBM-1 is fully trained, RBM-2 should be stacked on it, RBM-2's visible layer should be initialized by RBM-1's hidden layer, RBM-2 should be trained and then RBM-3 should be stacked on it, and so on. It has been demonstrated that this procedure provides an approximation of the maximum likelihood method that would ideally be applied for learning the weightings.

3 Data and Method

3.1 Data

The 1990-2016 500 hPa geopotential height data were obtained from the ERA-Interim reanalysis (Dee et al., 2011)produced by the European Centre for Medium-Range Weather Forecasts(ECMWF). The geographical range of the data was 0°-70°N,50°-146°E(Fig. 5). The horizontal spatial resolution and temporal resolution were 2.0°×2.0° and 6 h, respectively. The ERA-Interim dataset provided data at 0000,0600, 1200,and 1800 UTC. The grid size of the whole area was 36×49.

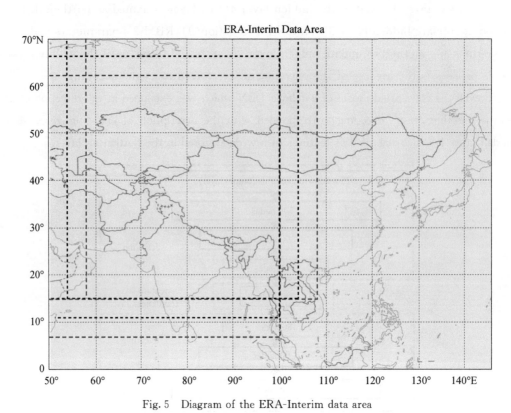

Fig. 5　Diagram of the ERA-Interim data area

The dashed box indicates the sliding windows. DBN extracts weather features from data in these sliding windows.
The sampling window slid along meridians/latitudinal directions in steps of two grids at a time

The 1990-2014 500 hPa geopotential height data were used to train the DBN. A total of 80% of this data was randomly selected as a training set, with the rest used as a validation set. To verify whether DBN could extract features effectively from the new weather process, 2015-2016 data was used as test data.

Many training cases are needed for the DBN to be fully trained, otherwise an over-fitting phenomenon will raise for the DBN have too many parameters(the 625-500-200-80-30 DBN had 433770 parameters). Thus, there were not enough training cases if the raw ERA-Interim data was used without any extra processing. A sliding window sampling technique was used to overcome this problem: a sampling window was prepared with 25×25 grid squares, and this sampling window was moved along the circle of longitude and latitude in steps of two grids at a time(see the dashed box in Fig. 5). Using this method, more than 2190000 training cases were generated from raw 1990-2014 ERA-Interim data, which was sufficient to meet the training needs.

The sliding window sampling method also avoided false features. The reconstructed 500 hPa geopotential height field would of had a permanent high-pressure center over South China(a false subtropical high) if we trained the DBN(the structure of raw-data DBN was 1764-1500-500-150-30 which had 3483388 parameters)with raw ERA-Interim data. This false subtropical high not only ignored the real weather process, but also interfered with the FE algorithm. There were two reasons for the false subtropical high:1)too many parameters in the DBN and too few training cases, which caused an over-fitting phenomenon; and 2)in most(but not all)training cases, there was indeed a subtropical high over South China.

For the probability of a visible node activation was used as input data, the geopotential height data had to be normalized to [0,1]. In the FE algorithm, we normalized each grid's value to [0,1], with its minimum and maximum value. A grid's minimum and maximum values were defined as the smallest and largest values on the grid from 1990 to 2014, respectively. These minimum and maximum values were also used to normalize data in 2015 and 2016. If a normalized value was >1, it was truncated to 1, whereas if a normalized value was <0, it was truncated to 0.

The CMA TC best track dataset (Ying et al., 2014) was used as the historical TC dataset. This best track dataset covered TCs that develop over the western North Pacific. The basin is to the north of the equator and to the west of 180°E, and includes the South China Sea (SCS). The present version of the dataset includes 6-hourly track and intensity analyses since 1949.

3.2 FE Method

The FE algorithm (step. 1, Fig. 1) used the encoding of DBN (Fig. 4) as the feature extractor. The DBN's structure was 625-500-200-80-30. This structure was an empirical result. In determining the structure of the DBN, we needed: 1) keep as much of the raw data as possible; 2) extract as few features as possible. A variety of structures were tested, and found the current structure was able to meet the above two requirements.

The detailed training process was given in section 2.3. Only the encoding and bottleneck layer (see Fig. 4) were retained after the DBN was fully trained. The extracted features were stored in the bottleneck layer, and were regarded as GPH-features.

3.3 Similarity Search Algorithm

A detailed statement of step 2 of SA_DBN (Fig. 1) is presented here. The retrieval method adopted the Chebyshev distance to compare the similarities of 500 hPa geopotential height field between the two moments:

$$t_f = \operatorname*{argmin}_{t_1}(d_{f,t_1}) = \operatorname*{argmin}_{t_1}\{\max_{i=1}^{30}[abs(feature_{i,t_0} - feature_{i,t_1})]\} \tag{7}$$

where $\operatorname*{argmin}_{t_1}(d_{f,t_1})$ means the value of t_1 is selected which minimizes the value of $d_{f,t1}$; $d_{f,t1}$ is the Chebyshev distance between t_0 and t_1; t_0 is the moment of the target TC, also known as the "target time"; t_1 traverses all the moments from 1990 to 2016 (excluding the year of t_0); $feather_{i,t}$ is the value of the i-th DBN feature at time t; and the most-similar-moment retrieved by this method is referred to as t_f. The smaller the value of $d_{f,t1}$, the more similar the two-moment data is.

In step 2, the search algorithm repeated Equation (7) 10 times, each time excluded the previous retrieval results, to get the top-10 most-similar-moments. The results were regarded as t_{min-1}, t_{min-2}, ⋯, t_{min-10}, the similarity was ranked as $t_{min-1} > t_{min-2} > ⋯ > t_{min-10}$.

3.4 Select the Most-Similar-TC Case

With the top 10 most-similar-moments, the algorithm retrieved historical TC data from the historical TC dataset. The retrieval process started from the moment with the highest similarity (i.e. t_{min-1}) to the moment with the lowest similarity (i.e. t_{min-10}), returned the first historical TC record (which met criterion 2 and 3) and its corresponding moment, as the most-similar-TC case and the most-similar-moment t_s, respectively.

4 Results

4.1 FE Algorithm Evaluation

The purpose of this section is to measure whether sufficient raw information is retained during the FE process(step 1, Fig. 1), since a very important evaluation indicator for a feature extractor is to retain the raw data information as much as possible. By comparing the differences between raw data and reconstructed data(reconstructed by the decoding in Fig. 4), whether the FE algorithm retains enough raw information can be verified.

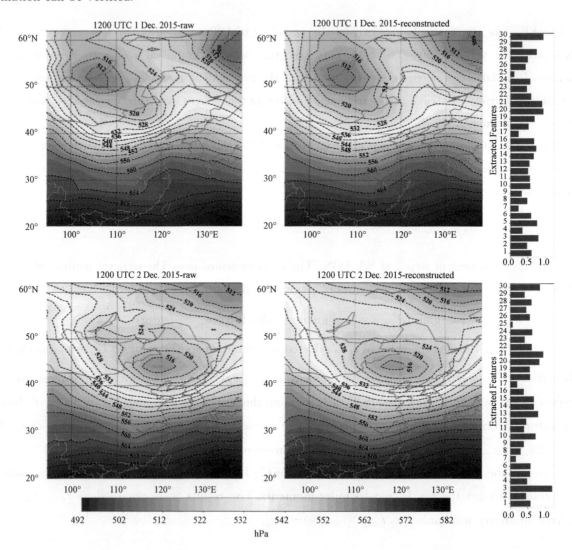

Fig. 6　The raw and reconstructed ERA-Interim 500 hPa geopotential height field over East Asia

The first column is the raw ERA-Interim 500 hPa geopotential height field, the second column is

the reconstructed field, and the third column gives the values of the 30 extracted feature nodes.

The reconstructed field(second column)was reconstructed based on the values of the 30 features(third column)

The mean error(ME)and root-mean-square error(RMSE)between raw ERA-Interim 500 hPa data and reconstructed data were −24.33 and 134.95 gpm, respectively. A cold vortex process over North Asia was shown in Fig. 6. The first column in Fig. 6 was the raw ERA-Interim 500 hPa field; the second column was the reconstructed field, which is reconstructed from the extracted 30 GPH-features; and the third column showed the values of 30 GPH-features. The major weather systems (a cold vortex over Mongolia and Northeast China and a subtropical high over the South China Sea and the Western Pacific) were well captured. The trend of geopotential height contours (dashed lines in Fig. 6) were also well preserved.

However, the strength of the low/high pressure centers of the reconstructed field were relatively weaker than the strength of the raw ERA-Interim 500 hPa geopotential height field. The reason for this was that in the DBN training procedure, the effect of extreme events on the parameters were smoothed by the effect of massive normal events. Training cases are fed into the DBN one after another to update its parameters. For local regions, a low-or high-pressure center is an extreme event. After feeding an extreme event, the DBN "remembered" this extreme event in the form of "update parameters"; however, the "remembered" extreme event will be overwritten if there are too many normal events after this extreme case.

4.2 Case 1-Tropical Cyclone Matmo(2014)

In this section and next section, two examples are presented to show the result of the similarity search algorithm (reference section 3.3). The 500 hPa geopotential height field of target-TC's moment and the most-similar-moment which was retrieved by Equation(7) were shown.

Typhoon Matmo(2014) (referred to as TC-1 hereinafter), the 10th typhoon to affect China in 2014, resulted in tremendous economic losses in the coastal provinces of China. By 0100 UTC 27 July 2014, the number of people affected by TC-1 in Liaoning, Jiangsu, Zhejiang, Anhui, Fujian, Jiangxi, Shandong, and Guangdong provinces reached 2.643 million; 13 people were killed; 289000 people were evacuated to resettlement locations; and 37000 people needed emergency supplies. Additionally, 190200 of hectares of crops were damaged, which resulted in a total crop failure of 13,000 hectares, with direct economic losses that reached ¥3.37 billion (US$524 million).

The ERA-Interim 500 hPa geopotential height field at 0000 UTC 22 July 2014 (regarded as $t_{0,1}$) was shown in Fig. 7a. At this time, TC-1's circulation was relatively weak (the 568 hPa closed isobar at 21°N, 123°E; this location was marked as L1). Northwest of TC-1, there was a strong subtropical high spread across central China (about 30°-40°N, 105°-125°E). Northeast of this subtropical high, there was a cold vortex located over the Stanovoy Range (about 57°N, 136°E).

To use the similarity search algorithm, the target time t_0 was set equal to $t_{0,1}$ in Equation(7). The search algorithm traversed all the moments from 1990 to 2016 (except 2014, which was the year containing the target time $t_{0,1}$) to obtain the case most similar to $t_{0,1}$. The search result was 0000 UTC 3 August 2003 (regarded as $t_{f,1}$). Fig. 7b showed the ERA-Interim 500 hPa geopotential height field at $t_{f,1}$.

There were several differences between the 500 hPa geopotential height fields of $t_{0,1}$ and $t_{f,1}$: 1) the shape of the subtropical high over central China was slightly different between $t_{0,1}$ and $t_{f,1}$; and 2) the intensity of the low pressure center (56°N, 134°E) at $t_{f,1}$ was weaker than that of $t_{0,1}$. Both fields had a high and a low pressure center over North Asia (45°-60°N), a subtropical high over central China (30°-40°N), and a TC over South China (20°-25°N). This indicated that the general weather circulation was almost the same.

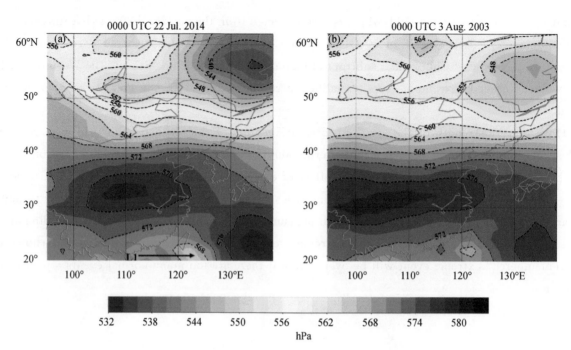

Fig. 7 The ERA-Interim 500 hPa geopotential height field at $t_{0,1}$
(0000 UTC 22 Jul. 2014) (a) and its most similar case ($t_{f,1}$) retrieved by the similarity search algorithm (b)

It was also found that the evolution of weather systems had a certain degree of similarity. Fig. 8 showed the 500 hPa geopotential height field at $t_{0,1}+30$ h (Fig. 8a) and $t_{f,1}+30$ h (Fig. 8b). TC-1 moved from location L1 to L2(24.5°N,119.5°E, the closed 564 hPa isobar in Fig. 8a). At the same time, around location L2, there was a TC(TC-2) at moment $t_{f,1}+30$ h (see Fig. 8b). We backtracked the 500 hPa geopotential height field from $t_{0,1}$ ($t_{f,1}$) to $t_{0,1}+30$ h ($t_{f,1}+30$ h), and found that although TC-2 was weaker than TC-1, they had almost the same track at the 500 hPa pressure level. The 500 hPa circulations at $t_{0,1}+30$ h (Fig. 8a) and $t_{f,1}+30$ h (Fig. 8b) also had a certain similarity.

4.3 Case 2-Tropical Cyclone Nepartak(2016)

Typhoon "Nepartak"(2016) (referred to as TC-3 hereinafter) had a great impact on China. By 0900 UTC 10 July 2016, the number of people affected by TC-3 across Fujian and Jiangxi provinces had reached 449000; six people were killed, and eight others were reported missing; 203000 people were evacuated to resettlement locations; and 2100 people needed emergency supplies. Almost 1900 buildings were demolished, and 11000 others were damaged to varying levels. Additionally, 15800 hectares(39000 acres) of crops were damaged, which resulted in a total crop failure of 1600 hectares; and total economic losses of ￥2.2 billion(US＄320 million).

The corresponding ERA-Interim 500 hPa geopotential height field at 1800 UTC 6 July 2016 (regarded as $t_{0,2}$) was shown in Fig. 9a. At this moment, TC-3 was located at L3 (see Fig. 9a, the closed 572 hPa isobar at 20°N,126°N). Northwest of TC-3, there was a weak and widespread area of high pressure over central and southern China (the high-pressure center was located at 30°N,110°E, the closed 576 hPa isobar in Fig. 9a). Northeast of this high pressure, there was a low-pressure center over the Stanovoy Range (58°N,136°E). The circulation at that moment was relatively weak, with no significant weather system except TC-3 over the South China Sea.

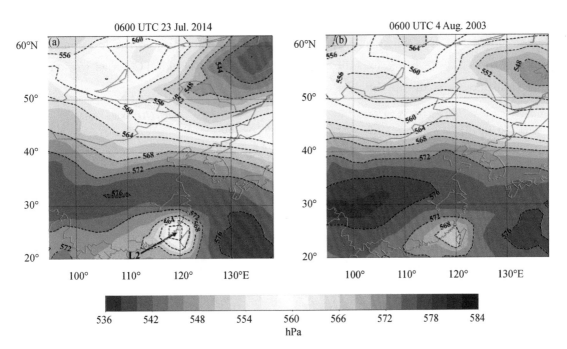

Fig. 8 The ERA-Interim 500 hPa geopotential height field at $t_{0,1}+30$ h (0600 UTC 23 Jul. 2014)(a) and at $t_{f,1}+30$ h (0600 UTC 4 Aug. 2003)(b)

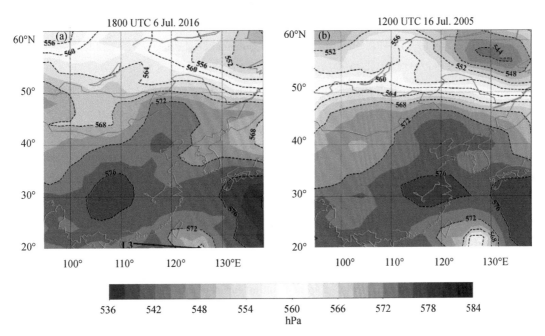

Fig. 9 The ERA-Interim 500 hPa geopotential height field at $t_{0,2}$ (1800 UTC 6 Jul. 2016) (a) and its most similar case ($t_{f,2}$) retrieved by the similarity search algorithm (b)

The most similar search algorithm's search result was 1200 UTC 16 July 2005 (regarded as $t_{f,2}$). As Fig. 9 showed, the 500 hPa circulation field had some similarities between the two times. At both times, the 500 hPa circulation field had a weak and wide spread area of high pressure in central and southern China (around 30°N), an area of low pressure over the Stanovoy Range (about 58°N, 136°E), and a TC over the South China Sea (around L3).

The TC appeared in $t_{f,2}$ was regarded as TC-4. Fig. 10 showed the 500 hPa geopotential height field at $t_{0,2}+48$ h(Fig. 10a)and $t_{f,2}+48$ h(Fig. 10b). TC-3 moved from location L3 to location L4(23°N, 119.5°E, the closed 564 hPa isobar in Fig. 10a). At the same time, around the location L4, TC-4 was present at moment $t_{f,2}+48$ h(see Fig. 10b). It was found that although TC-3 was weaker than TC-4, they had a similar track at the 500 hPa pressure level.

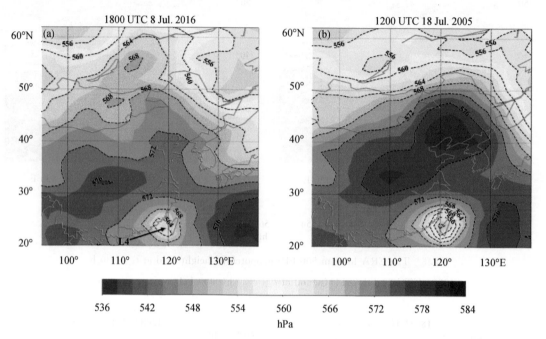

Fig. 10　The ERA-Interim 500 hPa geopotential height field at $t_{0,2}+48$ h (1800 UTC 8 Jul. 2016)(a)and at $t_{f,2}+48$ h(1200 UTC 18 Jul. 2005)(b)

4.4　TC Similarity Search Algorithm Evaluation

The general performance of SA_DBN is evaluated here. The assessment mainly examine the Δlon and Δlat between the most-similar-TC track and target-TC track at the corresponding moment.

From 1990 to 2016, there were a total of 330 TCs and 1410 records(each moment of a TC is one record)met criterion 1, and entered the statistical area which was defined as east of 130°E and north of 20°N. There were 1312 records that had the most-similar-TC, 98 records that failed to get the most-similar-TC.

The longitude and latitude differences were calculated as below:

$$\Delta \mathrm{lon}_i^{DBN} = \mathrm{lon}_{t_0+i}^{target-TC} - \mathrm{lon}_{t_s+i}^{similar-TC} \tag{8}$$

$$\Delta \mathrm{lat}_i^{DBN} = \mathrm{lat}_{t_0+i}^{target-TC} - \mathrm{lat}_{t_s+i}^{similar-TC} \tag{9}$$

where $\mathrm{lon}_{t_0+i}^{target-TC}$, resp. $\mathrm{lat}_{t_0+i}^{target-TC}$, is the longitude, resp. latitude, of target TC at moment t_0+i; $\mathrm{lon}_{t_s+i}^{similar-TC}$, resp. $\mathrm{lat}_{t_0+i}^{similar-TC}$, is the longitude, resp. latitude, of the most-similar-TC at moment t_s+i; $i = +6$ h, +12 h, …, +72 h, is the lead times. The superscript DBN indicates that these values correspond to the algorithm SA_DBN.

Table 2 showed the mean and standard deviation(STD)of $\Delta \mathrm{lon}_i^{DBN}$ and $\Delta \mathrm{lat}_i^{DBN}$. As table 2 showed, the lead time had very little effect on the mean of $\Delta \mathrm{lon}_i^{DBN}$, but the STD of $\Delta \mathrm{lon}_i^{DBN}$ increased significantly with the extension of lead time. The performance of $\Delta \mathrm{lat}_i^{DBN}$ was almost the same as that of $\Delta \mathrm{lon}_i^{DBN}$. This indicated that the distance between the most-similar-TC and target-TC might increase significantly with the extension of lead time.

Table 2　The mean and standard deviation(STD) of Δlon_i and Δlat_i. i is lead time(hour). For ease of writing, the superscript DBN is omitted

	mean	STD	mean	STD	mean	STD	mean	STD
i	6		12		18		24	
Δlon_i	−0.29	2.36	−0.41	2.68	−0.49	3.20	−0.49	3.92
Δlat_i	0.48	2.03	0.56	2.29	0.62	2.73	0.66	3.27
i	30		36		42		48	
Δlon_i	−0.52	4.70	−0.48	5.61	−0.43	6.63	−0.32	7.66
Δlat_i	0.74	3.88	0.80	4.56	0.90	5.30	0.99	6.02
i	54		60		66		72	
Δlon_i	−0.34	8.77	0.25	9.70	0.55	10.39	0.88	10.52
Δlat_i	1.07	6.73	1.33	7.24	1.38	7.44	1.58	7.54

4.5　A Simple Application

To show that the most-similar-TC retrieved by SA_DBN is able to improve TC forecasting, a simple TC track extrapolation method based on WPCLPR(western north pacific tropical cyclone prediction, Xu et al., 1985) was designed.

The TC historical dataset from 1990 to 2014 was used to fit the WPCLPR equations. Stepwise regression was used to retain the 32 most significant predictors. The predicted longitude and latitude values are named as $lon_{t_0+i}^{TC,WPCLPR}$ and $lat_{t_0+i}^{TC,WPCLPR}$, respectively. The superscript WPCLPR indicate that these values correspond to the algorithm WPCLPR.

Then several displacements were calculated:

$$\Delta lon_i^{target-TC} = lon_{t_0+i}^{target-TC} - lon_{t_0+i}^{target-TC,WPCLPR} \tag{10}$$

$$\Delta lat_i^{target-TC} = lat_{t_0+i}^{target-TC} - lat_{t_0+i}^{target-TC,WPCLPR} \tag{11}$$

$$\Delta lon_i^{similar-TC} = lon_{t_s+i}^{similar-TC} - lon_{t_s+i}^{similar-TC,WPCLPR} \tag{12}$$

$$\Delta lat_i^{similar-TC} = lat_{t_s+i}^{similar-TC} - lat_{t_s+i}^{similar-TC,WPCLPR} \tag{13}$$

where $\Delta lon_i^{target-TC}$, resp. $\Delta lat_i^{target-TC}$, is the longitude, resp. latitude, forecast error of target-TC; $\Delta lon_i^{similar-TC}$, resp. $\Delta lat_i^{similar-TC}$, is the longitude, resp. latitude, forecast error of the most-similar-TC.

Since a target-TC corresponded with a most-similar-TC, we used $\Delta lon_i^{similar-TC}$ and $\Delta lat_i^{similar-TC}$ to correct the value of $\Delta lon_i^{target-TC}$ and $\Delta lat_i^{target-TC}$. That was, based on historical TC data, retrieved each available target-TC and most-similar-TC pair, and then used a third-order polynomial to regress the forecast error equations:

$$\overline{\Delta lon_i^{target-TC}} = \alpha_{0,i} + \alpha_{1,i}\Delta lon_i^{similar-TC} + \alpha_{2,i}(\Delta lon_i^{similar-TC})^2 + \alpha_{3,i}(\Delta lon_i^{similar-TC})^3 \tag{14}$$

$$\overline{\Delta lat_i^{target-TC}} = \beta_{0,i} + \beta_{1,i}\Delta lat_i^{similar-TC} + \beta_{2,i}(\Delta lat_i^{similar-TC})^2 + \beta_{3,i}(\Delta lat_i^{similar-TC})^3 \tag{15}$$

Equations (14) and (15) were used to improve the target-TC's WPCLPR forecast, this method is regarded as modified WPCLPR, the symbol ⁀ indicated the modified WPCLPR method:

$$\overline{lon_{t_0+i}^{target-TC,WPCLPR}} = lon_{t_0+i}^{target-TC,WPCLPR} + \overline{\Delta lon_i^{target-TC}} \tag{16}$$

$$\overline{lat_{t_0+i}^{target-TC,WPCLPR}} = lat_{t_0+i}^{target-TC,WPCLPR} + \overline{\Delta lat_i^{target-TC}} \tag{17}$$

Since the WPCLPR method used the historical TC data (1990 to 2014) to fit extrapolation equations, the 2015-2016 TC data was used as a test dataset to show whether the modified WPCLPR method could improve the forecast skill. The forecast errors of standard and modified WPCLPR method were defined as:

$$e_i = \text{disHav}[(\text{lon}_{t_0+i}^{\text{target-TC}}, \text{lat}_{t_0+i}^{\text{target-TC}}), (\text{lon}_{t_0+i}^{\text{target-TC,WPCLPR}}, \text{lat}_{t_0+i}^{\text{target-TC,WPCLPR}})] \quad (18)$$

$$\hat{e}_i = \text{disHav}[(\text{lon}_{t_0+i}^{\text{target-TC}}, \text{lat}_{t_0+i}^{\text{target-TC}}), (\overline{\text{lon}_{t_0+i}^{\text{target-TC,WPCLPR}}}, \overline{\text{lat}_{t_0+i}^{\text{target-TC,WPCLPR}}})] \quad (19)$$

where the function disHav(p1,p2) is the great-circle distance between points p1 and p2. The mean values of e_i and $\hat{e}_i$ were shown in table 3. Besides the mean values, we did a two side t-test on e_i and $\hat{e}_i$. The null hypothesis of the t-test was that e_i and $\hat{e}_i$ had identical average values. The p-values of the t-test were shown in table 3, too. The smaller the p-value, the higher the probability of rejecting the null hypothesis. As shown in table 3, the p-value was smaller than 0.15 when the lead time was 54 h, 60 h or 66 h. That is, we can say that the modified WPCLPR method improve the forecast result (at confidence level 85%) when the lead time is 54 h, 60 h or 66 h.

Table 3   The mean of standard WPCLPR forecast errors e_i and modified WPCLPR forecast errors $\hat{e}_i$. i is the lead time

i	6	12	18	24	30	36
e_i	39.8	79.9	124.2	181.3	241.5	302.2
$\hat{e}_i$	40.0	80.3	124.8	184.7	233.9	290.5
p-value*	0.494	0.477	0.541	0.556	0.451	0.378
i	42	48	54	60	66	72
e_i	351.3	404.1	444.6	498.8	547.1	627.6
$\hat{e}_i$	339.3	372.8	396.9	436.1	473.7	585.5
p-value	0.340	0.190	0.110	0.080	0.092	0.258

* The p-value of a two side t-test. The null hypothesis of the t-test was that e_i and $\hat{e}_i$ had identical average values. The smaller the p-value, the higher the probability of rejecting the null hypothesis.

5 Discussion

A TC similarity search algorithm based on the features of 500 hPa geopotential height field was investigated. This algorithm retrieved the most-similar-TC case from the historical TC dataset, by comparing the features of 500 hPa geopotential height field. The features of 500 hpa geopotential height field were extracted by a DBN with structure 625-500-200-80-30, which was an unsupervised learning framework.

To show the differences between our algorithm and other similar TC search algorithms (Fraedrich et al., 1989; Alt et al., 2004; Pelekis et al., 2007; Yu et al., 2013), we tested two additional similar TC search algorithms. For ease of description, these two algorithms were denoted as SA_MA and SA_MD respectively. The principles of these two algorithms were briefly described as below:

(1) SA_MA: This algorithm was based on the principle of minimum area, which meant the polygon enclosed by the target TC's track and similar TC's track should be the smallest. The most-similar-TC and its corresponding moment, TC_{MA} and t_{MA}, were selected by the following equation:

$$\text{TC}_{\text{MA}}, t_{\text{MA}} = \underset{\text{TC},t}{\arg\min}[\text{polyArea}(p_{t_0}, p_{t_0-1}, p_{t_0-2}, p'_{t-2}, p'_{t-1}, p'_t)] \quad (20)$$

the function polyArea is the area of the polygon enclosed by points list $[p_{t_0}, p_{t_0-1}, p_{t_0-2}, p'_{t-2}, p'_{t-1}, p'_t]$, p_{t_0} is the location of target-TC at moment t_0, p'_t represents the location of any TC in the historical TC dataset at moment t, p_{t_0-1} is the location of target-TC at the previous moment of t_0, and so on; the argmin$_{TC,t}$ indicates selecting the TC and moment t which minimizes the value of function polyArea.

(2)SA_MD: This algorithm was based on the principle of minimum distance, which indicated the maximum distance between target-TC's track and similar TC's track should be the minimum. The most-similar-TC and its corresponding moment, TC$_{MD}$ and t$_{MD}$, were selected by the following equation:

$$TC_{MD}, t_{MD} = \underset{TC,t}{\arg\min}[\max_{i=0}^{2}(disHav(p_{t_0-i}, p'_{t-i}))] \quad (21)$$

the function disHav means the great-circle distance between p_{t_0-i} and p'_{t-i}, other definitions are consistent with Equation(20).

Like Equations(8)and(9), the same values for method SA_MA and SA_MD were calculated, superscription MA(MD)indicates that these values correspond to the algorithm SA_MA(SA_MD):

$$\begin{aligned}
\Delta lon_i^{MA} &= lon_{t_0+i}^{target-TC} - lon_{t_{MA}+i}^{TC_{MA}} \\
\Delta lat_i^{MA} &= lat_{t_0+i}^{target-TC} - lat_{t_{MA}+i}^{TC_{MA}} \\
\Delta lon_i^{MD} &= lon_{t_0+i}^{target-TC} - lon_{t_{MD}+i}^{TC_{MD}} \\
\Delta lat_i^{MD} &= lat_{t_0+i}^{target-TC} - lat_{t_{MD}+i}^{TC_{MD}}
\end{aligned} \quad (22)$$

Fig. 11 showed the space distribution of $[\Delta lon_i^{DBN}, \Delta lat_i^{DBN}]$ (red spots), $[\Delta lon_i^{MA}, \Delta lat_i^{MA}]$ (green spots)and $[\Delta lon_i^{MD}, \Delta lat_i^{MD}]$(blue spots)at different lead times. The performance of method SA_MD was better than the remaining two algorithms when lead time $i \leqslant 36$ h. This is because, in a short period of time, the TC's inertia has a more pronounced effect on its track. Therefore, the algorithm SA_MD which was based on the distance between TC tracks could achieve a better result. As the TC exchanges material with the surrounding atmosphere environment, the influence of the atmospheric environment on the TC gradually increases. Reflected in Fig. 11, as the lead time gradually increased, the performance of SA_DBN gradually caught up with SA_MD. On the other hand, compared with SA_MA, SA_DNB had better performance at all lead times. In additional, as the lead time increased, the correlation coefficient of Δlon_i^{DBN} and Δlat_i^{DBN} increased rapidly, which indicated the eccentricity of error ellipse decreased rapidly with the increased lead time.

Although in short term(the lead time$\leqslant$36 h), SA_DBN had no advantage over the method which directly compare TC's track(the method SA_MD), but for algorithm that retrieved similar track by comparing indirect indicators(for example, the method SA_MA, retrieve the most similar track through the minimum area), SA_DBN had obvious advantages.

Different search algorithms have its own advantages and disadvantages: in some cases the track of the most-similar-TC case retrieved by SA_DBN are more consistent with the target TC track than others, sometimes the situation is opposite. Therefore, if combine SA_DBN with other similar search algorithms, more diverse similar TC cases are available for forecasters to reference.

As we stated in section 2.1, the prominent characteristic of SA_DBN was that features were selected by DBN, which could exclude the influence of artificially specified thresholds(since the structure of the DBN was given by researchers, the artificially influence couldn't be completely ruled out).

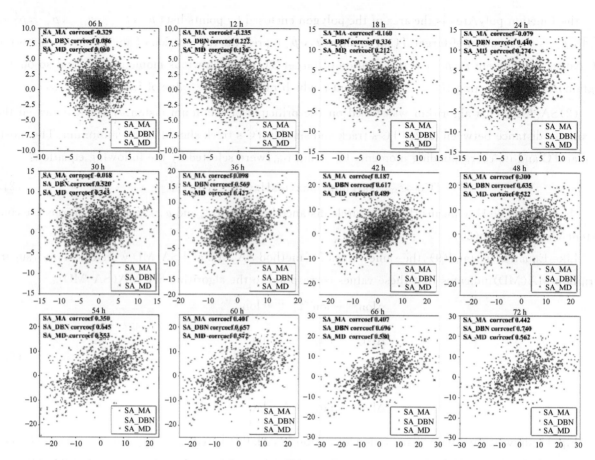

Fig. 11 The space distribution of $[\Delta \text{lon}_i^{DBN}, \Delta \text{lat}_i^{DBN}]$ (red), $[\Delta \text{lon}_i^{MA}, \Delta \text{lat}_i^{MA}]$ (green) and $[\Delta \text{lon}_i^{MD}, \Delta \text{lat}_i^{MD}]$ (blue) at different lead times. The corrcoef is an abbreviation for the correlation coefficient

Another reason for using DBN as a feature extractor was: we were not able to find a reliable human-consumable FE algorithm for our purpose. Therefore we used a NN(DBN is a kind of NN) to approximate the ideal FE algorithm.

As a tool, NNs can approximate functions that cannot be achieved explicitly. That is, when we cannot precisely implement a function $f(x)$ due to a lack of necessary knowledge, economic cost, or other reasons, we can try to find an approximation: $\tilde{f}(x)$. It should be as close to $f(x)$ as possible but does not strictly require $f(x) = \tilde{f}(x)$, which also means that a certain amount of error is allowed.

For example, any human driver can drive a car, but if we want to achieve the same function using a computer program, many questions must be addressed. It is almost impossible to achieve this function with a human-consumable format. However, it could be approximated by NN, e.g., Waymo(the Google self-driving car). In the training procedure, NN is updated to approximate the optimal solution gradually.

NN's black box cannot be opened with current technology. It is impossible to exhaust every possible situation. Even in this study, the algorithm extracted 30 characteristics, which corresponded to 2^{30} combinations of these features. Such situations are too complicated for a manual analysis.

This research showed that although these features were not human readable(the reconstructed field which set the value of a single feature to 1 and others to 0 was meaningless from a meteorological point of view, these figures were not shown here), we could still use them to measure the similarity between 500 hPa geopotential height fields. In order to make full use of these features, another supervised learning network, like a fully connected NN which the input data is the features extracted by DBN, can be trained to calculate the parameters we want. In this way, the DBN actually plays a similar role to the convolution part of CNN(extract features).

6 Conclusion

This study used DBN to extract ERA-Interim 500 hPa geopotential height field features from 25×25 geopotential height grid squares(these features are regarded as GPH-features). The RMSE and ME between the raw ERA-Interim 500 hPa geopotential height field and a reconstructed field(reconstructed based on GPH-features) were 134.95 and -24.33 gpm, respectively. The reconstructed field could restore the main weather systems at the 500 hPa geopotential height field, but the intensity of reconstructed low/high-pressure centers were weaker than the raw data suggested. This is because, for a single grid, a low-or high-pressure center is an extreme event, which would be smoothed by several normal events during the DBN training process.

A TC similarity search algorithm(SA_DBN) was designed. This algorithm used the GPH-features and Chebyshev distance between two moments to retrieve the most-similar-TC case. It was found that the retrieved moment's 500 hPa geopotential height field was very consistent with that of the target moment. Furthermore, the similarity was manifested in a time series, which meant that at the target moment the algorithm retrieved the most-similar-moment, and the 500 hPa geopotential height field's development trend had some degree of consistency. This phenomenon indicated that GPH-features had physically meaningful characteristics.

In this FE algorithm, only the 500 hPa geopotential height field were used when measuring the similarity between two moments. But the reality is that TC is a three-dimensional system. And not only the geopotential height field, but also the wind field, temperature field, water vapor filed, topography, et al., can affect the TC's development. Therefore our next jobs are:1)extend the FE method into 3-dimension field, including but not limited to geopotential height field, temperature field and water vapor field, design an algorithm to extract 3-dimensional features from these fields;2)find a suitable measure index to select the most-similar-TC automatically. This measure index will refer to not only the TC's environmental features but also the TC's forecasting parameters such as track and precipitation field.

Acknowledgement

This work was supported jointly by National Natural Science Foundation of China(grant numbers 41741013,41875049 and 41405110), the National Meteorological Center Youth Fund(grant number Q201620), and the Natural Science Foundation of Shandong Province(grant number ZR2016DM05). The authors thank ECMWF for providing the ERA-Interim data. The authors appreciates the reviewers' suggestions for this article. We gratefully acknowledge the support of NVIDIA Corporation with the donation of the Tesla K40 GPU used for this research.

References

ABBOT J,MAROHASY J,2014. Input selection and optimisation for monthly rainfall forecasting in Queensland, Australia, using artificial neural networks[J]. Atmos Res,138:166-178.

ABBOT J,MAROHASY J,2017. Skilful rainfall forecasts from artificial neural networks with long duration series and single-month optimization[J]. Atmos Res,197:289-299.

ALT H,KNAUER C,WENK C,2004. Comparison of distance measures for planar curves[J]. Algorithmica,38:45-58.

CARR L E,ELSBERRY R L,2000. Dynamical tropical cyclone track forecast errors. Part II:midlatitude circulation influences[J]. Weather Forecast,15:662-681.

CARREIRA-PERPINAN M A,HINTON G E,2005. Contrastive divergence learning[C/OL]//International conference on artificial intelligence and statistics. http://www.cs.utoronto.ca/~hinton/absps/cdmiguel.pdf.

CASADO M J,PASTOR M A,2016. Circulation types and winter precipitation in Spain[J]. Int J Climatol,36(7):2727-2742.

CHEN L S,DING Y H,1979. Introduction to the Western Pacific typhoon[M]. Beijing:Science Press.

CHEN Y,ZHAO X,JIA X,2015. Spectral-spatial classification of hyperspectral data based on deep belief network[J]. IEEE Journal of Selected Topics in Applied Earth Observations and Remote Sensing,8(6):2381-2392.

DEE D P,UPPALA S M,SIMMONS A J,et al.,2011. The ERA-Interim reanalysis:configuration and performance of the data assimilation system[J]. Q J Roy Meteor Soc,137(656):553-597.

DEMUZERE M,KASSOMENOS P,PHILIPP A,2011. The COST733 circulation type classification software:an example for surface ozone concentrations in Central Europe[J]. Theor Appl Climatol,105:143-166.

DIXON M,WIENER G,1993. TITAN:thunderstorm identification,tracking,analysis,and nowcasting-a radar-based methodology[J]. J Atmos Ocean Technol,10(6):785-797.

ELSBERRY R L,2014. Advances in research and forecasting of tropical cyclones from 1963-2013[J]. Asia-Pacific J Atmos Sci,50:3-16.

FENG B,LIU J N K,2005. Similarity retrieval from time-series tropical cyclone observations using a neural weighting generator for forecasting modeling[M]//KHOSLA R,HOWLETT R J,JAIN L C. Knowledge-Based Intelligent Information and Engineering Systems. KES 2005. Lecture Notes in Computer Science,vol 3683. Berlin:Springer.

FISCHER A,IGEL C,2014. Training restricted boltzmann machines[J]. Pattern Recongn,47(1):25-39.

FRAEDRICH K,LESLIE L M,1989. Estimates of cyclone track predictability. I:tropical cyclones in the Australian region [J]. Quarterly Journal of the Royal Meteorological Society,115:79-92.

GOERSS J S,SAMPSON C R,GROSS J M,2004. A history of Western North Pacific tropical cyclone track forecast skill[J]. Weather Forecast,19:633-638.

GOODFELLOW BENGIO I Y,COURVILLE A,2016. Deep learning[M]. Cambridge:MIT Press.

HAN L,FU S,ZHAO L,et al,2009. 3D convective storm identification,tracking,and forecasting-an enhanced TITAN algorithm[J]. J Atmos Ocean Technol,26(4):719-732.

HARISH S,COLIN P S,PAOLA P,2017. Multiresolution analysis of characteristic length scales with high-resolution topographic data[J]. J Geophys Res:Earth Surface,122:1296-1324.

HINTON G E,2009. Deep belief networks[J]. Scholarpedia,4(5):5947.

HINTON G E,2012. A practical guide to training restricted boltzmann machines[J]. Momentum,9(1):599-619.

HINTON G E,OSINDERO S,TEH Y W,2006a. A fast learning algorithm for deep belief nets[J]. Neural Comput,18(7):1527-1554.

HINTON G E,SALAKHUTDINOV R,2006b. Reducing the dimensionality of data with neural networks[J]. Science,313:504-507.

HUTH R,BECK C,PHILIPP A,et al,2008. Classifications of atmospheric circulation patterns:recent advances and applications[J]. Ann N Y Acad Sci,1146(1):105-152.

KOVORDÁNYI R,ROY C,2009. Cyclone track forecasting based on satellite images using artificial neural networks[J]. Isprs Journal of Photogrammetry and Remote Sensing,64(6):513-521.

LARSEN A B,SONDERBY S K,LAROCHELLE H,et al,2016. Autoencoding beyond pixels using a learned similarity metric[R]. International Conference On Machine Learning.

LAZRI M, AMEUR S, 2018. Combination of support vector machine, artificial neural network and random forest for improving the classification of convective and stratiform rain using spectral features of SEVIRI data[J]. Atmos Res, 203: 118-129.

LE J A, EL-ASKARY H M, ALLALI M, et al, 2017. Application of recurrent neural networks for drought projections in California[J]. Atmos Res, 188: 100-106.

LECUN Y, BOSER B E, DENKER J S, et al, 1990. Handwritten digit recognition with a back-propagation network[R]. Advances in Neural Information Processing Systems (NIPS 1989), Denver, CO (Vol. 2) Morgan Kaufmann.

LECUN Y, BENGIO Y, HINTON G E, 2015. Deep learning[J]. Nature, 521(7553): 436-444.

LEE Y, WAHBA G, ACKERMAN S A, 2004. Cloud classification of satellite radiance data by multicategory support vector machines[J]. J Atmos Ocean Technol, 21(2): 159-169.

MEYER H, KÜHNLEIN M, APPELHANS T, et al, 2016. Comparison of four machine learning algorithms for their applicability in satellite-based optical rainfall retrievals[J]. Atmos Res, 169: 424-433.

NASTOS P T, MOUSTRIS K P, LARISSI I K, et al, 2013. Rain intensity forecast using artificial neural networks in Athens, Greece[J]. Atmos Res, 119: 153-160.

NASTOS P T, PALIATSOS A G, KOUKOULETSOS K V, et al, 2014. Artificial neural networks modeling for forecasting the maximum daily total precipitation at Athens, Greece[J]. Atmos Res, 144: 141-150.

NEUMANN C J, 1972. An alternate to the HURRAN (Hurricane Analog) tropical cyclone forecast system[R/OL]. NOAA Tech. Memo. NWS SR-62, p. 34. https://repository.library.noaa.gov/view/noaa/3605.

PELEKIS N, KOPANAKIS I, MARKETOS G, et al, 2007. Similarity search in trajectory databases[R]. 14th International Symposium on Temporal Representation and Reasoning (TIME'07), pp. 129-140.

PENG X, FEI J, HUANG X, et al, 2017. Evaluation and error analysis of official forecasts of tropical cyclones during 2005-14 over the Western North Pacific. Part Ⅰ: storm tracks[J]. Weather and Forecasting, 32: 689-712.

PHILIPP A, BECK C, HUTH R, et al, 2014. Development and comparison of circulation type classifications using the COST 733 dataset and software[J]. Int J Climatol, 36(7): 2673-2691.

RIOS-BERRIOS R, TORN RD, DAVIS C A, 2016a. An ensemble approach to investigate tropical cyclone intensification in sheared environments. Part Ⅰ: Katia(2011)[J]. J Atmos Sci, 73(1): 71-93.

RIOS-BERRIOS R, TORN RD, DAVIS C A, 2016b. An ensemble approach to investigate tropical cyclone intensification in sheared environments. Part Ⅱ: Ophelia(2011)[J]. J Atmos Sci, 73(4): 1555-1575.

ROY C, KOVORDányi R, 2012. Tropical cyclone track forecasting techniques-a review[J]. Atmos Res, 104-105: 40-69.

SUTSKEVER I, TIELEMAN T, 2010. On the convergence properties of contrastive divergence[J]. Thirteenth International Conference on Artificial Intelligence and Statistics, PMLR, 9: 789-795.

TAO Y, GAO X, HSU K, et al, 2016. A deep neural network modeling framework to reduce Bias in satellite precipitation products[J]. J Hydrometeor, 17(3): 931-945.

TODD B K, MICHEAL J B, 2017. Global guide to tropical cyclone forecasting-chapter 3[R/OL]. WMO, Geneva, https://www.wmo.int/cycloneguide/pdf/Chapter-Three.pdf.

XIAO Y, CAO Z, ZHUO W, et al, 2015. mCLOUD: a multiview visual feature extraction mechanism for ground-based cloud image categorization[J]. J Atmos Ocean Technol, 33(4): 151209140713007.

XU Y M, NEUMANN C J, 1985. A statistical model for the prediction of Western North Pacific tropical cyclone motion[R/OL]. NOAA Tech Memo NWS NHC-28, p. 34. https://repository.library.noaa.gov/view/noaa/7117/Print.

XU Z, ZOU L, LU X, 2013. Fast prediction of typhoon tracks based on a similarity method and GIS[J]. Disaster Advances, 6: 45-51.

YING M, ZHANG W, YU H, et al, 2014: An overview of the China Meteorological Administration tropical cyclone database [J]. J Atmos Oceanic Technol, 31: 287-301.

YU H, CHEN G M, BROWN B, 2013. A new verification measure for tropical cyclone track forecasts and its experimental application[J]. Tropical Cyclone Research and Review,2(4):185-195.

ZHUO W,CAO Z,XIAO Y,2014. Cloud classification of ground-based images using texture-structure features[J]. J Atmos Ocean Technol,31(1):79-92.

Index of notation

b_i	the bias value of the i-th visible node
c_j	the bias value of the j-th hidden node
$d_{f,t1}$	the Chebyshev distance between t_0 and t_1
e_i	The distance between target-TC's best track data and its WPCLPR forecast position
h	$h = [h_1, h_2, \cdots, h_{n_h}]^T$, the state of hidden layer
h_j	the state of hidden node h_j
lat_t^{TC} (lon_t^{TC})	the latitude(longitude) of TC's best track at moment t
$\text{lat}_t^{TC,WPCLPR}$ ($\text{lon}_t^{TC,WPCLPR}$)	the WPCLPR forecast latitude(longitude) of TC's best track at moment t
n_h	the number of hidden layer nodes
n_v	the number of visible layer nodes
t_0	the moment to be retrieved by similarity search algorithm
$t_{0,1}$	0000 UTC 22 July 2014, the target time in case-1, section 4.2
$t_{0,2}$	1800 UTC 6 July 2016, the target time in case-2, section 4.3
t_1	in similarity search algorithm, traversed all the moments from 1990 to 2016 (excluding the year of t_0)
t_f	the most-similar-moment retrieved by similarity search algorithm
$t_{f,1}$	0000 UTC 3 August 2003, the most-similar-moment retrieved by similarity search algorithm in case-1, section 4.2
$t_{f,2}$	1200 UTC 16 July 2005, the most-similar-moment retrieved by similarity search algorithm in case-2, section 4.3
t_s	The moment which the most-similar TC corresponded with
v	$v = [v_1, v_2, \cdots, v_{n_v}]^T$, the state of visible layer
v_i	the state of visible node v_i
$w_{i,j}$	the weight between hidden layer node h_j and visible layer node v_i
λ	learning rate of RBM
θ	$\theta = \{b_i, c_j, w_{ij}\}, i = 1,2,\cdots,n_v, j = 1,2,\cdots,n_h$, all parameters in a RBM
Δlat_t^{TC} (Δlon_t^{TC})	TC's latitude(longitude) forecast error of WPCLPR method
$\Delta\text{lat}_t^{Method}$ ($\Delta\text{lon}_t^{Method}$)	the latitude (longitude) difference between the target-TC and the most-similar-TC retrieved by corresponding method. The method is one of DBN, MA and MD, corresponding with SA_DBN, SA_MA and SA_MD, respectively.

Any variable with ^ at the top, indicates that it is a variable of the modified WPCLPR.

Glossary

ANN	artificial neural network
CD	contrastive divergence
CNN	convolution neural network
DBN	deep belief network

ECMWF	European Centre for Medium-Range Weather Forecasts
FE	feature extraction
GPH-features	the 30 500 hPa geopotential height field's features extracted by DBN
h_j	the j-th hidden layer node
ME	mean error
NNs	neural networks
NWP	numerical weather prediction
RBMs	restricted Boltzmann machines
RMSE	root-mean-square error
RNNs	recurrent neural networks
SA_DBN	TC similarity search algorithm based on features extracted by DBN
SA_MA	TC similarity search algorithm based on minimum area principle
SA_MD	TC similarity search algorithm based on minimum distance principle
SML	stochastic maximum likelihood
STD	standard deviation
TC	tropical cyclones
V	the training set
v_i	the i-th visible layer node
WPCLPR	The western North Pacific climatology and persistence track forecast technique

利用高原积雪信号改进中国南方夏季降水预测的新方法及其在2014年降水预测中的应用试验

刘颖[1,2]　任宏利[1,*]　张培群[1]　贾小龙[1,2]　刘向文[1]　孙林海[1]

(1. 国家气候中心 中国气象局气候研究开放实验室，北京，100081；
2. 南京信息工程大学 气象灾害预报预警与评估协同创新中心，南京，210044)

摘　要：2014年夏季，中国南方出现严重洪涝、北方大部干旱，国内绝大多数预测模型在3月起报的汛期预测中均未能抓住位于南方地区的异常雨带，导致预测准确率明显偏低。基于模式对东亚地区夏季海平面气压场的高预报技巧和青藏高原冬季积雪与南方地区夏季降水的高相关性，本文提出一个针对中国夏季降水异常的组合统计降尺度预测新方法（Hybrid Statistical Downscaling Prediction，简称 HSDP）。该方法综合利用气候模式输出的高可预报性环流信息和前期观测的高原积雪异常信号，从而实现对中国南方夏季降水进行动力-统计相结合的改进预报，并据此方法建立一个基于国家气候中心气候预测模式的统计降尺度模型。对中国南方夏季降水进行跨季节预测的交叉检验结果显示，HSDP方法对南方地区多年平均空间距平相关系数从模式原始预报的-0.006提高到0.24，且在大多数年份均有改进。基于 HSDP 方法于3月制作的2014年夏季降水预测，能够很好地抓住南涝北旱的基本形势和中国南方的降水大值区，空间距平相关系数达到0.43。这表明，该方法对中国夏季降水预测具有较好的业务应用前景。

关键词：气候预测；夏季降水预测；2014南方洪涝；气候模式；统计降尺度

Improve the Prediction of Summer Precipitation in Southern China by a New Approach with the Tibetan Plateau Snow and the Applicable Experiment in 2014

LIU Ying[1,2]　REN Hongli[1]　ZHANG Peiqun[1]　JIA Xiaolong[1,2]　LIU Xiangwen[1]　SUN Linhai[1]

(1. Laboratory for Climate Studies, National Climate Center, China Meteorological Administration, Beijing, 100081; 2. Collaborative Innovation Center on Forecast and Evaluation of Meteorological Disasters, Nanjing University of Information Science & Technology, Nanjing, 210044)

Abstract: Sever floods occurred in southern China and droughts were prevalent in northern China in the

---

① 本文发表于《大气科学》2017年第2期。
资助项目：国家自然科学基金(41405080)；国家重点基础研究发展计划(973计划)项目(2015CB453203)；中国气象局气候研究开放实验室基金(2014年)。
第一作者：刘颖，副研究员，主要从事短期气候预测研究。E-mail: liuying@cma.gov.cn。
通讯作者：任宏利。E-mail: renhl@cma.gov.cn。

summer of 2014. Most predicted models in China missed the southern rain band in their flood season predictions conducted in March 2014, which led to relatively low prediction accuracy. Based on the higher prediction skill for summer sea level pressure of climate models and the significant relationship between the preceding winter Tibetan Plateau Snow and summer precipitation in the south, a new Hybrid Statistical Downscaling Prediction(abbreviated as HSDP) method for summer precipitation anomaly prediction in China was proposed in this paper. The method can integrate the information of the highly predictable circulation from climate models and the influential signal of Tibet Plateau Snow in the preceding winter to improve the dynamical-statistical combination prediction for summer precipitation in the south. Using this method, a statistical downscaling model was established based on the climate prediction model of National Climate Center of China. The cross validation of seasonal prediction for the summer precipitation in the south was performed and the results showed that the HSDP improved the multi-year average of anomaly correlation coefficient from -0.006 to 0.24, and it had a higher predicting skill than the original climate model in most years. Using HSDP, the precipitation prediction for the summer of 2014 could well capture the basic situations, i.e. floods in southern China and droughts in northern China, and the positive precipitation anomaly in the south. The anomaly correlation coefficient could reach 0.43. This result indicated that the HSDP has a great operational application prospect with regard to summer precipitation prediction in China.

Key words: climate prediction; summer precipitation; floods occurred in southern China; climate model; statistical downscaling

1 引言

中国地处东亚季风区，多因子的共同作用导致夏季降水年际变化复杂、预测难度较大。国家气候中心多年业务预报显示出一定技巧，但年际起伏大、总体上预报能力较低[1]。例如1998年的世纪大洪水，无论是动力模式还是统计模型均对其做出较成功的预测[2,3]，很大程度上得益于对厄尔尼诺-南方涛动(ENSO)现象发展演变的准确把握[4-6]。然而，对于2014年发生在中国南方的洪涝和北方的干旱状况，基于模式、统计和经验的绝大多数预测模型在3月的全国会商未能给出合理预测(图1)。这客观上反映出必须要在预报认知和方法论上进行深入探索。

目前针对中国夏季降水的影响因子和预测方法已有大量的研究成果。一方面，在影响中国夏季降水的众多信号中，ENSO一直是最为重要的预测参考，然而自2000年以来其信号减弱是不争的事实，尤其是2014年春夏季ENSO的变化不定可能是汛期降水预测不利的主要因素。相比之下，青藏高原冬春积雪是中国夏季降水预测另一个重要前期因子。它反映了高原的热状况，其异常可以通过积雪本身的持续性、土壤湿度异常影响后期大气环流、亚洲季风及夏季降水等[7-9]发现。进一步的研究表明，高原冬春积雪与中国夏季降水尤其是长江流域降水有显著的相关关系[10,11]。虽然高原积雪信息已经在经验预测、统计模型预测以及业务预测中得到应用，但由于资料时效性等原因高原积雪至今未能有效运用到汛期降水的统计降尺度客观定量实时预测中，亟待加强以其为基础的预报方法研究。

另一方面，现阶段气候模式对东亚季风区降水的模拟和预测效果仍不理想，短时期内实现模式大幅改进的难度较大。为此，统计方法与动力模式相结合已成为一种有效改进预报的方法[12-16]。其中，基于模式的统计降尺度方法能够从模式预报效果较好的大尺度环流场中提取信息推算出降水，较模式直接预测降水更加准确，中国学者近年来在这方面取得了较多研究进展[17-20]。为此，本文将基于已有研究，提出将高原积雪前期信号引入气候模式统计降尺度预报框架,发展一个动力-统计相结合的降尺度季节预测新方法,以期显著改善中国特别是南方地区夏季降水预测效果。

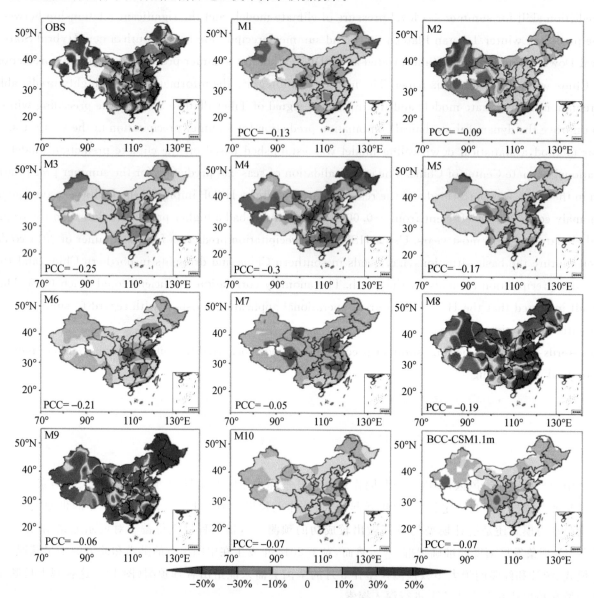

图1 2014夏季中国降水异常观测和3月全国汛期会商各参加单位预测的距平百分率空间分布

M1～M10表示会商单位的预测结果

2 数据和方法

观测数据为中国气象局国家气候中心整编的全国160个台站月平均降水资料;NCEP/NCAR再分析资料中的海平面气压场(SLP),水平分辨率为2.5°×2.5°[21];青海、西藏52个基本气象台站逐日积雪深度观测数据,该数据已剔除缺测较多的站点,并处理成月平均资料。

模式数据为国家气候中心第二代季节预测模式系统BCC-CSM 1.1 m模式(以下简称BCC模式)中1991—2014年的SLP场、降水场的月平均资料,水平分辨率为1°×1°。该模式每月初进行对未来1～13个月的预测,包含24个集合样本,文中使用24个样本的集合平均数据。由于全国汛期会商预测结果在3月下旬发布,本文选取3月初起报的6、7、8月SLP以及降水资料。

BCC 模式已经能够合理模拟出全球基本气候态、年平均降水以及热带降水年循环模态等大尺度气候要素的基本特征[22,23]，但由于模式自身不完善以及初始场等因素影响，BCC 模式针对中国夏季降水的直接预测结果还不够理想，对 2014 年夏季预测结果与实测相差较大（图1）。统计降尺度方法能够提取 BCC 模式最优预报信息、剔除系统性预报偏差，提高对中国夏季降水的预测能力。本文所要提出的方法是基于场信息耦合型技术的统计降尺度预测方法，其优点在于能够针对预测因子和预测量空间场的主要信息，通过提取两变量场之间的最优耦合变化型建立统计降尺度模型，详见文献[24-26]。在此基础上，我们提出将高原积雪前期信号引入这一模式统计降尺度预报框架，同样按照场信息耦合型技术形成新的预报因子，从而发展出一个组合统计降尺度预测新方法（HSDP）。

3 统计降尺度新方法检验与预测结果

由于东亚地区天气气候条件复杂，模式参数化方案估计降水的误差通常较大，导致气候模式对东亚大陆夏季降水预报技巧总体偏低。另外，气候模式对大尺度环流变量预测能力相对较好，例如，SLP 场能够很好地反映大气质量随着东亚夏季风环流的变化特征，可直接影响降水等气候要素变化。因此，充分利用耦合气候模式输出的具有较高可预报性的大尺度环流变量对较低可预报性的夏季降水进行统计降尺度预测，有望显著改善后者的预报性能。因此，本文选取 BCC 模式的 SLP 变量（区域为 5°N～55°N,60°E～160°E）作为夏季降水统计降尺度模型中的预测因子。图 2a 为 1991—2014 年夏季 BCC 模式与 NCEP/NCAR 再分析资料 SLP 的相关系数空间分布场。可以看到，中国大部分地区均为正相关，但高相关区集中在 30°N 以南，覆盖中国长江流域及以南地区。而且模式与再分析资料的 SLP 距平空间相关系数的多年平均为 0.17,24 年中超过 83% 的年份相关系数为正。由此可见，BCC 模式对东亚特别是中国长江流域以南地区的夏季 SLP 具有较好的预测能力。因此，本文选取东亚区域 SLP 场作为统计降尺度模型中预测因子。

中国长江流域及以南地区位于青藏高原下游，高原大地形的热力和动力作用对该地区的气候异常具有重要而直接的作用，前期冬季积雪与长江中下游夏季降水有着显著的正相关关系（图2b）。值得注意的是，春季高原积雪也是影响中国夏季降水异常的因素之一，但春季积雪一般是冬季积雪的延续且积雪异常持续时间较短，3 月以后积雪迅速减少[27]，同时考虑到实际降水预测的应用实效性，本文选取冬季高原积雪作为夏季降水预测因子。众多模式和统计模型均未准确预测出 2014 年中国夏季降水的空间分布，尤其是没有抓住长江及以南地区的降水大值区，很可能是未能有效刻画高原积雪的影响作用。HSDP 方法将前期冬季高原积雪影响引入模式统计降尺度预测，以期改进预测效果。

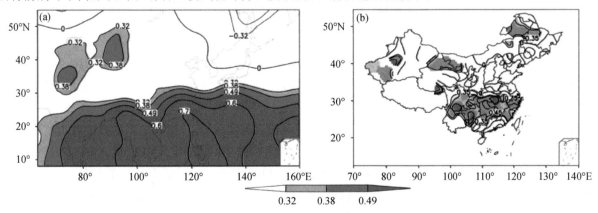

图 2 （a）夏季 BCC 模式与 NCEP/NCAR 再分析资料 SLP 的相关系数分析场；(b)冬季高原积雪指数与夏季中国降水相关阴影区颜色由浅到深为正/负相关系数通过 90%、95% 和 99% 显著性检验的区域

首先,基于 HSDP 模型对 1991—2013 年中国夏季降水进行回报试验,回报采取逐次去掉一年的交叉检验方法。图 3 为中国南方地区(31°N 以南,100°E 以东)空间距平相关系数的逐年变化曲线。可以看到,BCC 模式的多年距平相关系数平均值为 -0.006,而本文发展的 HSDP 方法距平相关系数的平均值提高到 0.24,其数值在大多数年份都大于 BCC 模式的值,所有年份的距平相关系数都在零以上,最高接近 0.4。为了进一步考察两个因子在 HSDP 模型中的各自贡献,我们也分别单独利用东亚地区 BCC 模式 SLP 和前冬高原积雪信息进行统计降尺度预测。交叉检验结果显示,南方地区夏季降水的多年平均距平相关系数分别达到 0.19(图 3 中 SD-SLP)和 0.21(图 3 中 SD-SNOW),均显著高于模式直接预报结果。这充分表明统计降尺度模型无论基于模式环流信息还是前期高原积雪信息均能有效提升预报性能,而综合运用两因子共同作用的 HSDP 模型更能发挥二者长处,进一步提高预报技巧。

我们也看到,统计降尺度模型在 1997 年和 2006 年夏季的回报技巧低于 BCC 模式结果,原因在于,这两年前期冬季青藏高原积雪偏多,而在接下来的实际夏季降水中长江流域均偏少,不符合两者之间存在的正相关统计规律,因此,青藏高原的作用在降尺度模型中效果不明显,导致空间距平相关不高。可以注意到,HSDP 统计降尺度模型对于南方地区夏季降水的距平相关系数表现比较平稳且略有上升,BCC 模式的距平相关系数预报技巧出现明显下降趋势,且年际变化较大,表现不稳定。

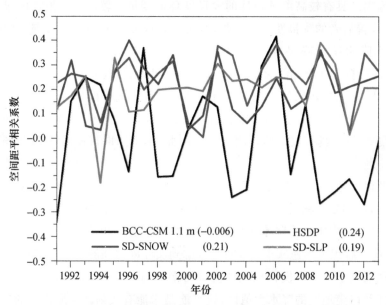

图 3　BCC 模式原始结果(黑色)、HSDP 结果(红色)、高原积雪预测结果(蓝色)和
SLP 降尺度预测结果(绿色)与观测之间的中国南方地区夏季降水空间距平相关系数

综合以上结果可以看到,经过 1991—2013 年的交叉检验,相对于 BCC 模式原始结果,HSDP 模型能够有效显著提高中国长江及以南地区预测技巧。由于青藏高原前期冬春积雪通过影响感热、垂直运动和环流场等,影响夏季风的强度,进而影响中国长江流域夏季降水[28],因此,高原前期积雪在 HSDP 模型中起到了至关重要的作用。

根据 1991—2013 年交叉检验结果可以看到,HSDP 模型确实较 BCC 模式结果具有较高的预测能力。那么,该模型是否能够提高 2014 年夏季中国降水的预测结果?图 4 给出了 BCC 模式和 HSDP 模型对 2014 年夏季中国降水距平的预测结果。对比前面图 1 的 OBS 结果,2014 夏季实测降水大值区主要集中在长江以南的江南地区,北方地区普遍降水偏少。BCC 模式预测的正降水大值区出现在西南地区,长江以南地区均预测成了降水负值区(图 4a)。相比来讲,HSDP 模型预测出了发生在江南地区的降水、黄淮地区的干旱(图 4b)。BCC 模式对 2014 夏季中国降水的距平相关系数为 -0.05,经过 HSDP 之后提高到 0.35,而均方根误差从 1.65 mm/d 下降到 1.51 mm/d。同时,长江及以南的南方地区的 49 个站点的距平相关系数可以达到0.43。

如果单纯利用气候模式因子SLP建立预测模型,其预测的2014年降水大值区集中在华南地区、华北以及东北地区,长江流域及以南大部分地区为少雨区,与观测结果有一定差距(南方地区距平相关系数为0.12);而单纯利用青藏高原积雪建立预测模型,其2014年夏季预测结果为:长江及以南地区、东北北部多雨,华南、江淮、华北以及东北大部分地区为少雨区,与观测结果更加接近(南方地区距平相关系数为0.4)(图略)。2013/2014年冬季青藏高原积雪面积总体为正距平,累积积雪深度在大部分地区也为正距平。2014夏季中国降水跨季节预测有效利用了前期青藏高原积雪的信号与长江及江南地区降水的密切关系,调整了前期预测方案,进而得到与观测事实更加接近的预测结果。值得注意的是,通常ENSO是影响中国夏季降水的重要年际信号,但在其信号不显著的年份(如2013—2014年冬春季),通过考虑与ENSO相对独立的高原积雪信号[29],仍有望取得良好预测效果。

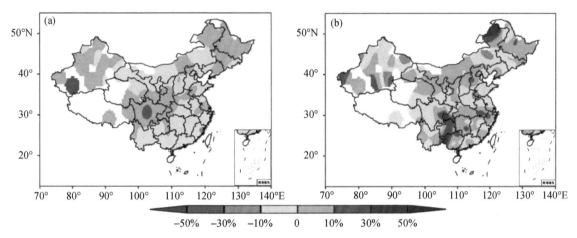

图4 预测的2014年夏季中国降水距平百分率空间分布型
(a)BCC模式结果;(b)HSDP结果

4 结语

青藏高原冬春积雪是影响中国夏季降水的重要外强迫因子之一,通过影响感热、大气环流以及东亚夏季风,进而影响到中国夏季降水。大部分模型对2014年夏季降水预测失效,尤其没有抓住南方地区的降水大值区,因此,本文提出将前期青藏高原积雪引入统计降尺度模型,发展了一个针对中国夏季降水异常的组合统计降尺度预报新方法(HSDP)。该HSDP模型充分利用BCC模式对SLP变量在中国南方地区的高预报技巧以及青藏高原冬季积雪与中国南方降水之间的显著关系,对夏季降水进行了1991—2013年交叉检验,并对2014年夏季降水进行预测试验。交叉检验结果显示,南方地区距平相关系数的平均值提高到0.24,相关系数值在大多数年份均大于BCC,最高可达0.4。2014年预测结果表明,该模型能够较好地将长江以南的降水大值区很好地预测出来,南方区域距平相关系数达到0.43,可能原因是充分利用了青藏高原前期冬季积雪与长江及以南的夏季降水具有显著的正相关关系,而观测到的2013/2014年冬季的青藏高原积雪面积与深度都是正距平值。

值得关注的是,2008/2009年冬季高原积雪显著偏少,HSDP模型较为准确地预测出了2009年夏季长江以南地区降水负值区以及华北、黄淮地区降水正值区,距平相关系数超过0.3,而BCC模式直接预报效果不佳(图略)。可能是由于模式对青藏高原热动力作用的模拟不够理想,因而未能将高原冬季积雪对夏季降水的影响有效表达出来,这需要进一步加以考察归因。当然,文中HSDP方法是交叉检验的结果,模型可靠性还需要更多实践检验。值得注意的是,由于中国南方和北方夏季降水异常分布的复杂性和非一致性[30,31],本文方法只针对有限个例中的江南地区夏季降水预测效果改进明显,结论仍存在一定局限性。由于台站观测的积雪面积和深度数据一般要滞后3个月以上才能供业务使用,因此如何使用时效性更好的卫星遥感替代资料将是关系到本文HSDP方法能否在春季会商及时做出预测的关键。另外,其他显著影响因子及高效预测方法,如影响中国南方降水异常的影响因素——欧亚大陆积雪[32,33],以及年际增量预测方法[34,35]的使用,也需要在下一步工作中重点加以考虑。

致谢:感谢王会军院士和李维京研究员在本文研究过程中给予的指导和帮助。

参考文献

[1] 钱维宏,陆波. 中国汛期季度降水预报得分和预报技巧[J]. 气象,36(10):1-7.

[2] 王会军. 关于中国几个大水年大气环流特征的几点思考[J]. 应用气象学报,2000,11(S):79-86.

[3] 王会军. 1998年夏季全球大气环流异常的预测研究[J]. 地球物理学报,2001,44:729-735.

[4] 黄荣辉,徐予红,王鹏飞,等. 1998年夏长江流域特大洪涝特征及其成因探讨[J]. 气候与环境研究,1998,3(4):300-313.

[5] 陶诗言,张庆云,张顺利. 1998年长江流域洪涝灾害的气候背景和大尺度环流条件[J]. 气候与环境研究,1998,3(4):290-299.

[6] GUO Y F,ZHAO Y,WANG J. Numerical simulation of the relationships between the 1998 Yangtze River Valley floods and SST anomalies[J]. Adv Atmos Sci,2002,19(3):391-404.

[7] 郭其蕴,王继琴. 青藏高原的积雪及其对东亚季风的影响[J]. 高原气象,1986,5(2):116-124.

[8] 陈兴芳,宋文玲. 冬季高原积雪和欧亚积雪对中国夏季旱涝不同影响关系的环流特征分析[J]. 大气科学,2000,24(5):585-592.

[9] 吴国雄,毛江玉,段安民,等. 青藏高原影响亚洲夏季气候研究的最新进展[J]. 气象学报,2004,62(5):528-540.

[10] 韦志刚,罗四维,董文杰,等. 青藏高原积雪资料分析及其与中国夏季降水的关系[J]. 应用气象学报,1998,9(S):39-46.

[11] WU T W,QIAN Z A. The relation between the tibetan winter snow and the Asian summer monsoon and rainfall:an observational investigation [J]. J Climate,2003,16(12):2038-2051.

[12] REN H L,CHOU J F. Strategy and methodology of dynamical analogue prediction [J]. Sci Chin Ser D Earth Sci,2007,50(10):1589-1599.

[13] 郑志海,任宏利,黄建平. 基于季节气候可预报分量的相似误差订正方法和数值实验[J]. 物理学报,2009,58(10):7359-7367.

[14] LANG X M. A hybrid dynamical-statistical approach for predicting winter precipitation over eastern China [J]. Acta Meteorola Sinica,2011,25(3):272-282.

[15] FAN K,LIU Y,CHEN H P. Improving the prediction of the East Asian summer monsoon:new approaches[J]. Wea Forecasting,2012,27(4):1017-1030.

[16] WANG H J,FAN K,SUN J Q,et al. A review of seasonal climate prediction research in China [J]. Adv Atmos Sci,2015,32(2):149-168.

[17] 贾小龙,陈丽娟,李维京,等. BP-CCA方法用于中国冬季温度和降水的可预报性研究和降尺度季节预测[J]. 气象学报,2010,68(3):398-410.

[18] SUN J Q,CHEN H P. A statistical downscaling scheme to improve global precipitation forecasting [J]. Meteor Atmos Phys,2012,117(3-4):87-102.

[19] LIU Y,FAN K. Improve the prediction of summer precipitation in the Southeastern China by a hybrid statistical downscaling model [J]. Meteor Atmos Phys,2012a,117(3-4):121-134.

[20] 刘颖,范可,张颖. 基于CFS模式的中国站点夏季降水的统计降尺度预测[J]. 大气科学,2013,37(6):1287-1299.

[21] KALNAY E,KANAMITSU M,KISTLER R,et al. The NCEP/NCAR 40-year reanalysis project [J]. Bull Amer Meteor

Soc,1996,77(3):437-471.

[22] 张莉,吴统文,辛晓歌,等. BCC_CSM 模式对热带降水年循环模态的模拟[J]. 大气科学,2013,37(5):994-1012.

[23] WU T W,LI W P,JI J J,et al. Global carbon budgets simulated by the Beijing climate center climate system model for the last century [J]. J Geophys Res,2013,118(10):4326-4347.

[24] LIU Y,FAN K. Prediction of spring precipitation in China using a downscaling approach [J]. Meteor Atmos Phys,2012b,118:79-93.

[25] LIU Y,FAN K. A new statistical downscaling model for autumn precipitation in China [J]. Int J Climatol,2013,33:1321-1336.

[26] LIU Y,REN H L. A hybrid statistical downscaling model for prediction of winter precipitation in China [J]. Int J Climatol,2015,35:1309-1321.

[27] 郑益群,钱永甫,苗曼倩,等. 青藏高原积雪对中国夏季风气候的影响[J]. 大气科学,2000,24(6):761-774.

[28] 张顺利,陶诗言. 青藏高原积雪对亚洲夏季风影响的诊断及数值研究[J]. 大气科学,2001,25(3):372-390.

[29] 陶亦为,孙照渤,李维京,等. ENSO 与青藏高原积雪的关系及其对中国夏季降水异常的影响[J]. 气象,2011,37(8):919-928.

[30] WU B Y,ZHANG R H,WANG B,et al. On the association between spring Arctic sea ice concentration and Chinese summer rainfall [J]. Geophys Res Lett,2009,36(9):L09501.

[31] WU B Y,ZHANG R H,D'ARRIGO R,et al. On the the relationship between winter sea ice and summer atmospheric circulation over Eurasia [J]. J Climate,2013,26(15):5523-5536.

[32] 许立言,武炳义. 欧亚大陆积雪与 2010 年中国春末夏初降水的关系[J]. 高原气象,2012a,31(3):706-714.

[33] 许立言,武炳义. 欧亚大陆积雪两种物理效应对 2010 年春末夏初华南降水的影响[J]. 大气科学,2012b,36(2):271-282.

[34] 范可,王会军,CHOI Y J. 一个长江中下游夏季降水的物理统计预测模型[J]. 科学通报,2007,52(24):2900-2905.

[35] 范可,林美静,高煜中. 用年际增量的方法预测华北汛期降水[J]. 中国科学 D 辑,2008,38(11):1452-1459.

多源气象数据融合格点实况产品研制进展[①]

师春香 潘旸 谷军霞 徐宾 韩帅 朱智 张雷 孙帅 姜志伟

(国家气象信息中心,北京,100081)

摘　要：阐述了中外主要的多源气象数据融合产品研究进展与趋势,重点介绍了中国气象局国家气象信息中心研制的陆面气象要素(包括气温、降水、湿度、风、气压、辐射等)、土壤温度与土壤湿度、洋面温度与洋面风、三维云等多源融合格点产品研发现状,以及中国气象局国家气象信息中心多源数据融合中试平台及统一质量检验评估系统的进展,并对未来多源气象数据融合产品研制进行了展望。

关键词：多源融合格点产品；降水融合；陆面气象驱动数据融合；陆面数据同化；海表气象要素融合；三维云量融合

A Review of Multi-source Meteorological Data Fusion Products

SHI Chunxiang　PAN Yang　GU Junxia　XU Bin　HAN Shuai
ZHU Zhi　ZHANG Lei　SUN Shuai　JIANG Zhiwei

(National Meteorological Information Center, Beijing, 100081)

Abstract: This paper is focused on the latest gridded fusion products developed by the National Meteorological Information Center(NMIC) of China Meteorological Administration(CMA), which include soil temperature and moisture, sea surface temperature, sea surface wind, three-dimensional cloud information and a series of meteorological elements data(such as air temperature, air pressure, precipitation, radiation, humidity, wind speed and direction, etc). The multi-source data fusion testbed and integrated quality evaluation system for these products are then introduced. Furthermore, recent progresses in multi-source data fusion in both foreign countries and China are reviewed in the paper, and the prospect of future development of multi-source meteorological data fusion is discussed.

Key words: multi-source gridded products; precipitation fusion; forcing data fusion; land data assimilation; SST and SSW fusion; three-dimensional cloud information fusion

1　引　言

随着气象观测系统的迅猛发展,利用地面自动气象站、雷达、卫星等获取的观测数据越来越多,多种

[①] 本文发表于《气象学报》2019年第4期。

资助课题：国家重点研发计划(2018YFC1506601)、国家自然科学基金项目(91437220)、公益性行业(气象)科研专项经费项目(GYHY201306045,GYHY201506002)、国家气象科技创新工程——"气象资料质量控制及多源数据融合与再分析"攻关任务。

第一作者：师春香,主要从事多源数据融合与再分析研究。E-mail:shicx@cma.gov.cn

数值模式模拟数据质量也在不断提高，同时，各行业对格点化的时空连续的气象数据产品要求越来越高。利用数据融合与数据同化技术，综合多种来源观测资料及多模式模拟数据，获得高精度、高质量、时空连续的多源数据融合气象格点产品是行之有效的手段。多源气象数据融合研究重点是，地面站点观测数据与卫星、雷达等遥感手段获取的面观测数据，不同分辨率面观测数据之间的时空匹配技术，以及不同观测之间系统性偏差订正技术，多源观测资料融合分析技术等。中外多源数据融合气象格点产品研究成果众多，涉及陆面、海洋、大气多个领域，已在天气、气候研究与业务，防灾、减灾等应用中发挥了重要作用。

中国多源气象数据融合研究起步相对较晚。中国气象局在2014年启动了国家气象科技创新工程"气象资料质量控制及多源数据融合与再分析"（以下简称"创新工程"），其攻关任务目标之一是研制高质量的陆面、海洋与三维云雨多源数据融合产品及相关技术。依托创新工程，中国气象局国家气象信息中心在引进国际先进融合技术的基础上，消化吸收并自主创新，建成了业务化的亚洲区域中国气象局陆面数据同化系统（CMA Land Data Assimilation System，CLDAS）和中国区域融合降水分析系统（CMA Multi-source Precipitation Analysis System，CMPAS），以及全球海表温度融合分析系统（CMA Ocean Data Analysis System-SST，CODAS-SST）和中国区域三维云融合分析系统（3D Cloud Analysis System，3DCloudAS）。2017年，中国气象局天气预报业务由原来的站点预报升级为智能网格预报，一系列多源数据融合产品（包括气温、降水、湿度、风、总云量、能见度等）通过优化产品时效、调整网格，已提供给智能网格预报业务应用。此外，包括地面气象要素及土壤温湿度、径流、蒸散发等在内的由中国气象局陆面数据同化系统（CLDAS）生产的系列产品也提供给中国气象局智慧农业业务应用。

文中首先阐述了中外主要的多源气象数据融合产品研究进展与趋势，重点介绍中国国家气象信息中心研制的陆面气象要素、土壤温度与土壤湿度、海洋洋面温度与洋面风、大气三维云信息等多源融合格点产品研发现状，之后介绍中国气象局国家气象信息中心多源数据融合中试平台及统一质量检验评估系统的建设进展，最后对未来多源气象数据融合产品研制进行了展望。

2 降水融合进展

多源降水融合分析起始于20世纪90年代，为了综合利用静止气象卫星红外探测时空连续分辨率高和极轨气象卫星被动微波降水精度高的优势，对多颗卫星不同类型探测资料反演的降水进行校正和融合，形成多卫星集成降水产品，再采用地面观测对卫星降水的系统偏差进行订正，形成最终的融合降水产品。初期由于卫星数量较少，卫星降水产品如GPCP（Global Precipitation Climatology Project，Huffman，et al.，1997）、CMAP（Climate Prediction Center（CPC）Merged Analysis of Precipitation，Xie，et al.，1997）的时空分辨率低，时间序列从1979年开始，适于气候研究。随着卫星遥感探测技术的发展，发展了概率密度函数（PDF）匹配等卫星资料校正及偏差订正技术，且最优插值（OI）、卡尔曼滤波（KF）等融合技术在降水融合领域也有了成熟广泛的应用，卫星集成及融合降水产品的时空分辨率显著提高，如美国的TMPA（TRMM Multi-satellite Precipitation Analysis，Huffman，et al.，2007）、CMORPH（CPC Morphing technique，Joyce，et al.，2004）、日本GsMAP（East Asian Multi Satellite Integrated Precipitation，Ushio，et al.，2009）等产品的时空分辨率已不低于3 h、0.25°，为全球大范围降水监测提供了可能。

在区域高分辨率降水监测领域，雷达定量降水估测（QPE）产品仍是主流（表1）。雷达探测的是与降水直接的结构信息，精度较卫星明显要高。针对雷达估测降水的偏差，以雨量计为基准发展了基于卡尔曼滤波、最优插值、距离反比加权（IDW）等方法的平均场系统误差订正和局部偏差订正技术（Seo，et al，2002），已在美国国家环境预报中心（NCEP）的StageⅣ系统和美国国家强风暴实验室（NSSL）的MRMS（Multi-Radar Multi-Sensor）系统中成熟应用。目前，美国MRMS产品的最高时空分辨率达到2 min、

1 km,在强降水灾害天气监测和临近预报方面有重要应用价值。

与国际主流高分辨率的降水产品以雷达资料应用为主的情况不同,中国复杂地形和雷达型号差异造成雷达组网的技术困难,再加上超折射、负折射、电磁干扰、海浪回波、地形遮挡、地物杂波、0℃层亮带、衰减、未完全充塞、扫描方式局限等雷达回波的基本质量问题尚未完全解决,中国雷达定量降水估测产品质量与美国相比有很大差距。如中国雷达定量降水估测产品夏季的相关系数不超过0.6,均方根误差(RMSE)大于1.3 mm/h(潘旸 等,2018),而美国本土雷达定量降水估测暖季相关系数大于0.75,均方根误差小于1 mm/h(Wu et al.,2012)。目前,中国区域业务化的高分辨率融合降水产品多是模式预报或卫星与地面观测资料的融合产品。其中,中国气象局国家气象信息中心引进吸收美国国家海洋和大气管理局(NOAA)气候预测中心(CPC)的"概率密度函数+最优插值"两步融合法(Xie,et al.,2011),研制逐时、10 km地面和卫星二源降水融合产品(潘旸 等,2012)。2014年,潘旸等提出"概率密度函数+贝叶斯模型平均(BMA)+最优插值"方法,引入中国气象局气象探测中心的雷达定量降水估测产品,研制了逐时、5 km的地面、卫星、雷达三源融合降水产品,2016年将产品的空间分辨率提高至1 km(潘旸 等,2018)。

表1 国外主要高分辨率降水业务产品

国家	机构	产品名称	主要技术	产品要素	空间覆盖	起始时间	时空分辨率
美国	NCEP	Stage Ⅱ	地面杂波、异常回波、固定地面目标去除;0℃层亮带检测;静态Z-R关系	累计降水量	美国本土	1994年	1 h,4 km
	NCEP	Stage Ⅳ (Kitzmiller, et al.,2013)	地面雨量计偏差订正、人工质量控制、卫星填补	累计降水量	美国本土	2011年	1 h,4 km
	NSSL	MRMS (Zhang,et al.,2016)	dpQC、无缝混合扫描、波束阻挡订正、VPR订正、降水类型分类、Mountain Mapper、基于降水类型Z-R关系、地面雨量计局部偏差订正	降水率、累计降水量、误差	美国本土	2006年	1 h,1 km,2 min 更新
法国	法国气象局	业务雷达降水 第1代	地物杂波抑制	降水	单部雷达	1997年	15 min,1 km
		第2代 (Tabary,2007)	地物杂波抑制、VPR订正、波束阻挡订正	降水	法国组网	2007年	5 min,1 km
		第3代 (Jordi,et al.,2013)	地物杂波抑制、VPR订正、波束阻挡订正、地面雨量计校准、双偏振晴空回波去除及降水估计	降水	法国组网	2013年	5 min,1 km
德国	德国气象局	RADOLAN (Radar-Online-Aneichung)	地物杂波抑制、地形遮挡订正、气候态订正	降水	德国	2001年	1 h,1 km
澳大利亚		雷达定量降水估测产品 (http://www.bom.gov.au/australia/radar/about/radar_coverage_national.shtml)	去除地面杂波、海量杂波;降水类型分类;基于卡尔曼滤波方法的地面雨量计校准	累计降水量	单部雷达		6 min

3 陆面数据融合同化进展

目前,国外陆面数据融合产品主要包括陆面大气驱动场(如气温、气压、湿度、风速、降水、辐射)和陆面要素数据融合分析(如土壤湿度、土壤温度、地表温度、地表热通量、径流、积雪等)两大类产品。

陆面大气驱动场是陆面过程模式的输入数据,其数据来源可以是大气再分析资料中的近地面要素场,也可以是在此基础上通过数据融合订正等技术对再分析场进行优化获取的更高质量陆面大气驱动场。主要包括 Sheffield 大气驱动数据场、Qian 大气驱动数据和美国全球陆面数据同化系统(Global Land Data Assimilation System,GLDAS)大气驱动场数据等,其研制技术主要采用数值模式预报、卫星资料反演、多源数据融合等(Sheffield,et al.,2004;Qian,et al.,2006)。以上 3 种陆面大气驱动场覆盖区域均为全球,时间分辨率为 3 h,但空间分辨率各有不同,Sheffield 大气驱动数据场与 Qian 大气驱动数据已经回算到 1948 年,具有较长的时间序列,适宜开展长时间的陆面模拟分析和气候评估研究等;美国全球陆面数据同化系统大气驱动场数据作为美国 NOAA 业务产品,实时向用户提供最新时次产品。

国外上陆面要素数据融合分析产品主要来自于各业务和科研单位构建的陆面数据同化系统,如欧洲陆面数据同化系统(European Land Data Assimi-lation System,ELDAS)(Albergel,et al.,2013)、美国全球陆面数据同化系统(Rodell,et al.,2004)北美陆面数据同化系统(North American Land Data Assimi-lation System,NLDAS)(Xia,et al.,2012)、东京大学陆面数据同化系统(LDAS-UT)(Rasmy,et al.,2011)等。多数陆面数据同化系统尚未真正同化陆面状态变量,仍然局限于提高陆面模式驱动及参数数据等,例如全球陆面数据同化系统和北美陆面数据同化系统致力于完善反照率、土地覆盖/土地利用类型、植被绿度和叶面积指数等。也有部分系统开展了同化研究及业务应用,如欧洲陆面数据同化系统采用最优插值和扩展卡尔曼滤波(EKF)方法,可同时同化 2 m 高气温和湿度以及微波反演土壤湿度资料。总体来看,尽管陆面数据同化是当前的一个研究热点,但在业务天气预报模式中的应用仍处于起步阶段。

美国 NCEP 准业务运行的北美陆面数据同化系统虽然没有真正同化陆面状态变量,但是已在美国干旱监测等领域发挥了重要作用。其下一阶段的目标是引入最新版本的 LIS(Land Information Sys-tem)系统,以真正实现业务化的陆面数据同化,陆面要素产品的空间分辨率将达到 3.125 km。另外,美国国家大气研究中心(NCAR)也已开展 4 km 分辨率陆面数据同化(HRLDAS)研究(Chen,et al.,2007),通过同化观测数据,为天气研究预报模式(WRF)提供高质量、高时空分辨率陆面分析数据。

中国陆面数据融合与同化研究起步相对较晚,但发展较快。表 2 给出了中国主要陆面数据融合与同化产品。其中,中国气象局国家气象信息中心研发的中国气象局陆面数据同化系统第 1 版(CLDAS V1.0)于 2013 年率先在中国实现了国家级陆面要素融合产品的业务化生产和发布,重点解决了东亚(尤其是中国)区域陆面大气驱动场的多源融合技术难题。引进和改进的美国 NOAA 地球系统研究实验室(ESRL)的时、空多尺度分析系统(ST-MAS),改进了中国气象局国家卫星气象中心基于离散坐标法物理模型(Hybrid)的短波辐射遥感反演业务算法(刘军建 等,2018),采用基于"概率密度函数+最优插值"的融合降水,发展基于 CLM3.5、CoLM、Noah-MP(4 套参数化方案)多陆面模式集合模拟技术,于 2015 年研制了中国气象局陆面数据同化系统第 2 版(CLDAS V2.0),实时发布亚洲区域逐时和逐日的 0.0625°分辨率的大气驱动场和陆面要素集合分析产品。同时,在 CLDAS V2.0 关键技术研发基础上,高分辨率中国气象局陆面数据同化系统第 1 版(HRCLDAS V1.0)投入试运行,产品分辨率提高至 1 km(韩帅 等,2018)。中、外同类产品比较分析结果表明,中国气象局陆面数据同化系统系列大气驱动场产品、陆面要素融合分析产品在中国区域的时空分辨率和质量更高(韩帅 等,2017;孙帅 等,2017)。2017 年年底,中国气象局国家气象信息中心研制完成了中国气象局陆面数据同化系统第 3 版(CLDAS V3.0),实现了中国 FY-3C 卫星反演土壤湿度资料同化、地表温度同化以及地表温度与微波亮温资料协同同化;开展了 FY-3 积雪覆盖率以及雪深资料的同化研究,进一步提高了 CLDAS 产品精度,积雪同化的效果明显,尤其在东北、新疆及青藏高原地区(张帅 等,2018;师春香 等,2018)。

表 2 中国主要陆面数据融合与同化产品

机构	产品名称	主要技术	产品要素	空间覆盖	起止时间	时空分辨率	下载地址
中国气象局国家气象信息中心	CLDAS 陆面数据同化产品	多重网格变分;空间格点拼接、离散纵坐标短波辐射遥感反演;CLM、Noah-MP、CoLM 多陆面模式集合模拟	气温、气压、湿度、风速、降水、短波辐射、土壤湿度、土壤温度、地表温度、土壤相对湿度	亚洲	2008—	1 h,0.0625°	中国气象数据网(http://data.cma.cn/)
中国气象局国家气象信息中心	HRCLDAS 高分辨率陆面数据同化产品	多重网格变分;离散纵坐标短波辐射遥感反演和融合、多卫星集成降水融合;CLM 陆面模式模拟	气温、气压、湿度、U/V 风、风速、降水、短波辐射、土壤湿度、土壤温度、地表温度、土壤相对湿度	中国	2015—	1 h,0.01°	全国综合气象信息共享平台(CIMISS)
中国科学院寒区旱区环境研究所	中国西部陆面数据同化产品	陆面模式模拟	土壤水分、土壤温度、积雪、冻土	中国西部	2002—	3 h,0.25°	寒区旱区科学数据中心 http://westdc.westgis.ac.cn/
中国科学院青藏高原研究所	中国区域地面气象要素驱动数据集	双线性空间插值;薄板样条插值;Hybrid Model 辐射估算	气温、气压、湿度、风速、降水、太阳辐射	中国	1981—2008	3 h,0.1°	

4 海表要素融合进展

海表要素的融合思路和技术与降水融合类似,采用实测数据订正多卫星反演要素,再将卫星反演与船舶、浮标等实测数据融合,诸如泊松方程、最优插值、概率密度函数匹配、二维变分等偏差订正和融合技术的应用也较降水应用更早、更成熟。目前,海表要素融合研究主要集中在海表温度(SST)、洋面风、海冰等融合产品的研发。

海表温度是全球海洋大气系统中最为重要和基础的海洋要素之一,各国研究机构和业务部门陆续研制出了高质量的海表温度融合产品(表 3),如逐月哈得来中心(Hedley Center)海冰和海温产品(HadISST,Rayner,et al.,2003),逐日、0.25°的最优插值海表温度(Optimum Interpolation Sea Surface Temperature,OISST,Reynolds,et al.,2002)以及逐日、(1/12)°的 RTG-HR(Daily Real-time Global sea Surface Temperature-high Resolution analysis,Gemmill,et al.,2007)等。最近,英国气象局基于 GHRSST-PP(Group for High Resolution Sea Sur-face Temperature Pilot Project)提供的多卫星数据和实测数据,研制了 0.05°的全球逐日海温-海冰融合产品 OSTIA(Operational Sea Surface Tempera-ture and Sea Ice A-nalysis,Donlon,et al.,2012)。

表3 国际主要海表温度融合产品列表

国家	机构	产品名称	主要技术	产品要素	空间覆盖	起始时间	时、空分辨率	下载网址
美国	NOAA	OISST	Poisson方法校正卫星资料大尺度偏差,最优插值方法融合	海表温度	全球	1981年	周,1° 月,1°	https://www.ncdc.noaa.gov/oisst
		OISST V2.0	月平均资料由周资料线性插值得到该月每日值,再经累计得到	海表温度	全球	1981年	日,0.25°	
		RTG-HR	二维变分方法	海表温度	全球	2005年	日,0.5°~(1/12)°	http://polar.ncep.noaa.gov/sst/rtg_high_res/
英国	Hadley Center	HadISST1	两步约化空间最优插值(RSOI),叠加高质量的格点观测(RDB)	海表温度 海冰覆盖度	全球	1870年	月,1°	https://www.metoffice.gov.uk/hadobs/hadisst/data/download.html
		OSTIA	多尺度最优插值技术	海表温度	全球	2006年	日,0.05°	http://ghrsst-pp.metoffice.com/pages/latest_analysis/ostia.html

海冰的高反照率、隔绝海-气热量和动量交换作用对区域乃至全球气候变化均有重要的调节作用。HadISST是国际上应用最为广泛的全球海冰密集度融合资料之一(Rayner,et al.,2003),利用数字化图表订正微波数据,保证了融合产品的均一性。还有一些区域的海冰融合资料,如美国的IMS(the Interactive Multisensor Snow and Ice Mapping System,Ramsay,2000)和MASIE(Multisensor Analyzed Sea Ice Extent project for the Northern Hem-isphere,Fetterer,2006)北半球逐日海冰覆盖资料,也常用于天气、气候实时监测业务。

洋面风场影响大气-海洋的相互作用,也影响船只航行、海上工程等活动。国际上应用较多的洋面风融合有CCMP(Cross-Calibrated Multi-Platform Ocean Surface Wind,Atlas,et al.,2011)和BSW(Blended Sea Winds,Zhang,et al.,2006a,2006b)。

中国也开展了相关研究,王际朝(2014)分别以QSCAT(QuikSCAT satellite)/NCEP和CCMP再分析风场为背景场,利用最优插值方法对研究区域内浮标的风速和风向进行融合研究。中国国家海洋环境预报中心在西北太平洋海域利用最优插值方法将船舶报资料和全球海洋观测网(ARGO)海温数据同化到数值模拟中,有效地改进了三维海温模拟的结果(李云 等,2008)。中国科学院大气物理研究所的全球海洋资料同化系统(ZFL-GODAS)能够同化包括卫星高度计资料、卫星海表温度资料以及AR-GO、投弃式温度剖面测量系统(XBT)、热带大气海洋观测阵列(TAO)等各种不同来源的温盐廓线资料(路泽廷等,2014)。但是,目前业务工作中所使用到的海表温度、海冰和洋面风资料主要依靠国际卫星及融合产品。中国气象局国家气象信息中心2016年采用时空多尺度分析方法实现了中国风云3B卫星(FY-3B)、日本全球变化观测任务卫星(GCOM-W1)、欧洲气象业务化卫星(MetOP-B)等反演海温产品与浮标、船舶观测海表温度与欧洲中期天气预报中心(ECMWF)精细化预报海温产品的融合,能够更好地反映出台风路径上的海表温度变化(徐宾 等,2018),基于此方法建立的全球海表温度融合分析系统于2018年年底实现了业务化运行。同时,中国气象局国家气象信息中心还开展了基于欧洲气象卫星应用组织(EUMETSAT/OSI-SAF)制作的全球海冰密集度分析日产品、美国NISE(Near-Real-Time SSM/I EASE-Grid

Daily Global Ice Concentration and Snow)产品、风云卫星微波成像仪(MWRI)反演产品、IMS(the Interactive Multi-sensor Snow and Ice Mapping System)等海冰覆盖度产品的全球多卫星海冰覆盖度融合试验。

5　三维云融合进展

美国NOAA地球系统研究实验室发展了局地分析预报系统(LAPS),能够通过融合数值预报产品、地面、探空、雷达、静止气象卫星、GPS/MET、风廓线雷达、飞机等多源观测数据,获得三维云融合格点数据,为数值预报模式提供更好的初始场,改进数值预报模式的短期预报水平。其他机构也发展了类似的三维云融合系统,例如美国的ARPS(Advanced Regional Prediction System)、RUC(Rapid Update Cycle)、奥地利的INCA(Integrated Nowcasting through Comprehensive Analysis)等(表4)。近年来,美国通过在目前的业务同化系统(GSI)中嵌入ARPS和RUC,研发了GSI-Cloud云分析模块,实现了云分析的功能,可以为数值预报模式提供包含更高精度云信息的初始场,意味着三维云融合将成为同化系统的重要模块。

表4　国际三维云融合系统概况

国家	研发机构	系统名称	系统介绍
美国	NOAA/ESRL	LAPS	LAPS系统能够将多种观测数据(气象地面观测网、雷达、卫星、垂直探测器、飞机等)进行融合分析,得到高分辨率的三维分析结果(Albergel,et al.,2013)
美国	俄克拉何马大学/风暴分析和预测中心	ARPS	ARPS的云分析模块主要从LAPS发展而来,也采用了逐步订正的方法进行三维云分析(Xue,et al.,2001;Wang,et al.,2003;朱立娟 等,2017)
美国	NOAA/ESRL	RUC	云分析模块采用了逐步订正的方法(Benjamin,et al.,2015)
美国	俄克拉何马大学/风暴分析和预测中心,NOAA/ESRL/全球系统部	GSI-Cloud	结合了RUC和ARPS的优势,通过融合卫星的云产品、地面云观测、雷达和闪电定位等资料,形成较为准确的三维云融合产品,并应用于北美快速天气模拟业务中(Hu,et al.,2007)
奥地利	奥地利国家气象局	INCA	通过融合数值预报产品、卫星、雷达和地面观测数据,并结合精细化下垫面信息,对数值预报产品进行订正后,形成高分辨率的分析场与外推预报场;每15 min计算1次(Haiden,et al.,2011)

中国在三维云融合方面也开展了一些研究工作,主要是引进国际的融合系统,并基于中国的业务环境开展本地化研发(表5),如中国气象局国家卫星气象中心基于中国区域业务环境搭建的三维云信息融合分析系统;中国气象局国家气象信息中心针对中国区域特点以及中国现有资料情况,基于局地分析预报系统和全国综合气象信息共享平台(CIMISS)数据环境,建成了适用于中国区域的3DCloudAS系统,实现了中国区域的地面观测、机场观测、探空观测、飞机观测、风云-2号/风云-4号/葵花-8静止气象卫星、多普勒雷达、风廓线雷达等多种观测资料的融合。该系统于2018年实现业务运行,可实时逐时输出三维云量,以及三维温度场、湿度场、风场等的网格化产品,空间覆盖为中国区域,分辨率为0.05°。

6　多源数据融合中试平台设计

中试平台是完善气象科技成果转化应用体制机制、推动气象科技成果转化应用的重要途径。多源数据融合中试平台的设计目的即推动多源数据融合关键技术快速地转化为业务应用能力。

多源数据融合中试平台作为气象信息中试基地的有机组成部分,在其框架下,发展多源数据融合科研成果迅速转化为业务,主要用于实现多源数据融合产品的研发和业务转化。中试平台主要包括:依托气象信息中试基地提供的软硬件环境及数据环境,构建多源数据融合的资源环境;打造包括降水融合分析、陆面数据融合同化、海表要素融合、三维云大气融合等多源数据融合分析功能于一体的中试系统环境;重点研发融合格点实况产品检验评估的统一规范,开发评估工具;建立统一的准入规则、中试系统开发、检验评估、业务转化流程等规范。

表 5 中国三维云融合系统概况

机构	系统平台	区域	系统介绍
上海市气象局	LAPS	华东	引进并业务化,能够融合自动站、雷达径向风和回波、风云卫星云图等资料(刘寿东 等,2012)
武汉暴雨研究所	LAPS	华中	根据本地探测资料的种类和特点开展LAPS本地化移植及二次开发(李红莉 等,2009)
北京市气象局	LAPS	北京	能够融合雷达、卫星云导风、常规探空及自动气象站等观测资料(高华 等,2009)
中国气象局国家卫星气象中心	LAPS	中国	搭建三维云信息融合分析系统,可生成集多颗卫星、多部雷达、地面常规和非常规等多种观测资料优势为一体的三维大气信息产品(刘瑞霞 等,2013)
中国气象局公共气象服务中心	LAPS	中国	引进并业务化,使用GRAPES-Meso预报场作为背景场,融合风云卫星、葵花卫星、探空、地面、多普勒雷达等多种资料,产生每小时1次的三维分析场(李超 等,2017)
中国气象局国家气象信息中心	LAPS	中国	建成三维云融合业务系统,能够融合包括地面观测、探空观测、雷达、GPS/MET、风云-2号卫星、葵花-8静止气象卫星等多种观测资料,实时生成三维大气融合格点产品与三维云融合格点产品

7 未来发展

多源数据融合格点分析中需要解决的科学问题主要集中在以下两个方面:

第一,融合分析的时空尺度代表性问题,即在不同观测数据条件,如空间上站网疏密变化、时间上观测频次的增减等,以及在不同地形条件、不同气候背景、不同天气系统条件下如何合理制定多源融合格点分析产品时空分辨率,从而得到最优分析尺度上时序列一致的多源融合产品。如青藏高原典型区域,站点稀疏且受高大地形、积雪覆盖、复杂下垫面条件等多种因素影响,成为融合分析中的难点。

第二,由于多源数据融合格点分析产品常被用来对数值预报产品等进行检验评估,以及在智能网格气象预报业务中被作为实况产品应用,因而如何对融合数据的"真实性"进行评判,需要利用多源协同观测和设计外场科学试验进行独立检验,建设多源数据融合格点分析产品质量评价体系,并将其视作真值的质量标准。

在多源数据融合格点分析产品研制中,一些关键的技术问题尚需要深入研究,包括地面、探空、卫星、雷达、模式等多种来源资料的协同质量控制,针对不同气候背景、复杂地形、下垫面条件下各来源资料的误差分析系统及偏差订正技术,时空多尺度分析系统、最优插值、EnKF、Hybrid等方法的局地优化应用技术等。引进机器学习、大数据挖掘等人工智能方法在降水、陆面、海洋以及三维云融合中的应用也是值得探索的研究内容。

未来,将在多源数据融合科学问题与关键技术研究基础上,在业务单位形成多源数据融合格点分析产品业务体系。图1是目前中国气象局国家气象信息中心多源数据融合多个业务系统成熟度及产品特性。在保证产品质量的同时,产品空间分辨率由现在的千米级提高到米级,时间分辨率由小时级提高到分钟级,产品时效逐步提高到分钟级,产品覆盖范围从中国扩展到全球(重点是"一带一路"区域)。同时对历史数据进行回算,建成长时间序列的多源数据融合格点分析产品数据集,提供应用。

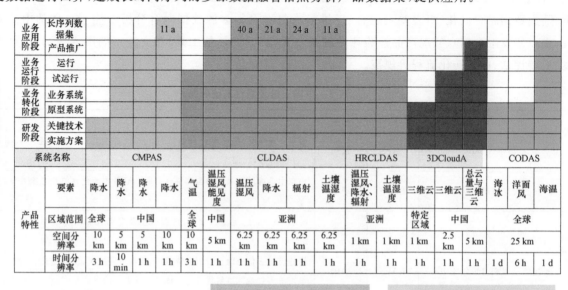

图1 中国气象局国家气象信息中心多源数据融合业务系统成熟度及产品特性

参考文献

高华,谭旭光,李英华,等,2009. 局地分析和预报系统(LAPS)在北京市气象局移植与应用[C]//第26届中国气象学会年会论文集. 杭州:中国气象学会.

韩帅,师春香,姜立鹏,等,2017. CLDAS 土壤湿度模拟结果及评估[J]. 应用气象学报,28(3):369-378.

韩帅,师春香,姜志伟,等,2018. CMA 高分辨率陆面数据同化系统(HRCLDAS-V1.0)研发及进展[J]. 气象科技进展,8(1):102-108,116.

李超,唐千红,陈宇,等,2017. 多源数据融合系统 LAPS 的研究进展及其在实况数据服务中的应用[J]. 气象科技进展,7(2):32-38.

李红莉,崔春光,王志斌,2009. LAPS 的设计原理、模块功能与产品应用[J]. 暴雨灾害,28(1):64-70.

李云,刘钦政,张建华,等,2008. 最优插值方法在西北太平洋海温同化中的应用研究[J]. 海洋预报,25(2):25-32.

刘军建,师春香,贾炳浩,等,2018. FY-2E 地面太阳辐射反演及数据集评估[J]. 遥感信息,33(1):104-110.

刘瑞霞,徐祥德,刘玉洁,2013. JICA 综合观测与卫星数据在高原地区三维云和水汽场构建中的应用[J]. 高原气象,32(6):1589-1596.

刘寿东,唐玉琪,邵玲玲,等,2012. LAPS 分析场在一次强对流天气过程尺度分析中的应用[J]. 大气科学学报,35(4):391-403.

路泽廷,朱江,符伟伟,等,2014. 全球海洋资料同化系统 ZFL_GO-DAS 的研制和初步评估试验[J]. 气候与环境研究,19(3):321-331.

潘旸,沈艳,宇婧婧,等,2012. 基于最优插值方法分析的中国区域地面观测与卫星反演逐时降水融合试验[J]. 气象学报,70(6):1381-1389.

潘旸,谷军霞,宇婧婧,等,2018. 中国区域高分辨率多源降水观测产品的融合方法试验[J]. 气象学报,76(5):755-766.

师春香,张帅,孙帅,等,2018. 改进的 CLDAS 降水驱动对中国区域积雪模拟的影响评估[J]. 气象,44(8):985-997.

孙帅,师春香,梁晓,等,2017. 不同陆面模式对中国地表温度模拟的适用性评估[J]. 应用气象学报,28(6):737-749.

王际朝,2014. 基于风场和海浪同步观测的海浪同化模式构建[D]. 青岛:中国科学院研究生院(海洋研究所).
徐宾,宇婧婧,张雷,等,2018. 全球海表温度融合研究进展[J]. 气象科技进展,8(1):164-170.
张帅,师春香,梁晓,等,2018. 风云三号积雪覆盖产品评估[J]. 遥感技术与应用,2018,33(1):35-46.
朱立娟,龚建东,黄丽萍,等,2017. GRAPES三维云初始场形成及在短临预报中的应用[J]. 应用气象学报,28(1):38-51.
ALBERGEL C,DORIGO W,BALSAMOG,et al,2013. Monitoring multi-decadal satellite earth observation of soil moisture products through land surface reanalyses[J]. Remote Sens Environ,138:77-89.
ATLAS R,HOFFMAN R N,ARDIZZONE J,et al,2011. A cross-calibrated,multiplatform ocean surface wind velocity product for meteorological and oceanographic applications[J]. Bull Amer Meteor Soc,92(2):157-174.
BENJAMIN S G,WEYGANDT S S,BROWN J M,et al,2015. A North American hourly assimilation and model forecast cycle:The rapid refresh[J]. Mon Wea Rev,144(4):1669-1694.
CHEN F,MANNING K M,LEMONE M A,et al,2007. Description and evaluation of the characteristics of the NCAR high-resolution land data assimilation system[J]. J Appl Meteor Climatol,46(6):694-713.
DONLON C J,MARTIN M,STARK J,et al,2012. The operational sea surface temperature and sea ice analysis (OSTIA) system[J]. Remote Sens Environ,116:140-158.
FETTERER F,2006. A selection of documentation related to National Ice Center sea ice charts indigital format[R]. Boulder,CO,USA:National Snowand Ice Data Center.
JORDI F V I,TABARY P,2013. The new French operational polarimetric radar rainfall rate product[J]. J Appl Meteor Climatol,52(8):1817-1835.
GEMMILL W,KATZ K,LI X,2007. Daily real-time global sea surface temperature-high resolution analysis at NOAA/NCEP[R]. NOAA/NWS/NCEP/MMAB Office Note,260,39.
HAIDEN T,KANN A,WITTMANN C,et al,2011. The integrated now-casting through comprehensive analysis (INCA) system and its validation over the eastern Alpine region[J]. Wea Forecasting,26(2):166-183.
HU M,XUE M,2007. Implementation and evaluation of cloud analysis with WSR-88D reflectivity data for GSI and WRF-ARW[J]. Geophys Res Lett,34(7):L07808.
HUFFMAN G J,ADLER R F,ARKIN P,et al,1997. The global precipitation climatology project(GPCP)combined precipitation dataset[J]. Bull Amer Meteor Soc,78(1):6-20.
HUFFMAN G J,BOLVIN D T,NELKIN E J,et al,2007. The TRMM multisatellite precipitation analysis(TMPA):quasi-global, multiyear, combined-sensor precipitation estimates at fine scales[J]. J Hydrometeorol,8(1):38-55.
JOYCE R J,JANOWIAK J E,ARKIN P A,et al,2004. CMORPH:a method that produces global precipitation estimates from passive microwave and infrared data at high spatial and temporal resolution[J]. J Hydrometeorol,5(3):487-503.
KITZMILLER D,MILLER D,FULTON R,et al,2013. Radar and multisensor precipitation estimation techniques in national weather service hydrologic operations[J]. J Hydrol Eng,18(2):133-142.
QIAN T T,DAI A G,TRENBERTH K E,et al,2006. Simulation of global land surface conditions from 1948 to 2004. Part Ⅰ:Forcing data and evaluations[J]. J Hydrometeorol,7(5):953-975.
RAMSAY B H,2000. Prospects for the interactive multisensor snow and ice mapping system(IMS)[C]//Proceedings of the 57th Eastern Snow Conference. Syracuse,NY.
RASMY M,KOIKE T,BOUSSETTA S,et al,2011. Development of a satellitel and data assimilation system coupled with a mesoscale model in the Tibetan Plateau[J]. IEEE Trans Geosci Remote Sens,49(8):2847-2862.
RAYNER N A,PARKER D E,HORTON E B,et al,2003. Global analyses of sea surface temperature,sea ice,and night marine air temperature since the late nineteenth century[J]. J Geophys Res,108(D14):4407.
REYNOLDS R W,RAYNER N A,SMITHT M,et al,2002. An improved ins itu and satellite SST analysis for climate[J]. J Climate,15(13):1609-1625.
RODELL M,HOUSER R,JAMBOR U,et al,2004. The global land data assimilation system[J]. Bull Amer Meteor Soc,85(3):381-394.
SEO D J,BREIDENBACH J P,2002. Real-time correction of spatially nonuniform bias in radar rainfall data using rain gauge

measurements[J]. J Hydrometeorol, 3(2): 93-111.

SHEFFIELD J, ZIEGLER A D, WOOD E F, et al, 2004. Correction of the high-latitude rain day anomaly in the NCEP-NCAR reanalysis for land surface hydrological modeling[J]. J Climate, 17(19): 3814-3828.

TABARY P. 2007. The new French operational radar rainfall product. Part Ⅰ: Methodology[J]. Wea Forecasting, 22(3): 393-408.

USHIO T, KUBOTA T, SHIGE S, et al, 2009. A Kalman filter approach to the global satellite mapping of precipitation (GSMaP) from combined passive microwave and infrared radiometric data[J]. J Meteor Soc Japan, 87A: 137-151.

WANG D, GAO J, BREWSTER, et al, 2003. The advanced regional prodiction system (ARPS), storm-scale numerical weather prediction and data assimilation[J]. Meteor Atmos Phys, 82(1): 139-170.

WU W R, KITZMILLER D, WU S R, 2012. Evaluation of radar precipitation estimates from the national mosaic and multisensor quantitative precipitation estimation system and the WSR-88D precipitation processing system over the conterminous United States[J]. J Hydrometeorol, 13(3): 1080-1093.

XIA Y L, MITCHELL K, EK M, et al, 2012. Continental-scale water and energy flux analysis and validation for North American land data assimilation system project phase2 (NLDAS-2): 2. Validation of model-simulated streamflow[J]. J Geophys Res, 117(D3): D03110.

XIE P, ARKINP A, 1997. Global precipitation: a 17-year monthly analysis based on gauge observations, satellite estimates, and numerical model outputs[J]. Bull Amer Meteor Soc, 78(11): 2539-2558.

XIE P P, XIONG A Y, 2011. A conceptual model for constructing high-resolution gaugesatellite merged precipitation analyses[J]. J Geophys Res, 116(D2): D21106.

XUE M, DROEGEMEIER K K, WONG V, et al, 2001. The advanced regional prediction system (ARPS): a multi-scale non-hydrostatic atmospheric simulation and prediction tool. Part Ⅱ: Model physics and applications[J]. Meteor Atmos Phys, 76(3-4): 143-165.

XUE M, WANG D H, GAOJ D, et al, 2003. The advanced regional prediction system (ARPS), storm-scale numerical weather prediction and data assimilation[J]. Meteor Atmos Phys, 82(1-4): 139-170.

ZHANG H M, BATES J J, REYNOLDS R W, 2006a. Assessment of composite global sampling: sea surface wind speed[J]. Geophys Res Lett, 33(17): L17714.

ZHANG H M, REYNOLDS R W, BATES J J, 2006b. Blended and gridded high resolution global sea surface wind speed and climatology from multiple satellites: 1987-present//Proceedings of the 14th Conference on Satellite Meteorology and Oceanography[R]. Atlanta, GA: American Meteorological Society.

ZHANG J, HOWARD K, LANGSTO N C, et al, 2016. Multiradar multisensor (MRMS) quantitative precipitation estimation: Initial operating capabilities[J]. Bull Amer MeteorSoc, 97(4): 621-638.

Roles of Synoptic to Quasi-Monthly Disturbances in Generating Two Pre-Summer Heavy Rainfall Episodes over South China

JIANG Zhina[1]   ZHANG Da-Lin[1,2]   LIU Hongbo[3]

(1. State Key Laboratory of Severe Weather, Chinese Academy of Meteorological Sciences, Beijing, 100081, China; 2. Department of Atmospheric and Oceanic Science, University of Maryland, College Park, Maryland, 20742-2425, USA; 3. LASG, Institute of Atmospheric Physics, Chinese Academy of Sciences, Beijing, 100029, China)

Abstract: In this study, power spectral analysis and bandpass filtering of daily meteorological fields are performed to explore the roles of synoptic to quasi-monthly disturbances in influencing the generation of pre-summer heavy rainfall over South China. Two heavy rainfall episodes are selected during the months of April-June 2008-2015, which represent the collaboration between the synoptic and quasi-biweekly disturbances and the synoptic and quasi-monthly disturbances, respectively. Results show that the first heavy rainfall episode takes place in a southwesterly anomalous flow associated with an anticyclonic anomaly over the South China Sea (SCS) at the quasi-biweekly scale with 15.1% variance contributions, and at the synoptic scale in a convergence zone between southwesterly and northeasterly anomalous flows associated with a southeastward-moving anticyclonic anomaly on the leeside of the Yungui Plateau and an eastward-propagating anticyclonic anomaly from higher latitudes with 35.2% variance contribution. In contrast, the second heavy rainfall episode takes place in southwest-to-westerly anomalies converging with northwest-to-westerly anomalies at the quasi-monthly scale with 23.2% variance contributions to the total rainfall variance, which are associated with an anticyclonic anomaly over the SCS and an eastward-propagating cyclonic anomaly over North China, respectively. At the synoptic scale, it occurs in south-to-southwesterly anomalies converging with a cyclonic anomaly on the downstream of the Yungui Plateau with 49.3% variance contributions. In both cases, the lower-tropospheric mean south-to-southwesterly flows provide ample moisture supply and potentially unstable conditions; it is the above synoptic, quasi-biweekly or quasi-monthly disturbances that determine the general period and distribution of persistent heavy rainfall over South China.

Key words: synoptic scale; pre-summer rainfall; quasi-biweekly scale; quasi-monthly disturbances

Article Highlights

• During the pre-summer periods of 2008-2015, the most dominant low-frequency contributions to persistent heavy rainfall production are the synoptic frequency, and then the quasi-biweekly and quasi-monthly frequency, based on the daily mean rainfall averaged over South China by spectral analysis.

• Although the lower-tropospheric south-to-southwesterly mean flows provide a convectively favorable environment, it is the synoptic, quasi-biweekly or quasi-monthly disturbances that determine the general period and distribution of persistent heavy rainfall.

• The origins of these disturbances influencing the south-to-southwesterly supply and lifting of warm and moist air over South China can be traced back to midlatitudes or the South China Sea and Indian Ocean.

1 Introduction

South China is situated to the east of the Yunnan-Guizhou Plateau and bordered with the South China Sea(SCS), roughly covering the continental area of 104°-120°E, south of 27°N. It is one of the most centralized regions for heavy rainfall over China. The pre-summer rainfall season, which is the first rainy period, usually referring to April to June, accounts for 40%-50% of the total annual rainfall over South China[1].

Huang found that a subtropical jet would "jump" northward twice each year, with the first jump of about 7° latitude during March-April and the second one of about 4° during July-August[1]. Accompanying this phenomenon, the ridge of the western North Pacific subtropical high(WNPSH) moves northward by about 7° latitude during spring and 5° during June-July. The pre-summer rainfall period coincides with the first simultaneous northward movements of the subtropical jet and WNPSH. This period is dominated by southwesterly winds, which can transport ample warm and moist air of tropical origin for the generation of rainfall over South China.

Previous studies have indicated that warm-season rainfall over South China exhibits substantial intraseasonal variability, with low-frequency modes in both rainfall and atmospheric circulations[2-5]. A 30-60-day periodic variation appears to occur under the influences of the Madden-Julian Oscillation[6,7] and Mascarene and Australian highs[8], while a 10-20-day periodic variation can be traced to atmospheric perturbations over middle to higher latitudes and tropical regions[9-12]. Cao et al. attributed the low-frequency rainfall over South China to the juxtaposition of a lower-level west-to-northwestward propagating low-frequency anticyclone near 150°E(with an accompanying cyclone over the SCS) with the westward extension of the WNPSH[13]. Pan et al. emphasized the roles of a lower-tropospheric anomalous anticyclone encircling the Tibetan Plateau interacting with a southeastward-propagating wave train in the middle to upper troposphere[11].

Recently, some studies have begun to note the importance of 3-8-day synoptic disturbances in generating heavy rainfall. Liu et al. claimed that Rossby wave trains at higher latitudes, cyclonic and anticyclonic anomalies downstream of the Tibetan Plateau, and the WNPSH, played different roles in producing the summer 2003 heavy rainfall over the Yangtze-Huai River Basin(YHRB) during the mei-yu season[14]. Li et al. indicated that the location and strength of the intraseasonal oscillation and the synoptic disturbances play a decisive role in controlling the severity and duration of the heavy precipitation events over South China during summer, in which wind variations contribute more than humidity variations to changes in moisture divergence[15]. However, few studies have focused on the relationship between pre-summer rainfall and synoptic distances until a very recent study by Huang et al.[16], who found a dominant frequency mode at the synoptic scale(i.e., 3-8 days) with well-defined positive rainfall anomalies over South China. They traced some disturbances accounting for extreme rainfall events over South China to the development of cyclonic anomalies downstream of the Tibetan Plateau. Given the frequent passage of tropical and higher-latitude disturbances at different scales(likely up to monthly) across South China and the upstream influences of high topography and the SCS[17], we may hypothesize their different roles in determining the general generation and persistence of the pre-summer heavy rainfall over South China. Therefore, the major objectives of this study are to (1) investigate what low frequencies of disturbances contribute more significantly to the pre-summer heavy rainfall over South China on the daily mean rainfall basis, and (2) examine where these disturbances originate and what their roles are. The above objectives are achieved by performing a spectral analysis of pre-summer rainfall events occurring during the years of 2008-2015, and then exploring the roles of different frequency modes in two representative heavy rainfall episodes over South China.

The next section describes the data and methodology. The spectral characteristics of pre-summer rainfall over South China during 2008-2015 are also presented. Section 3 shows the selection of two heavy rainfall episodes over South China and provides a brief overview of the associated rainfall events. Section 4 analyzes the temporal evolution of mean flow structures and the different roles of synoptic, quasi-weekly and quasi-monthly disturbances in influencing the two heavy rainfall episodes. A summary and concluding remarks are given in the final section.

2 Data and methodology

In this study, merged rain gauge-satellite 0.1°×0.1° resolution gridded hourly precipitation data across China from 2008 onward are used[18], archived by the National Meteorological Information Center of the China Meteorological Administration(CMA). This product is developed with the optimum interpolation technique by combining the CMA's hourly rain-gauge network data with a satellite-retrieved precipitation product of the National Oceanic and Atmospheric Administration(NOAA)/Climate Prediction Center's morphing technique(CMORPH) dataset[19]. Daily rainfall data used herein are obtained by accumulating hourly rainfall records.

Large-scale meteorological conditions associated with the pre-summer rainfall events are analyzed using the National Centers for Environmental Prediction Final Global Analysis dataset with 1°× 1° grid resolution at six-hourly intervals. For our study, the daily mean fields are calculated by averaging the original six-hourly data. Besides, the daily mean interpolated outgoing longwave radiation(OLR) data from NOAA are used to represent large-scale tropical convective activity, with a horizontal resolution of 2.5° latitude ×2.5° longitude.

To identify the prominent temporal characteristics of rainfall events over South China, a spectral analysis of the above-mentioned daily averaged data is performed with the fast Fourier Transform, following [14]. A temporal mean is first removed, and then a smoothing with a 10% tapered window is applied to the perturbation time series. The statistical significance of power spectra is tested, based on the power spectrum of the red noise according to [20]. In addition, a bandpass filter [21] is used to extract the periodical disturbances from the raw daily rainfall and circulation field, so as to better identify the temporal evolution of their structures corresponding to the generation of heavy rainfall over South China.

The pre-summer rainfall season over South China is one of the most prominent rainfall periods in China. The spatial distribution of the daily mean rainfall during April to June of 2008 to 2015 is given in Fig. 1, showing three local maxima: one in North Guangxi, and the other two in the north-central region and southeastern coastal region of Guangdong Province, respectively. Given the dominant rainfall amounts associated with the latter two local maxima, we select the region(21°-25°N, 112°-117°E) shown by a dashed box in Fig. 1 as the target area, which includes most of Guangdong Province and its coastal waters. Then, the power spectra of the area-averaged daily mean rainfall during the pre-summer rainy season(i. e. , from 1 April to 30 June)for the years 2008-2015 are calculated and Markov's red-noise spectrum and 95% upper confidence bound are applied, respectively(Fig. 2), in order to understand the frequency characteristics of rainfall over the target region. In [9,22,23], they all take the red-noise spectrum as the confidence limit. In addition, the 95% upper confidence bound is another higher confidence limit.

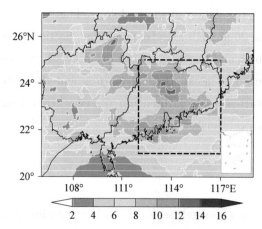

Fig. 1 Horizontal distribution of the daily mean rainfall(shading; units: mm) during the pre-summer months(April-June) of 2008-2015

The dashed box(21°-25°N, 112°-117°E) is used to define the target domain of interest-similarly for the rest of figures

Despite their different magnitudes and frequency characteristics, 3-8 days(i. e. , the synoptic scale) appears to be the dominant frequency band, as compared to the 95% upper confidence bound for all the eight pre-summer periods. In addition, 10-20 days(i. e. , the quasi-biweekly scale) can be identified as a secondary frequency band, which is also significant against the red-noise background for five pre-summer periods(i. e. ,2008,2010,2012,2014,2015). The above analysis indicates that the synoptic and quasi-biweekly disturbances are the two most dominant modes in these years' pre-summer rainfall over South China on the daily mean basis. A further examination of Fig. 2 also reveals that the frequency band at 15-40 days (i. e. , the quasi-monthly scale) are significant during the pre-summer periods of 2009 and 2013. This scale of disturbance(15-35 days) has been found to be related to Yangtze River rainfall over eastern China during the summer of 1991 by [24], and later further explored by [25], who pointed out

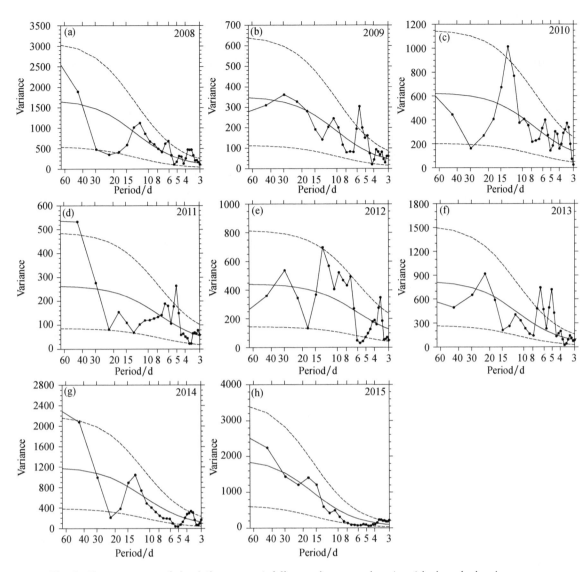

Fig. 2 Power spectra of the daily mean rainfall over the target domain with the calculated spectrum (solid lines with dots), Markov red-noise spectrum (solid lines), 95% upper and 5% lower (dashed lines) confidence bounds during the months of April–June 2008–2015

Note that different scales in ordinates are used for different years

that this time scale of summer Yangtze River rainfall arises in response to intraseasonal variations in the WNPSH. We should note that the 10-20-day and 15-40-day modes overlap with each other and thus cannot be examined at the same time. To our knowledge, the impact of quasi-monthly disturbances on heavy rainfall over South China has not been examined in the literature. Thus, in the next two sections, we examine how the three dominant modes contribute to the development of pre-summer rainfall over South China through two representative case studies: one is mainly contributed by synoptic and quasi-biweekly disturbances, and the other is mainly influenced by synoptic and quasi-monthly disturbances.

3 Selection and overview of two typical rainfall events

To address the objectives raised in section 1, it is desirable to select two representative rainfall episodes that could include the synoptic and quasi-biweekly disturbances or the synoptic and quasi-monthly disturbances. To this end, we first calculate the area-averaged daily mean rainfall over Southeast China, and the 10-20-day and 3-8-day bandpass-filtered daily mean rainfall anomalies in the pre-summer seasons of 2008,2010,2012,2014 and 2015, and the 15-40-day and 3-8-day bandpass-filtered daily mean rainfall anomalies in the pre-summer seasons of 2009 and 2013. Then, we check the above two kinds of persistent heavy rainfall events, in which the wettest rainfall peaks coincide well with the peak amplitudes of synoptic and quasi-biweekly disturbances or synoptic and quasi-monthly disturbances. As a result, the persistent and comparatively heavier rainfall periods of 6-10 May 2014 and 19-22 May 2013 are found to better reveal the general features of the above-mentioned disturbances than the other periods.

To intuitively examine the contributions of synoptic, quasi-biweekly and quasi-monthly disturbances to the heavy rainfall events, the area-averaged daily rainfall series against the 3-8-day, 10-20-day or 15-40-day bandpass-filtered rainfall anomalies over the target domain are shown in Figs. 3a and c for the above two rainfall episodes. For the first rainfall episode of interest, the synoptic-scale and quasi-biweekly filtered rainfall variances contribute 35.2% and 15.1% to the total rainfall variance during the period of 6-10 May 2014(Fig. 3a), respectively. Moreover, one can see prominent synoptic-scale(i. e. ,6-10 May) and quasi-biweekly(i. e. ,3-15 May)processes involved in this rainfall episode, whose wettest phase matches well with the heaviest rainfall on 8 May(cf. Figs. 3a and e). For the second rainfall episode of interest, the persistent rainfall events occur during 19-23 May 2013, and they are dominated by synoptic-scale disturbances with about 49.3% variance contributions to the total rainfall variance, and quasi-monthly disturbances with 23.2% variance contributions. Prominent synoptic-scale(i. e. ,19-23 May) and quasi-monthly(i. e. ,11-30 May)processes are involved in this rainfall episode, whose wettest phase matches well with the heaviest rainfall on 21 May 2013(cf. Figs. 3 c and e).

Fig. 3b and d show the spatial distribution of the temporally averaged daily rainfall peaked on 8 May 2014 and 21 May 2013, respectively. One can see that the 6-10 May 2014 rainfall episode is mainly distributed over the Pearl River Delta of Guangdong Province, whose local maximum daily mean rainfall amount exceeds 70 mm(Fig. 3b). In contrast, the 19-22 May 2013 rainfall episode is linearly distributed along the coastline over Southeast China with another branch extending to the SCS, whose rainfall region with local maximum rainfall amount over 70 mm is comparatively small(Fig. 3 d). Thus, the above two rainfall episodes of different geographical and morphological characteristics are selected herein to examine what their dominating circulation anomalies are, and then trace where they originate from.

It is important to realize that the above two rainfall episodes must be produced by different types of mesoscale convective systems(MCSs) associated with different larger-scale circulations(e. g. , Fig. 1 in [26]). The lower-frequency disturbances under study provide favorable large-scale forcing that influences the general period and distribution of regional persistent heavy rainfall. In fact, many previous studies have indicated that coastal precipitation over South China is closely related to land and sea breezes, low-level jets, coastal orography and slow-propagating weak baroclinic disturbances[27-30]. For example, Chen et al. [27] illustrated that the nocturnal offshore rainfall over the South China coastal region is induced by a convergence line between the low-level monsoonal wind and the land breezes. Du et al. [29] pointed out that the two propagation modes(onshore and offshore)of the diurnal rainfall cycle near the coastline are associated with inertia-gravity waves, in terms of speed and phase. It is the background wind that changes the pattern of the inertia-gravity waves and further affects the diurnal propagation. Du et al. [30] showed that the initiation of a heavy-rain-producing MCS that occurred over the southern coast of China on 10-11

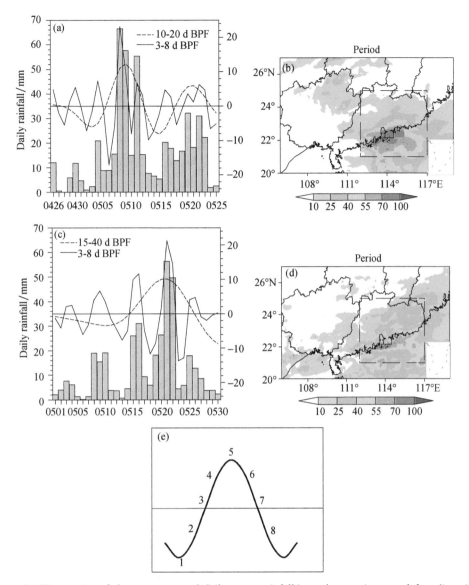

Fig. 3 (a) Time series of the area-averaged daily mean rainfall (gray bars; units: mm; left ordinate) and 10-20-day (dashed lines) and 3-8-day (solid lines) bandpass-filtered daily mean rainfall anomalies (units: mm; right ordinate) during the one-month period of 26 April to 25 May 2014. (b) Spatial distribution of the temporally averaged daily rainfall (units: mm) during the period 6-10 May 2014. (c) As in (a) but for 15-40-day (dashed lines) and 3-8-day (solid lines) bandpass-filtered daily mean rainfall anomalies during the period 1-30 May 2013. (d) As in (b) but during the period 19-23 May 2013. (e) Diagram of eight phases during one life cycle of the phase analysis. The phase 3(7) is crossed by a horizontal line, which represents transition from a dry (wet) state to a wet (dry) state

May 2014 was associated with the convergence between a synoptic-scale low-level jet and a boundary-layer jet. This means that most of the remaining variances in the selected two rainfall episodes may be explained by mesoscale processes that could not be resolved herein and the interactions across different scales.

4 Mean flows and filtered circulations

In this section, we examine the temporal evolution of daily mean flow fields, and the bandpass-filtered anomalous circulation systems from dry to wet phases, i.e., phase 1(P1) to phase 5(P5) given in

Fig. 3e, in order to gain insight into the lower-frequency background in which each heavy rainfall episode occurs and the origins of disturbances at different scales that influence the generation of the heavy rainfall episode.

4.1 The heavy rainfall episode that peaked on 8 May 2014

Fig. 4 shows the daily mean flows at 850 hPa, and the daily mean geopotential height field and precipitable water during 6-8 May 2014. It is evident that on the heavy rainfall day of 8 May, South China is dominated by warm and moist southwesterly to southerly flows of tropical origin (even from the Indian Ocean) in the lower troposphere (Fig. 4 c). The southwesterly flow is associated with the WNPSH, whose axis extends from the central North Pacific to midwest Indian Ocean. This southwesterly flow turns cyclonically in the north of the target domain to a more southerly flow at about 25°N and southeasterly flow at higher latitudes, as an anticyclone (denoted by "A") moves to the middle-east of China, which strengthens the supply of warm and moist air by southwesterlies to the target domain (cf. Figs. 4b

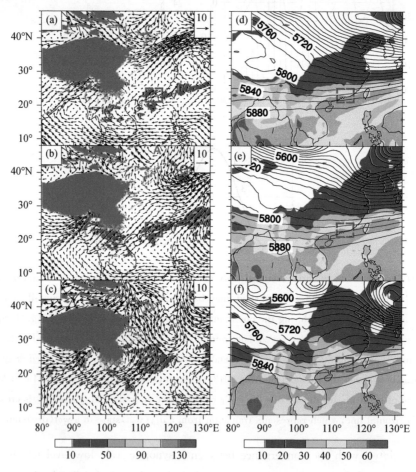

Fig. 4 Temporal evolution of (a-c) the daily mean 850 hPa wind vectors (units: m/s), superimposed with daily mean rainfall (shaded; units: mm), and (d-f) the daily mean 500 hPa geopotential height (solid lines; units: gpm) and precipitable water (shaded; units: mm) during the period 6-8 May 2014. The thick solid line refers to the 5880-m geopotential isoline at 500 hPa. The gray shading represents topography greater than 1600 m. (a,d) 6 May; (b,e) 7 May; (c,f) 8 May. The inner box dashed in red denotes the target area, similarly for the rest of figures. The capital letter "A" ("C") represents anticyclone (cyclone), similarly for the rest of figures

and c). At 500 hPa, South China is located ahead of a weak trough axis with prevailing west-southwesterly winds extending from the southeast of the Tibetan Plateau (Fig. 4f). Clearly, the juxtaposition of midlevel quasi-geostrophic lifting, albeit weak, with the lower-level southwesterly warm and moist air from the SCS and Indian Ocean helps pre-condition a favorable environment for heavy rainfall production. The favorable environment, though relatively weak, appears to account for the generation of a near-zonally oriented rainfall belt, which is consistent with the mean flow at 500 hPa (cf. Figs. 4a, b and 4 d, e). The rainfall pattern occurring on 8 May looks more like a warm-sector rainfall episode. In particular, precipitable water in the upstream regions reaches more than 50 mm, implying the presence of considerable water vapor available for the generation of heavy rainfall. Note that although the precipitable water pattern in the vicinity looks similar during the period of May 7-8, the moisture flux, the product of water vapor and horizontal winds, increases with increasing southwesterly flows (Figs. 4a-c), with its peak flux on 8 May. Ultimately, it is the moisture flux convergence that determines the amount of rainfall over the region.

However, the above favorable mean flow conditions, resulting from the interaction of several atmospheric weather systems, are significantly different when they are traced back two days earlier, i.e., 6 May (Fig. 4a). That is, the anticyclone, which accounts for the cyclonic turning of the southwesterly flow, originates from Mongolia, west of an extratropical cyclone (denoted by "C") (Fig. 4b). It grows in amplitude as it intrudes southeastward (Figs. 4b and c). Meanwhile, the WNPSH retreats equatorward, as indicated by the 5880-m geopotential isoline at 500 hPa (Figs. 4 d-f), but strengthens, as it extends more into the Bay of Bengal-similarly for the strengthened southwesterly flow, as indicated by the horizontal wind vectors. At 500 hPa, South China is governed by westerly flow, with the southeastward intrusion of colder and drier air on the northern side of the Tibetan Plateau on 6 May (Fig. 4 d). This southeastward intrusion, as indicated by a dry pocket (shown in blue shades) of precipitable water, helps hydrostatically strengthen the midlevel trough (cf. Figs. 4e and f).

After seeing the evolution of large-scale mean flows in relation to the generation of the heavy rainfall episode that peaked on 8 May 2014, we can now examine what circulation anomalies at the quasi-biweekly and synoptic scales are involved. For this purpose, Figs. 5 and 6 show the filtered quasi-biweekly (10-20-day) and synoptic scale (3-8-day) disturbances, respectively, in different phases at 850 hPa and 200 hPa in association with the filtered rainfall. In the driest phase (i.e., phase 1, or P1) of the quasi-biweekly scale (Fig. 5, left column), South China is dominated by northeasterly anomalies, with an anticyclonic anomaly (A1) on the southeast side of the Tibetan Plateau, and a cyclonic anomaly (C) and an anticyclonic ridge (A2) over the SCS. Although the anticyclonic (A1) and cyclonic (C) anomalies gradually move eastward to eastern-central China and the Bashi Strait, respectively, from P2 to P3, the northeasterly anomaly still dominates South China. Only after the midlatitude anticyclonic anomaly (A1) migrates into the Sea of Japan and the anticyclonic ridge (A2) strengthens into an anticyclonic anomaly in P4 does a large-scale southwesterly anomaly associated with A2 begin to prevail on the southeast side of the Tibetan Plateau. As a result, South China is influenced by the weak northeasterly and weak southwesterly anomalies, which are associated with the above-mentioned anticyclonic anomaly (A1) and summer monsoon, respectively. The anticyclonic anomaly (A2) over the SCS helps strengthen the southwesterly monsoonal flow across South China and generate a confluence zone along the coastline.

In the wettest phase (i.e., P5), the southwesterly anomaly reaches a larger intensity than before,

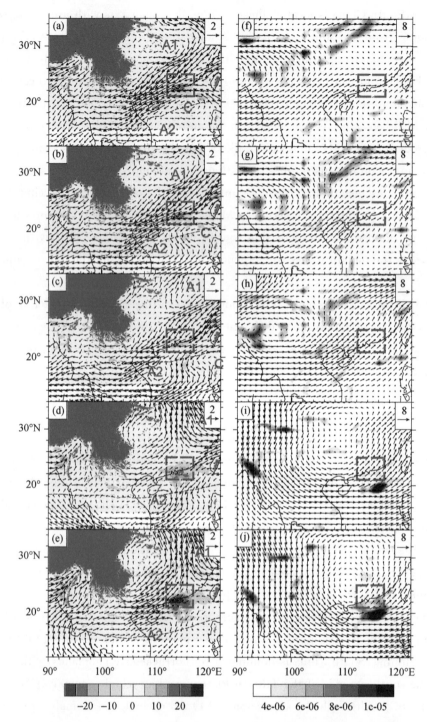

Fig. 5 (a-e) Phase evolution of the quasi-biweekly bandpass-filtered daily mean 850 hPa wind vector(units: m/s) superimposed with quasi-biweekly rainfall rate anomalies(shaded; units: mm/d). Black-dashed lines refer to the 5880-m geopotential isoline at 500 hPa. (f-j) As in(a-e) except for the quasi-biweekly bandpass-filtered daily mean 200 hPa wind vector(units: m/s) and the corresponding divergence(shaded; units: s^{-1}). (a,f)P1 on 3 May; (b,g)P2 on 4 May; (c,h)P3 on 5 May; (d,i)P4 on 7 May and (e,j)P5 on 8 May 2014

but gradually weakens from the southwestern to northeastern part in the target domain. Furthermore, this anomalous flow together with the trailing anomalous easterly flow of the anticyclonic anomaly(A1)

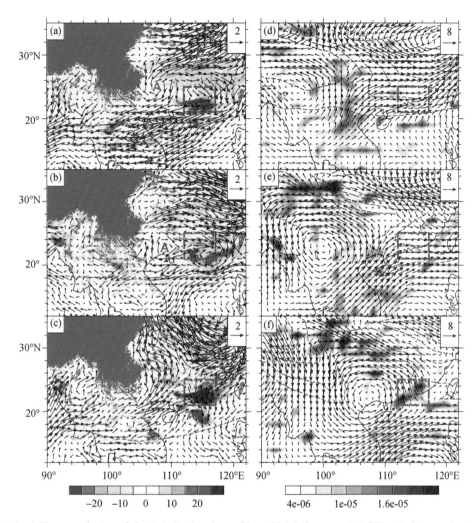

Fig. 6 (a-c) Phase evolution of the 3-8-day bandpass-filtered (a) daily mean 850 hPa wind vector (units: m/s) superimposed with 3-8-day bandpass-filtered rainfall rate anomalies (shaded; units: mm/d). Black-dashed lines refer to the 5880-m geopotential isoline at 500 hPa. (d-f) Temporal evolution of the 3-8-day bandpass-filtered daily mean 200 hPa wind vector (units: m/s), the total wind speeds (dashed line; units: 1/s) and divergence (shaded; units: 1/s). (a,d) P1 on 6 May; (b,e) P3 on 7 May; (c,f) P5 on 8 May 2014

leads to an intense southeast-northwest-oriented convergence zone where heavy rainfall is distributed on its southern side (see P5 in Fig. 5). It follows that the converging warm and moist air masses at the quasi-biweekly scale contribute positively to the generation of this heavy rainfall episode.

The quasi-biweekly filtered fields in the upper troposphere show the migration of a cyclonic anomaly into the study area from the west during P1-P5 (Fig. 5, right column), which gives rise to the presence of south-to-southwesterly anomalies over South China. In P4, a pronounced divergence region is observed between Hainan and Taiwan, and in P5 it expands toward the southern coastal region, where diffluent horizontal wind vectors are evident. This upper-level diffluence is closely collocated with the lower-tropospheric convergence area. Thus, such a tropospheric juxtaposition tends to generate a favorable upward motion for the generation of the present heavy rainfall episode.

On the synoptic time scale, the lower troposphere from P1 to P3 exhibits a southeastward propagating anticyclonic anomaly (i. e. , A), which originates from the lee side of the Yungui Plateau at P1 and dominates South China with two centers in P3 (Fig. 6, left column). This scenario is different from that of a pre-summer heavy rainfall mode found by[16]. In P5, however, this anticyclonic anomaly moves southeastward into the SCS. This displacement allows a cyclonical turning of northeasterly anomalous flows(with likely cold air), associated with a higher-latitude anticyclonic anomaly, over the target domain, thereby assisting in lifting the southwesterly warm and moist air for the generation of the heavy rainfall episode that peaked on 8 May 2014.

The upper troposphere at the synoptic time scale shows the southeastward propagation of a cyclonic anomaly, which is similar to but much better organized than that at the quasi-biweekly scale, with the presence of an upper-level jet stream around 35°N. South China is located over the southern entrance region of the jet stream, where favorable divergence can be seen as expected[31]. In P5, the cyclonic anomaly moves close to the heavy rainfall region, where significant diffluence is present, like that on the quasi-biweekly time scale(cf. Figs. 6 and 5).

To further see the contribution of quasi-biweekly and synoptic disturbances to the heavy rainfall production, the latitude-height cross sections of divergence along 115°E in P5 for the two different scales of disturbances are given in Fig. 7. One can see evident convergence in the 850-400 hPa layer with two convergence centers in the vertical direction, and divergence aloft, implying the presence of favorable deep local upward motion for the rainfall generation. This should also be expected from Figs. 5 and 6, showing isentropic lifting as a warm and moist air parcel moves northward in the southwesterly flow[32,33]. In general, the amplitudes of convergence and divergence of the 3-8-day filtered disturbances are twice more than those of the 10-20-day filtered disturbances. This suggests the more important dynamical forcing of the synoptic scale disturbances than the quasi-biweekly ones in influencing the persistent heavy rainfall event.

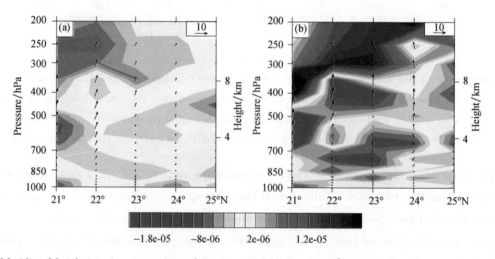

Fig. 7　Meridional-height(y,z)cross sections of divergence(shaded;units:s^{-1})and in-plane flow vectors(units:m/s) with the vertical motion magnified by 500, taken along 115°E in P5 for(a)the 10-20-day bandpass-filtered disturbances and(b)the 3-8-day bandpass-filtered disturbances

Fig. 8a and b show the temporal evolution of the respective 10-20-day and 3-8-day bandpass-filtered area-averaged moisture and potential temperature tendencies, superimposed with the area-averaged filtered vertical motion disturbances (w). The above area-averaged tendencies are calculated from their corresponding flux convergence across the target area:

$$\int \frac{\partial \pi}{\partial t} ds = -\int \nabla \cdot (\pi V) \, ds - \int \left[\frac{\partial (\pi \omega)}{\partial P}\right] ds$$

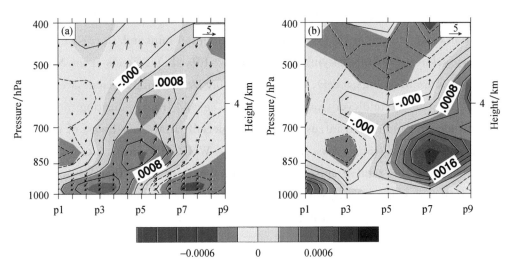

Fig. 8 (a) Time-pressure diagram of the 10-20-day bandpass-filtered anomalies of moisture flux convergence (shaded; units: g/(kg · s)), thermal flux convergence (contours; units: K/s) and meridional-vertical (y,z) flow vectors (units: m/s) with vertical motion magnified by 500 averaged over the area (21°~25°N, 112°~117°E) on 3(P1), 5(P3), 8(P5), 12(P7) and 15(P9) May 2014. (b) As in (a) but for the 3-8-day bandpass-filtered fields on 6(P1), 7(P3), 8(P5), 9(P7) and 10(P9) May 2014

where V denotes the horizontal wind vector, ω is the vertical velocity in P-coordinates; s is the target domain, and π denotes a conserved variable (i.e., $\frac{d\pi}{dt}=0$), which is the specific humidity or potential temperature herein. This equation is derived by combining $\frac{d\pi}{dt}=0$ and the mass continuity equation.

At the quasi-biweekly scale (Fig. 8a), we can see favorable upward motion from P4 to P6. From P1 to P3, increased warm and moist air prevails below 800 hPa, and extends upward to about 700 (500) hPa during P4 (P5), which provides sufficient moisture supply and potential instability (i.e., with warm air below cold air) to be favorable for the convective development associated with the heavy rainfall episode. The cold and dry air above 800 hPa from P1 to P3 is consistent with the presence of dominant northerly flow shown in Fig. 5. After P5, dry and cold air tendency appears, as can be expected. By comparison, the synoptic-scale fields show the presence of warm and moist disturbances in the lower troposphere, mostly below 850 hPa prior to P3, and they gradually weaken with time (Fig. 8b). Clearly, the quasi-biweekly and synoptic-scale disturbances both provide the moisture and energy needed for the generation of heavy rainfall. Certainly, the quasi-biweekly disturbances appear to generate more significant thermal and moisture transport than the synoptic disturbances to the heavy rainfall production, especially during P4 to P5.

4.2 The heavy rainfall episode that peaked on 21 May 2013

Fig. 9 shows the temporal evolution of large-scale mean flows at 850 and 500 hPa associated with a heavy rainfall episode that peaked on 21 May 2013 (see Fig. 3c). Results show that this episode occurs ahead of a mesoscale cyclone (denoted by "C") to the north-to-northwest under the influences of a southwesterly monsoonal flow originating from the Indian Ocean in the lower troposphere (Figs. 9a-c). Of relevance here is that the southwesterly flow strengthens during the period of 19-21 May, in association with the generation of an elongated southwest-northeast-oriented rainband. Corresponding to the lower-tropospheric mesocyclone is an intensifying trough at 500 hPa that extends from Northeast China to North Bay from 19 to 21 May(Figs. 9 d-f). Unlike the 8 May 2014 rainfall episode, we see greater amplitude and spatial extent of the midlevel trough with a surface cold front, indicating more important roles of the higher-latitude cold and dry air intrusion behind, especially in the northwest of the target area, and favorable lifting ahead of the trough axis. In addition, the WNPSH retreats more eastward to the Indochina Peninsula during this rainfall episode. Precipitable water of more than 60 mm is distributed along the coastal land and water regions, where the heavy rainfall event takes place. On average, this precipitable water amount is at least 10 mm higher than that associated with the episode that peaked on 8 May 2014(cf. Figs. 9 and 4).

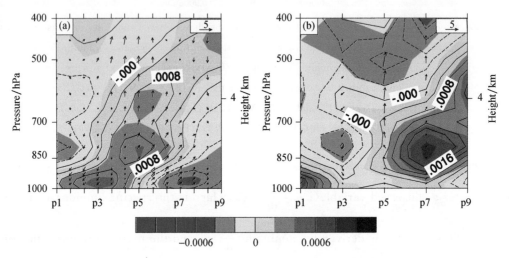

Fig. 9　As in Fig. 4 but during the period 19-21 May 2013: (a,d)19 May; (b,e)20 May; (c,f)21 May

Fig. 10 shows the roles of the band-filtered quasi-monthly disturbances in determining the heavy rainfall episode that peaked on 21 May 2013. It is evident that this episode occurs in southwest-to-westerly anomalies converging with northwest-to-westerly anomalies located on the south of a cyclonic anomaly at 850 hPa in P5(Fig. 10, left column). This cyclonic anomaly can be traced back to North China (denoted by "C") in P1. The southeastward movement of this cyclonic perturbation from P1 to P5 facilitates the influences of westerly to northwesterly anomalies over South China. The associated southwesterly anomaly accounts for the advection of warm and moist air into the heavy rainfall region in P5. In particular, the heavy rainfall zone coincides well with the confluence of an anomalous northwesterly flow associated with the eastward-moving cyclonic anomaly and anomalous west-to-southwesterly flows associated with a quasi-stationary anticyclonic anomaly over the SCS during P4 and P5. This suggests the presence of cold frontal lifting at this time scale, which facilitates the generation of favorable conditions for the heavy rainfall production. Again, the heavy rainfall generation occurs during the eastward retreat of the WNPSH from the Bay of Bengal to the east side of the Indochina Peninsula(cf. Figs. 10 and 5). In the upper troposphere, a Rossby wave train with an anticyclonic anomaly is seen propagating northeastward into the western North Pacific during P1 to P5(Fig. 10, right column). A distinct divergence zone associated with the propagating anticyclonic anomaly is observed to move southeastward across the target area, which facilitates a favorable vertical circulation for the occurrence of heavy rainfall.

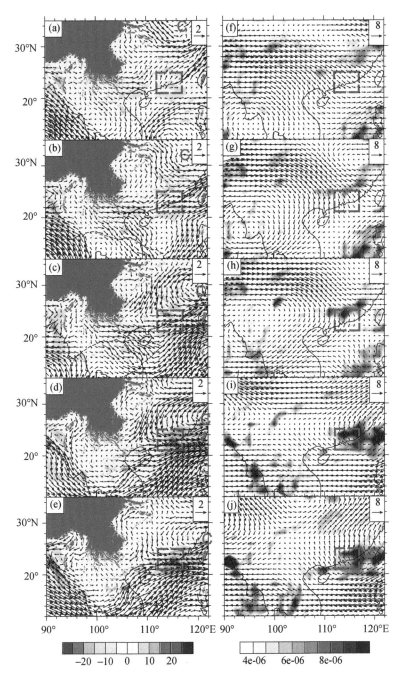

Fig. 10 As in Fig. 5 but with 15-40-day bandpass-filtered fields: (a,f) P1 on 11 May; (b,g) P2 on 13 May; (c,h) P3 on 15 May; (d,i) P4 on 18 May; (e,j) P5 on 21 May 2013

To further illustrate the origins of quasi-monthly disturbances, Fig. 11 shows the evolution of the 15-40-day filtered OLR field. In the driest phase (P1), a weak positive (with depressed convection) OLR anomaly over South China is accompanied to its southeast by a strong negative OLR anomaly around the Philippine islands. This means that the quasi-monthly oscillation of heavy rainfall over South China may have a connection with the anomalous convective system in the western North Pacific and the SCS. This seesaw structure weakens in intensity as it migrates northwestward from P1 to P3. Its northwestward migration is more evident from P3 to P5, during which period a negative OLR anomaly (i. e., with active convective development) approaches the edge of South China in P4 and covers the rainfall center in

P5. At the same time, another positive OLR anomaly is seen on its southeast, which represents an eastward shift of the WNPSH, as also shown in Fig. 10. This implies that the 15-40-day oscillation of pre-summer rainfall over South China may also correspond to variations in the WNPSH, in which the wave-like anomalous convective system originates from the equatorial western Pacific. In fact, this kind of opposite-phase variation in convective anomalies between South China and the SCS-Philippine islands is very similar to that found between the Yangtze River basin and the SCS-Philippine islands, as found by Mao and Wu(2006) and Mao et al. (2010).

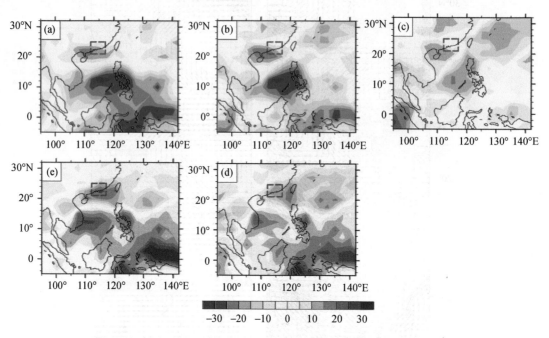

Fig. 11   As in Fig. 10 but for the 15-40-day filtered OLR(units: W/m²)

On the synoptic time scale(Fig. 12, left column), the filtered disturbances at 850 hPa show the presence of a cyclonic anomaly(i. e. , "C") in P5 that corresponds to a cyclone in the northwest(cf. Figs. 12 and 9 c). Like that shown in Figs. 9a-c, this anomaly results from the southwestward extension of a cyclonic disturbance in P1. This disturbance grows in both amplitude and coverage from P1 to P5, due partly to the continued latent heat release associated with the heavy rainfall event. The southwestward extension of this cyclonic anomaly is facilitated by the eastward movement of an anticyclonic anomaly (i. e. , "A") out of the target area from P3 to P5. As the cyclonic anomaly approaches the target area, South China is under the influence of south-to-southwesterly anomalous warm and moist currents. In the upper troposphere(Fig. 12, right column), a higher-latitude Rossby wave train with a cyclonic anomaly is seen propagating southeastward. South China is located over the southern entrance region of an upper-level jet stream, where favorable diffluence facilitates the generation of favorable vertical motion for pre-conditioning a convectively unstable environment.

Fig. 13 summarizes the temporal evolution of the area-averaged quasi-monthly and synoptic band-pass-filtered moisture flux convergence, thermal flux convergence and vertical motion fields, which differs somewhat from those associated with the episode that peaked on 8 May 2014. Specifically, at the quasi-monthly scale(Fig. 13a), upward motion with cold and dry anomalous air appears from P3 to P6, which corresponds to the northwest-to-westerly anomalous advection from higher latitudes(cf. Figs. 13a

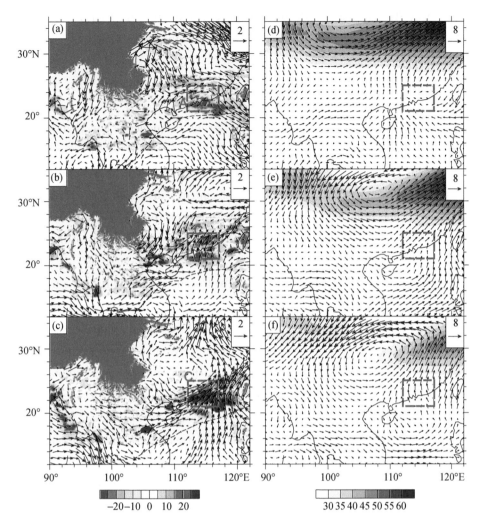

Fig. 12 As in Fig. 6 but in (a,d) P1 on 19 May, (b,e) P3 on 20 May, and (c,f) P5 on 21 May 2013.

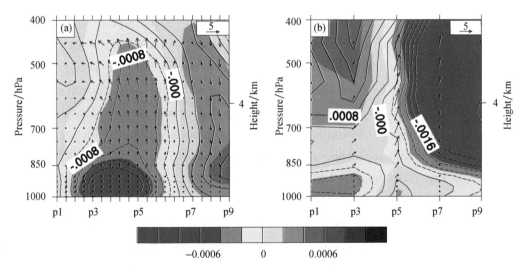

Fig. 13 As in Fig. 8 but with (a) 15-40-day on 11(P1), 15(P3), 21(P5), 27(P7) and 31(P9) May 2013 and (b) 3-8-day on 19(P1), 20(P3), 21(P5), 22(P7) and 23(P9) May 2014

and 10). The opposite is true both prior to P2 and after P6. However, large moisture and heat flux divergence (convergence) prevails above (below) 850 hPa prior to P3. Subsequently, both the moisture flux divergence and the heat flux divergence decrease in magnitude first and then switch sign prior to P5, which is consistent with the southerly warm and moist flows into the target region(cf. Figs. 13b and 12f).

5 Summary and concluding remarks

In this study, we investigate the roles of three different frequencies(i.e., synoptic, quasi-biweekly and quasi-monthly)of meteorological disturbances in influencing two pre-summer heavy rainfall episodes over South China. Results reveal the most dominant contributions are the synoptic frequency disturbances, and then the quasi-biweekly and quasi-monthly frequency disturbances, to heavy rainfall production during the pre-summer months of 2008-15. Two representative heavy rainfall episodes, i.e., that peaked on 8 May 2014 and 21 May 2013, are selected to study the contributions of the three different frequencies of disturbances.

The heavy rainfall episode that peaked on 8 May 2014 occurs as a result of the interaction of several perturbations in the lower and upper troposphere at both the quasi-biweekly and synoptic time scales, with 15.1% and 35.2% variance contributions to the total rainfall variance, respectively. The upper troposphere at both time scales is controlled by approaching cyclonic anomalies around 24°N: one with south-to-southwesterly anomalies and the other with south-to-southeasterly anomalies in the wettest phase across South China. In the lower troposphere, it takes place mostly in a southwesterly anomaly converging with a trailing anomalous easterly flow at the quasi-biweekly scale. They are associated more with an anticyclonic anomaly over the SCS and an eastward traveling anticyclonic anomaly from downstream of the Tibetan Plateau, respectively. The heavy rainfall episode also takes place in a southwesterly converging with northwest-to-northeasterly anomalies at the synoptic time scale, which are associated with a southeastward migration of an anticyclonic anomaly on the lee side of the Yungui Plateau and an eastward-propagating anticyclonic anomaly at higher latitudes, respectively. Positive warm and moist air supplied at both the synoptic and the quasi-biweekly scale, which is consistent with the lower-tropospheric southwesterly flow at both time scales, contributes to the heavy rainfall production during P4-P5. Thus, we may state that the lower-tropospheric southwesterly(warm and moist) monsoonal air at both time scales accounts for the generation of the heavy rainfall episode that peaked on 8 May 2014. This result appears to differ from that found by[14], in which the strong northeasterly component with a deep cold and dry layer at the synoptic scale helped generate the heavy rainfall over southern YHRB, although this process is also dominated by the quasi-biweekly and synoptic scale disturbances.

The anomalous flow fields in the heavy rainfall episode that peaked on 21 May 2013 appear to differ from those associated with the episode that peaked on 8 May 2014. The quasi-monthly and synoptic disturbances contribute to the total rainfall with 23.2% and 49.3% variances, respectively. That is, the upper troposphere experiences the passage of an anticyclonic anomaly around 25°N at the quasi-monthly scale, but a cyclonic anomaly at the synoptic scale centered around 30°N. Furthermore, the heavy rainfall zone at the quasi-monthly scale coincides with the confluence of anomalous west-to-northwesterly flows associated with an eastward-moving cyclonic anomaly and anomalous west-to-southwesterly flows associated with a quasi-stationary anticyclonic anomaly over the SCS. The presence of the anomalous west-to-northwesterly flows indicates the possible roles of higher-latitude cold and dry air intrusion in triggering deep convection and organizing the development of the associated MCSs. By comparison, the synoptic-scale heavy rainfall region is collocated with the confluence of south-to-southwesterly anomalies associated with a cyclonic anomaly that can be traced back to a cyclonic perturbation over North China in the very early phase. This perturbation grows in both amplitude and coverage from the driest to the wettest phase as a consequence of continued latent heat release. The south-to-southwesterly flow confluence and the presence of higher precipitable water than that associated with the 8 May 2014 rainfall episode account for the synoptic-scale heavy rainfall production over the target region.

Based on the above results, we may conclude that although the lower-tropospheric south-to-southwesterly mean flows provide a convectively favorable environment, it is the passage or interaction of the above-mentioned synoptic, quasi-biweekly and quasi-monthly disturbances that determines the general period and distribution of regional heavy rainfall production. Results show the important roles of these disturbances, which can be traced back to their higher latitude origins (i. e., on the downstream side of the Tibetan/Yungui Plateau or in North China) or the SCS and Indian Ocean, in enhancing the south-to-southwesterly supply of warm and moist air or in providing persistent lifting of warm and moist air along a cold front. This is more or less consistent with the findings of [15], who also emphasized the deterministic role of intraseasonal and synoptic components rather than the low-frequency background state in regulating the overall changes in moisture convergence and deep convection. Our study could not reveal the intraseasonal contributions because of the limited number of the pre-summer rainfall days used for the power spectral analysis. Certainly, our results are very preliminary. What frequencies of disturbances dominate pre-summer heavy rainfall production over South China, and where these disturbances originate, require more case studies and statistical analyses in order to help improve our understanding of the relationship between various low-frequency disturbances and the pre-summer heavy rainfall production over South China. In addition, hourly rainfall observations could be used to analyze the high-frequency characteristics of the heavy rainfall events in order to explain the residual variances shown herein.

Acknowledgments

We are grateful to the two anonymous reviewers for their valuable comments that have helped improve the quality of this work. This work was jointly supported by the National Department of Public Benefit Research Foundation (Grant No. GYHY201406003), the National Key R&D Program of China (Grant No. 2018YFC1507403), the National Natural Science Foundation of China (Grant No. 41475043), and the National Basic Research (973) Program of China (Grant Nos. 2014CB441402 and 2015CB954102).

References

[1] HUANG S S. The heavy rain during the pre-summer period over Southern China[M]. Guangzhou: Guangdong Technology Press, 1986. (in Chinese)

[2] CHEN G H, SUI C H. Characteristics and origin of quasi-biweekly oscillation over the western North Pacific during boreal summer[J]. J Geophys Res, 2010, 115: D14113.

[3] GU D J, JI Z P, GAO X R, et al. The relationship between the rainfall during the annually first rainy season in Guangdong and the quasi-biweekly oscillation of wind field in the north of South China Sea[J]. Journal of Tropical Meteorology, 2013, 29: 189-197. (in Chinese)

[4] HONG W,REN X J. Persistent heavy rainfall over South China during May-August:subseasonal anomalies of circulation and sea surface temperature[J]. Acta Meteorologica Sinica,2013,27:769-787.

[5] YANG J,BAO Q,WANG B,et al. Distinct quasi-biweekly features of the subtropical East Asian monsoon during early and late summers[J]. Climate Dyn,2014,42:1469-1486.

[6] LIN L H,HUANG X L,LAU N C. Winter-to-spring transition in East Asia:a planetary-scale perspective of the south China spring rain onset[J]. J Climate,2008,21:3081-3096.

[7] ZHANG L N,WANG B Z,ZENG Q C. Impact of the Madden-Julian Oscillation on summer rainfall in southeast China [J]. J Climate,2009,22:201-216.

[8] ZHANG T,WEI F Y,HAN X. Low frequency oscillations of southern hemispheric critical systems and precipitation during flood season in south China[J]. Journal of Applied Meteorological Science,2011,22:265-274. (in Chinese)

[9] CAO X,REN X J,SUN X G. Low-frequency oscillations of persistent heavy rainfall over Yangtze-Huaihe River Basin[J]. Journal of the Meteorological Sciences,2013,33:362-370. (in Chinese)

[10] CHEN S,JIAN M Q. Propagation of the quasi-biweekly oscillations related to rainfall abnormity in the first rainy season over Southern China[J]. Acta Scientiarum Naturalium Universitatis Sunyatseni,2015,54:130-137. (in Chinese)

[11] PAN W J,MAO J Y,WU G X. Characteristics and mechanism of the 10-20-day oscillation of spring rainfall over Southern China[J]. J Climate,2013,26:5072-5087.

[12] TONG T N,WU C S,WANG A Y,et al. An observational study of intraseasonal variations over Guangdong Province China during the rainy season of 1999[J]. Journal of Tropical Meteorology,2007,23:683-689. (in Chinese)

[13] CAO X,REN X J,YANG X Q,et al. The quasi-biweekly oscillation characteristics of persistent severe rain and its general circulation anomaly over southeast China from May to August[J]. Acta Meteorologica Sinica,2012,70:766-778. (in Chinese)

[14] LIU H B,YANG J,ZHANG D L,et al. Roles of synoptic to quasi-biweekly disturbances in generating the summer 2003 heavy rainfall in east China[J]. Mon Wea Rev,2014,142:886-904.

[15] LI R C Y,ZHOU W. Multiscale control of summertime persistent heavy precipitation events over South China in association with synoptic,intraseasonal,and low-frequency background[J]. Climate Dyn,2015,45:1043-1057.

[16] HUANG L,LUO Y L,ZHANG D L. The relationship between anomalous presummer extreme rainfall over South China and synoptic disturbances[J]. J Geophys Res,2018,123:3395-3413.

[17] JIANG Z N,ZHANG D L,XIA R D,et al. Diurnal variations of presummer rainfall over southern China[J]. J Climate,2017,30:755-773.

[18] PAN Y,SHEN Y,YU J J,et al. Analysis of the combined gauge-satellite hourly precipitation over China based on the OI technique[J]. Acta Meteorologica Sinica,2012,70:1381-1389. (in Chinese)

[19] JOYCE R J,JANOWIAK J E,ARKIN P A,et al. CMORPH:a method that produces global precipitation estimates from passive microwave and infrared data at high spatial and temporal resolution[J]. Journal of Hydrometeorology, 2004,5:487-503.

[20] GILMAN D L,FUGLISTER F J,MITCHELL J R J M. On the power spectrum of "red noise"[J]. J Atmos Sci,1963, 20:182-184.

[21] DUCHON C E. Lanczos filtering in one and two dimensions[J]. J Appl Meteorol,1979,18:1016-1022.

[22] WANG L J,PANG Y,YU B,et al. The characteristics of persistent heavy rain events and 10-30 day low-frequency circulation in Yangtze-Huaihe River Basin during Meiyu period[J]. Journal of Tropical Meteorology,2014,30:851-860. (in Chinese)

[23] WEI L,FANG J B,YANG X Q. Low frequency oscillation characteristics of 12-30 d persistent heavy rainfall over South China[J]. Acta Meteorologica Sinica,2017,75:80-97. (in Chinese)

[24] MAO J Y,WU G X. Intraseasonal variations of the Yangtze rainfall and its related atmospheric circulation features during the 1991 summer[J]. Climate Dyn,2006,27:815-830.

[25] MAO J Y,SUN Z,WU G X. 20-50-day oscillation of summer Yangtze rainfall in response to intraseasonal variations in

the subtropical high over the western North Pacific and South China Sea[J]. Climate Dyn,2010,34:747-761.

[26] MIAO C S,YANG Y Y,WANG J H,et al. A comparative study on characteristics and thermo-dynamic development mechanisms of two types of warm-sector heavy rainfall along the South China coast[J]. Journal of Tropical Meteorology,2018,24(4):494-507.

[27] CHEN X C,ZHANG F Q,ZHAO K. Diurnal variations of the land-sea breeze and its related precipitation over South China[J]. J Atmos Sci,2016,73:4793-4815.

[28] CHEN X C,ZHANG F Q,ZHAO K. Influence of Monsoonal wind speed and moisture content on intensity and diurnal variations of the Mei-Yu season coastal rainfall over South China[J]. J Atmos Sci,2017,74:2835-2856.

[29] DU Y,ROTUNNO R. Diurnal cycle of rainfall and winds near the south coast of China[J]. J Atmos Sci,2018,75:2065-2082.

[30] DU Y,CHEN G X. Heavy rainfall associated with double low-level jets over Southern China. Part II:Convection initiation[J]. Mon Wea Rev,2019,147:543-565.

[31] UCCELLINI L W,JOHNSON D R. The coupling of upper and lower tropospheric jet streaks and implications for the development of severe convective storms[J]. Mon Wea Rev,1979,107:682-703.

[32] RAYMOND D J,JIANG H. A theory for long-lived mesoscale convective systems[J]. J Atmos Sci,1990,47:3067-3077.

[33] ZHANG M,ZHANG D L. Subkilometer simulation of a torrential-rain-producing mesoscale convective system in East China. Part I:Model verification and convective organization[J]. Mon Wea Rev,2012,140:184-201.

最优集合预报订正方法在客观温度预报中的应用

郝 翠[1,2]　张迎新[2,*]　王在文[1]　付宗钰[2]　DELLE MONACHE Luca[3]

(1. 北京城市气象研究所,北京,中国,100089;2. 北京市气象台,
北京,中国,100089;3. 国家大气研究中心,博尔德,美国)

摘　要:数值模式的直接输出预报在实际应用时常与实况产生一定的偏差,对模式预报进行有效的本地化订正是提高预报准确率的重要手段。本文采用欧洲中期天气预报中心(ECMWF)模式细网格资料,应用最优集合(Anolog Ensemble,简称AnEn)预报订正方法对北京市各站1～7 d的日最高气温和日最低气温进行订正,并对相关参数进行本地化。采用滑动训练期、优化变量权重两种方案进行训练。检验评估结果表明:①滑动训练期采用60 d时能同时保证计算效率和预报准确率;采用最优变量权重方案后,与预报员主观预报准确率对比,AnEn的最低气温优于预报员主观预报,最高气温基本相当;增加训练期的长度(引入多年的历史资料)相比优化变量权重方案能更有效地提高预报准确率。②AnEn预报订正方法改善数值模式预报的系统性偏差(如由数值模式对局地地形、边界层日变化等形成的误差)效果显著,有较好的应用价值;在局地天气(如霾、降水、大风等)影响的情况下,AnEn的温度预报准确率虽优于ECMWF,但不如主观预报,未来还有改进空间。本文还对检验结果进行时间和空间验证,确保在以后的业务,尤其是智能网格预报业务中的运行效果。

关键词:最优集合;客观温度预报;滑动训练期;优化变量权重

Application of Analog Ensemble Rectifying Method in Objective Temperature Prediction

HAO Cui[1,2]　ZHANG Yingxin[2,*]　WANG Zaiwen[1]　FU Zongyu[2]　DELLE MONACHE Luca[3]

(1. Institute of Urban Meteorology,Beijing,China,100089;2. Beijing Meteorological Observatory,
Beijing,China,100089;3. National Center for Atmospheric Research,Boulder,USA)

Abstract:Model-based numerical prediction is often affected by bias when compared to local observations. In this study, the European Center for Medium-Range Weather Forecasting (ECMWF) data were used to generate the Analog Ensemble (AnEn) prediction over the 15 national weather stations and 274 automatic stations of Beijing, with a focus on correcting ECMWF prediction of the daily maximum and minimum temperature, 1-7 day ahead, twice a day. The analog of a forecast for a given location and time is defined as the observation that corresponds to a past prediction matching selected features of the

---

① 本文发表于《气象》2019 年第 8 期。
资助项目:国家重点研发计划(2018YFC1507305 和 2018YFF0300104)、中央级公益性科研院所基本科研业务费专项(IUMKY201719)、中国气象局发展预报业务检验评估关键技术项目[YBGJXM(2017)06-4]。
第一作者:郝翠,主要从事天气预报客观方法研究。E-mail:haocui2015@163.com。
通信作者:张迎新,主要从事天气预报技术方法研究。E-mail:zhangyx9@sina.com。

current forecast. The best analogs form AnEn, which produces accurate predictions and a reliable quantification of their uncertainty with similar or superior skill compared to traditional ensemble methods while requiring considerably less real-time computational resources. An analysis of the performance of ECMWF and AnEn in space and time was presented. The results demonstrate that a short training period of 60 days may be a good compromise for the computational efficiency and the quality of deterministic predictions. Extending the training periods would further increase the prediction quality than optimizing the environmental parameters, no matter 1-month, 3-month or 6-month optimizations. AnEn correction results are better than the predictions generated by the forecasters, particularly for daily minimum temperatures. AnEn effectively reduces the bias of ECMWF predictions, resulting in a skilled downscaled prediction at the observation location, consistently over time and space. However, AnEn is not very effective in improving predictions of haze, precipitation, and strong winds, which may require a much longer training data set. Furthermore, this study tests the results over time and space to make sure the method's reliability for the future smart grid forecast operation.

Key words: Analog Ensember(AnEn); objective temperature prediction; slide training period; optimizing the environmental parameters weight

引 言

数值预报模式的直接输出结果受很多因素影响进而产生不确定性。这些不确定性可能来源于初始条件(丑纪范,2007)、模型建立过程中对多物理过程描述或在多个模型集合时缺乏细节或未描述地形、相关地理信息等(康红文 等,2012),也存在随机参数化过程等引入的不确定性;侧边界条件和下边界条件设定时都会引入不确定性,比如下垫面条件,城市、非城市地区,以及海拔高度等在不同地区有不同问题(桑建国 等,1992)。模式的预报结果与实况观测结果相比存在很大误差,因此对模式预报进行有效订正、提高产品的业务应用性能是一项重要的研究课题。

目前,数值预报的订正方法有很多,主要有模式输出统计(MOS)、完全预报法(PP)、卡尔曼滤波(KLM)、人工神经网络(ANN)、相似预报法等(赵声蓉,2006;吴君 等,2007;张庆奎 等,2008;Delle Monache et al.,2011;吴启树 等,2016)。每种方法都有其优势的地方,而相似预报法从过去历史资料中寻找相似天气形势或相似个例,来对当前的预报进行订正,是气象学中的经典方法,是较接近预报业务思路的方法。Lorenz(1965,1969)认为,大气可预测性是由于自然发生的相似的天气形势决定的,认为在全球范围内选取完全一样的天气形势需要1030年,然而在大范围区域(例如整个北半球)找到观测误差范围内的2个相似流体(天气形势)是可以在相对短的时间内实现的。那么在10~100年的观测时段内,针对某一有限区域,假设2~3个自由度,寻找相似个例的概率会很大。基于这样的理论背景,研究者们对相似方法在气象和气候学上的应用也较为广泛。Bergen 等(1982)运用多个相似预报的加权平均;Shabbar 等(1986)使用n天大气状态综合选相似。上述工作均是使用相似来做长期预报,对于短期天气预报的应用也很多。Van den Dool(1987,1989,1991)系统地研究了相似和反相似的预报性能和使用前景,并指出可用区域取代全球找相似来克服资料少的困难。区域性相似的问题也被Robber 等(1998)详细讨论。鉴于单个取样相似的不确定性,可通过选取一组相似来实现集合概率预报(Kruizinga et al.,1983;Gruza et al.,1993)。基于统计学方法的相似预报在短期预报中的效果也较好。到目前为止,美国已实现了相似预报技术在业务上的应用,中国也有很多相关的研究(鲍名 等,2004;陈凯,2014;周海,2009;任宏利,2006)。鉴于对相似天气形势的判断有难度,Delle Monache 等(2011)利用寻找相似的气象要素,即寻找当前预报与过去预报相似的气象要素,再用这些相似的历史个例的观测值来校正当前预报,利用最优集合(Analog Ensemble,AnEn)预报订正方法对美国西部地区的风和温度预报进行订正并取得了较好的效果,模式预报的偏差得到有效订正。

目前,中国业务采用的温度预报订正方法主要基于MOS,包括一元和多元线性回归等,对温度的订正效果较好,为业务应用提供了较好的参考。这些方法是根据温度在时间上的连续性以及与其他气象要素的相关性对模式结果进行误差订正的。就应用效果而言,各方法均有其占优势的地方,很难找出一个完美的、在所有时段所有站点都应用效果很好的方法,因此,开发不同的温度预报订正方法给业务应用提供参考有一定的研究意义。针对AnEn方法在偏差订正方面的研究,本文拟采用该订正方法对北京地区基于ECMWF预报的日最高气温和日最低气温进行订正,测试训练日期长度、环境变量权重、相似天气个例数等变量,寻找优化的参数变量值,完成参数的本地化,并分析AnEn的实际应用效果,尤其是在山区的订正效果。

1 资 料

本文采用2012年1月1日至2017年8月31日ECMWF细网格逐3 h(或6 h)的资料(0~72 h时效内的时间分辨率为3 h,72~168 h时间分辨率为6 h),水平分辨率为0.125°(2015年1月15日之前为0.25°),每日08:00和20:00(北京时)两时次,预报时效为0~168 h。选取2 m气温、10 m风速、10 m风向、2 m露点温度等要素预报作为环境变量。

为确保实况资料的可靠及完整性,本文采用北京地区2012年1月1日至2017年9月6日国家基本气象站(15个站)及自动站(274个站)逐5 min的地面气温资料作为实况。

预报员的主观预报数据采用与ECMWF同时间段内的资料,每日两次,即08:00和20:00起报的0~168 h内的日最高气温、日最低气温预报。为了准确衡量AnEn在实际应用中的预报订正效果,本研究采用与业务应用时效一致的方案,即前一日20:00的数值模式预报资料对应当日08:00的主观预报,而当日08:00的数值模式预报对应当日20:00的主观预报,并将起报时刻开始至12 h内的温度预报剔除不做订正和检验(预报员无法参考该时段资料)。

2 方 法

2.1 最优集合预报订正技术

AnEn旨在改进由于数值模式产生的不确定性引起的误差,尤其是误差中的系统误差(偏差)部分。它采用寻找过去一段时间内的相似个例集合,并对这些相似个例进行加权平均来实现对误差的订正,达到减小模式预报误差的目的,获得一个更接近实况观测的结果。本研究采用下述方法进行最优的选择(Delle Monache et al.,2011):

$$d_t = |f_t - g_t| = \sum_{v=1}^{N_v} \frac{w_v}{\sigma_{fv}} \sqrt{\sum_{k=-t}^{t} (f_{t+k}^v - g_{t+k}^v)^2} \tag{1}$$

式中,d_t为当前预报与过去预报的接近程度(即相似程度);f_t为t时刻对应的预报值;g_t为t时刻对应过去的预报值;w_v为环境变量对预测要素的权重,可以通过历史资料进行优化;σ_{fv}在固定权重配置时设为1,为第v个环境变量的预报值f的方差;N_v为环境变量的总数;$t+k$时刻对应过去数据的预测值;k为预测窗的一半(用整数表示),可以自行设定,如$t=1$,则k为-1,0,1。

本文针对2 m气温的预报,先确定与温度相关的环境变量:过去的2 m气温、10 m风速、10 m风向、2 m露点温度等;对这些环境变量的过去的预报和当前的预报根据式(1)计算相似距离d_t,即对每个环境变量求当前预报和过去预报之间的差,再将不同变量按照权重求和,最后得出当前预报与过去预报的差别程度,即距离d_t,根据d_t排序,筛选出d_t最小的那些值,即为最优的相似个例集。

针对选取的最优预报个例集,进行集合计算(Delle Monache et al.,2011):

$$F'_t = \sum_{i=1}^{N} r_i O_{i,t} \tag{2}$$

$$r_i = \frac{d_{t_i}}{\sum_{i=1}^{N} d_{t_i}} \tag{3}$$

式中,F'_t为t时刻的预报订正值;N为选取的最优集合的数量;$O_{i,t}$为t时刻预报的最优个例的观测值;r_i为第i个观测的权重,距离t时刻越近,则权重越大;d_{t_i}为t时刻第i个距离值,由式(1)计算。

将日最高气温和日最低气温分别进行上述计算,寻找历史资料中ECMWF模式资料的相似个例(式1),并将这些最优个例的实况值加权平均[(式(2)和式(3)],作为对当前时刻的ECMWF预报的订正值,即AnEn的预报。从原理上看,AnEn利用历史资料的观测来校正当前数值预报的偏差。

2.2 数据处理方法

为提高计算效率,同时不降低预报质量,训练集数据采用滑动方式(吴启树 等,2016),即将预报日的前n天与前一年的后n天资料作为训练期,随预报日滑动取样。检验日即为预报日。

实况2 m气温采用5 min数据提取成日最高气温和最低气温(根据实际业务应用标准,日最高气温取02:00—02:00时的最高温度,日最低气温取20:00—20:00的最低温度),数据缺失量大于20%时将该站该日数据从训练集剔出,数据量缺失小于20%时采用内插方法补足数据。

综合考虑插值准确程度和计算效率,ECMWF模式的格点场资料插值到气象站时,本研究采用最近邻距离插值方法插值到气象站上,同时利用15 m的数字高程模型(DEM)资料对气象站邻近的四个ECMWF格点的高程进行对比,如果该站与周围格点高度相差100 m以上,则该格点不参与插值,以此确保插值到站点后的值接近气象站的实况值。

2.3 检验方法

为了评估AnEn的预报效果,本研究对2016年9月1日至2017年8月31日的效果用均方根误差(RMSE)、偏差(BIAS)、温度预报准确率(F_a)和温度预报技巧评分(TSS)来检验效果;同时对2016年(训练集为2014—2015年)和2015年(训练集为2013—2014年)的主客观预报效果进行时间上的检验分析;对2016年9月1日至2017年8月31日的276个自动站进行空间效果的检验。其中检验计算公式如下:

$$\text{RMSE} = \frac{1}{n} \sum_{i=1}^{n} (O_i - A_i)^2 \tag{4}$$

$$\text{BIAS} = \bar{A} - \bar{O} \tag{5}$$

其中,

$$\bar{A} = \frac{1}{n} \sum_{i=1}^{n} A_i, \bar{O} = \frac{1}{n} \sum_{i=1}^{n} O_i \tag{6}$$

$$F_2 = \frac{n_2}{n} \times 100\% \tag{7}$$

$$\text{TSS} = \frac{T_{\text{mean}} - T_{\text{meat}}}{T_{\text{mean}}} \times 100\% \tag{8}$$

式中，O_i 为 i 时刻的实况值；n 为样本量；A_i 为 i 时刻的 AnEn 估计值；$\bar{A}$ 和 $\bar{O}$ 分别为 AnEn 的均值和实况的均值；F_2 为温度预报与实况值的误差不超过 2 ℃ 的百分率；n_2 为不超过 2 ℃ 的样本量；TSS 为预报技巧评分；T_{mean} 和 T_{meaf} 分别为 ECMWF 和对比预报的平均误差。

3 方案对比与改进

为了提高算法的计算效率，首先对训练日期长度进行试验，结果表明，60 d 以后结果波动变小（图 1），为了在保证预报准确率的基础上提高效率，本文以 60 d 为训练期，即采用预报日前 60 d，预报日去年的后 60 d（以此类推）的资料作为训练样本，对环境变量和权重参数进行优化：

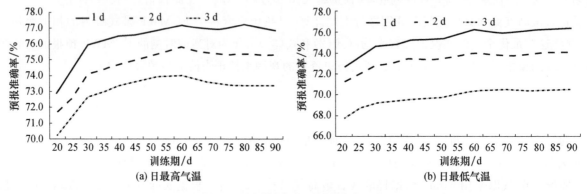

图 1　不同训练期的预报准确率变化

方案 1：统一选用 4 个环境变量（Delle Monache et al.，2011），选取 20 个相似值，固定权重，即设置 2 m 气温、10 m 风速、10 m 风向和 2 m 露点温度的权重分别为 0.7、0.1、0.1、0.1。

方案 2：由于 6—11 月霾和降水天气相对较多，针对这些天气的相似案例较少，因此，对 12 月—次年 5 月选取 20 个相似值，6—11 月选取 15 个相似值，即选用公式（3）中对 d_i 排序后的前 20 或 15 个历史个例，固定权重配置（如方案 1）。

方案 3：优化权重选取上一年同期的 1、3、6 个月，结果表明 1 个月的优化权重不稳定，综合效果较差，而 3 和 6 个月的优化差距不大，但 6 个月的优化更稳定。本文仅列出 6 个月的优化效果。

方案 4：在方案 2 的基础上延长训练期时间，将 2012 年 1 月 1 日至 2016 年 8 月 31 日的资料作为训练集。

4 检验效果

4.1 几种方案的结果对比

对 2016 年 9 月 1 日至 2017 年 8 月 31 日的逐日结果进行检验，从固定权重下两种方案的对比（表 1）可以看出，不论日最高气温还是日最低气温，AnEn 的效果比 ECMWF 原始预报的均方根误差和预报偏差都有不同程度的改进，其中方案 2 的改进效果更好一些，相对于 ECMWF 的预报技巧提高了 10% 以上。而在优化环境变量的权重之后，三种参数方案有所差距，以月为周期的权重配置波动性较大，稳定性最差，而以 3 和 6 个月为周期的权重配置较为稳定，结果相差不大。表 2 仅列出 6 个月下的优化效果。与优化环境变量权重相比,延长训练期的时间对预报质量的提高更为显著(表2),日最高气温的订正效果好于日最低气温,预报技巧总体分别提高了15.4%和12.4%,且预报时效越短订正效果越好。方案4对日最低气温的RMSE改进效果不如方案2的效果好,但BIAS比方案2普遍低,可能是2015年1月15日之后ECMWF模式有所调整,分辨率不同造成的,因此虽然历史资料的延长可以减小预报偏差,但由于资料前后不一致导致的随机误差也较大,日最高气温的RMSE和BIAS在方案4中是最小的。相比预报员的主观预报,不论是日最高气温还是日最低气温,在4~7 d的中期预报中AnEn的预报技巧比预报员分别提高了3.5%和6.3%;1~3 d的日最高气温的预报AnEn比ECMWF提高18.4%,相比预报员有2.0%的差距;日最低气温在1 d的AnEn预报比ECMWF提高12.5%,与预报员相比有一定的差距。

表1 不同方案之间的日最高气温与日最低气温预报检验

要素	方案	检验	1 d	2 d	3 d	4 d	5 d	6 d	7 d
日最高气温	ECMWF	RMSE	2.087	2.178	2.369	2.688	2.941	3.236	3.549
		BIAS	−1.148	−1.183	−1.179	−1.26	−1.251	−1.367	−1.372
	方案1	RMSE	1.836	1.899	2.093	2.339	2.622	2.846	3.198
		BIAS	−0.305	−0.222	−0.176	−0.231	−0.281	−0.35	−0.383
		预报技巧	0.16	0.17	0.15	0.15	0.13	0.10	0.10
	方案2	RMSE	1.775	1.882	2.079	2.158	2.493	2.776	2.909
		BIAS	−0.221	−0.174	−0.217	−0.188	−0.281	−0.176	−0.144
		预报技巧	0.18	0.18	0.16	0.15	0.13	0.12	0.12
日最低气温	ECMWF	RMSE	2.05	1.959	2.04	2.38	2.500	2.587	2.802
		BIAS	−0.095	−0.108	−0.058	0.828	0.836	0.808	0.835
	方案1	RMSE	1.882	1.813	1.912	2.022	2.189	2.288	2.14
		BIAS	−0.034	−0.043	0.095	0.049	0.089	−0.005	0.028
		预报技巧	0.09	0.10	0.07	0.15	0.12	0.11	0.11
	方案2	RMSE	1.853	1.754	1.832	1.969	2.109	2.213	2.430
		BIAS	−0.017	−0.027	0.125	0.103	0.122	0.007	0.006
		预报技巧	0.11	0.11	0.10	0.16	0.14	0.13	0.13

表2 不同优化权重配置下的日最高气温与日最低气温预报检验及与主观预报的对比

要素	方案	检验	1 d	2 d	3 d	4 d	5 d	6 d	7 d
日最高气温	主观预报	RMSE	1.726	1.77	1.996	2.502	2.64	2.833	3.177
		BIAS	−0.172	−0.13	−0.141	−0.317	−0.302	−0.288	−0.321
		预报技巧	0.25	0.22	0.18	0.07	0.08	0.12	0.10
	方案3	RMSE	1.749	1.850	2.066	2.320	2.626	2.810	3.286
		BIAS	−0.025	−0.158	−0.165	−0.168	−0.236	−0.249	−0.420
		预报技巧	0.20	0.19	0.16	0.17	0.14	0.13	0.10
	方案4	RMSE	1.699	1.829	2.044	2.119	2.504	2.726	2.857
		BIAS	−0.059	−0.069	−0.09	−0.13	−0.275	−0.158	−0.082
		预报技巧	0.22	0.20	0.17	0.15	0.13	0.12	0.14

续表

要素	方案	检验	1 d	2 d	3 d	4 d	5 d	6 d	7 d
日最低气温	主观预报	RMSE	1.691	1.846	1.946	2.079	2.289	2.474	2.597
		BIAS	−0.182	−0.113	−0.06	0.08	0.113	0.148	0.138
		预报技巧	0.13	0.06	0.05	0.12	0.08	0.04	0.07
	方案 3	RMSE	1.930	1.785	1.880	2.001	2.154	2.232	2.484
		BIAS	−0.072	−0.025	0.113	0.073	0.106	0.074	0.039
		预报技巧	0.08	0.10	0.08	0.15	0.12	0.13	0.11
	方案 4	RMSE	1.862	1.787	1.827	1.988	2.118	2.227	2.431
		BIAS	0.014	0.102	0.112	0.088	0.083	−0.036	−0.006
		预报技巧	0.11	0.10	0.10	0.16	0.14	0.13	0.13

4.2 时间和空间检验评估

为验证 AnEn 方法对客观温度预报的稳定性,本文从时间和空间上对该方法进行了进一步检验(图2和图3)。图2为2016年的预报准确率检验,AnEn 的效果都比预报员高出0.1%~4.5%,在时间上表现出良好的稳定性。

从空间上看,北京地区274个自动站的预报准确率 AnEn 都比 ECMWF 有了较大提高,1~3 d 的日最高气温预报准确率分别提高了19.0%、19.8%和16.5%,对应的日最低气温预报准确率分别提高了14.0%、13.3%和10.2%。日最高气温的 RMSE 和 BIAS 分别由2.526和−1.616减至2.016和−0.212,日最低气温的 RMSE 和 BIAS 分别由2.474和0.266减至1.931和0.009,预报偏差几乎减至无偏,表明 AnEn 对预报偏差的订正效果较好。图3表明,在 ECMWF 的预报质量较差(预报准确率通常低于50.0%)的地区,大部分都是地形较为复杂和受边界层影响较大的地区,AnEn 体现了较好的订正能力。未来采用合适的插值方案,AnEn 可以从基于站点的预报向智能网格预报业务应用,提高智能网格客观预报质量。

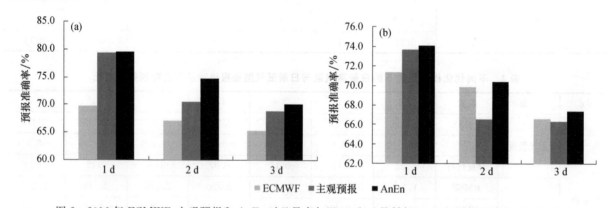

图 2　2016年 ECMWF、主观预报和 AnEn 对日最高气温(a)和日最低气温(b)的预报准确率对比分析

4.3 最优方案的效果及与主观预报的对比

根据方案 4 对 1~3 d 的气温预报进行分月详细分析,虽然预报技巧评分比 ECMWF 直接输出高,但是总体看无论最高还是最低气温在5、6和10月的预报技巧较差,这可能跟大气环流的季节性调整有关。但逐日分析发现,在雾霾、阴雨等天气发生的时间段内,AnEn 的订正效果虽比 ECMWF 直接输出提高不少,但预报技巧还不尽人意。这也是下一步要解决的问题。

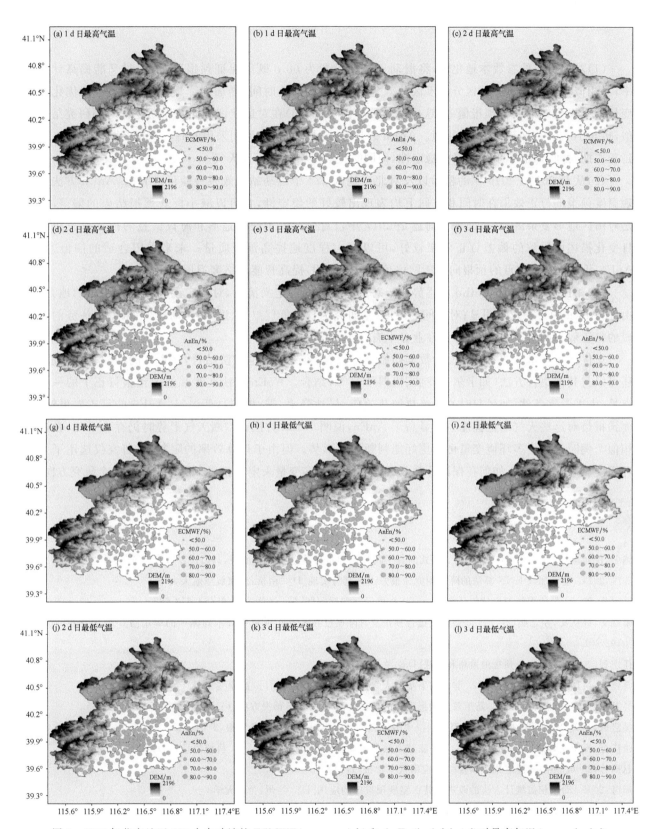

图 3 2017 年北京地区 274 个自动站的 ECMWF(a、c、e;g、i、k)和 AnEn(b、d、f;h、j、l)对最高气温(a、c、e;b、d、f)和最低气温(g、i、k;h、j、l)的预报准确率对比(DEM 为高程信息)

5 结论和讨论

(1)对 AnEn 的参数本地化中,将滑动训练期设置为 60 d 既能保证预报质量,同时又能提高计算效率;对相似个数的选取进行区分比不区分效果好,而延长训练时间(增加训练集的历史资料)比优化环境变量的效果更好地减少预报偏差,这一点与吴启树等(2016)在对最佳训练期的周期优化时的研究结论相似,也与 Junk 等(2015)的结论相似。AnEn 的环境变量权重优化周期采用的时间越短,结果越不稳定;而延长周期时间,保证了预报质量的稳定性,但同时也会平滑掉转折性天气。

(2)从对 2015—2016 年的检验以及对北京地区 274 个自动站的效果检验可以看出,AnEn 方法对客观温度预报的订正效果在时间和空间上均表现出较好的稳定性,它可以减小模式预报结果的偏差,尤其是对山区地形复杂区域(空间评分前述是 2017 年),对模式由于缺乏地形和海拔信息的描述以及边界层日变化描述而导致的偏差订正效果较好,可以较大程度地提高预报质量。未来采用合适的插值方案,AnEn 可以从基于站点的预报向智能网格预报业务应用,提高智能网格客观预报质量。

(3)相比 MOS 方法,AnEn 的优势在于对误差中的偏差定义清晰,对固定误差即偏差明显的地方,如山区,AnEn 的订正效果明显;对夜间的温度(本研究为日最低气温)来说,受扰动较少,偏差占优势的时间段的订正效果明显。因此,该方法在业务上有一定的应用前景。

最优预报技术方法不考虑原始数据初始分布状态,容易与其他方法结合,除本文的加权平均方法外,还可以与其他集合方法(如卡尔曼滤波、机器学习等)结合。AnEn 算法在客观温度预报订正上的一个缺点是,由于天气条件,如风起时的非绝热加热效应、局地降水、霾、极端高低温等,AnEn 的订正效果较差,而预报员对这些天气的把握能力明显高于 AnEn,说明 AnEn 方法在判断天气形势时仍存在不足,在选取相似个例时采用更多环境变量可以更好地判断天气形势。但由于计算效率的限制,本研究仅选用了 4 个环境变量来判断相似,如何在保证计算效率的同时采用多变量来定义相似个例是未来的一个研究方向。

参考文献

鲍名,倪允琪,丑纪范,2004. 相似-动力模式的月平均环流预报试验[J]. 科学通报,49(11):1112-1115.

陈凯,2014. 基于加权 KNN 算法的降水相似预报方法研究与实现[D]. 南京:南京航空航天大学.

丑纪范,2007. 数值天气预报的创新之路——从初值问题到反问题[J]. 气象学报,65(5):673-682.

康红文,祝从文,左志燕,等,2012. 多模式集合预报及其降尺度技术在东亚夏季降水预测中的应用[J]. 气象学报,70(2):192-201.

任宏利,2006. 动力相似预报的策略和方法[D]. 兰州:兰州大学.

桑建国,吴熠丹,刘辉志,等,1992. 非均匀下垫面大气边界层的数值模拟[J]. 高原气象,11(4):400-410.

吴君,裴洪芹,石莹,等,2007. 基于数值预报产品的地面气温 BP-MOS 预报方法[J]. 气象科学,27(4):430-435.

吴启树,韩美,郭宏,等,2016. MOS 温度预报中最优训练期方案[J]. 应用气象学报,27(4):426-434.

张庆奎,寿绍文,陆汉城,等,2008. 卡尔曼滤波方法在极端温度预报中的应用[J]. 科技信息(35):811-812.

赵声蓉,2006. 多模式温度集成预报[J]. 应用气象学报,17(1):52-58.

周海,2009. 动态相似统计方法的改进及其在温度预报中的应用[D]. 兰州:兰州大学.

BERGEN ROBERT E,ROBERT P H,1982. Long-range temperature prediction using a simple analog approach[J]. Mon Wea Rev,110:1083-1099.

DELLE MONACHE L,NIPEN T,LIU Y B,et al,2011. Kalman filter and analog schemes to postprocess numerical weather predictions[J]. Mon Wea Rev,139(11):3554-3570.

GRUZA G V,ESTHER Y R,1993. Potential predictability assessment for analog long-term forecasting[R]. Russian Meteor-

ology and Hydorlogy,9:1-7.
JUNK C,DELLE MONACHE L,ALESSANDRINI S,et al,2015. Predictor-weighting strategies for probabilistic wind power forecasting with an analog ensemble[J]. Meteor Z,24(4):361-379.
KRUIZINGA S,ALLAN H. M,1983:Use of an analogue procedure to formulate objective probabilistic temperature forecasts in the Netherlands[J]. Mon Wea Rev,111:2244-2254.
LORENZ E N,1965. A study of the predictability of a 28-variable atmospheric model[J]. Tellus,17:321-333.
LORENZ E N,1969. Atmospheric predictability as revealed by naturally occurring analogues[J]. J Atmos Sci,26:636-646.
ROBBER P J,LANCE F B,1998:The sensitivity of precipitation to circulation details. Part Ⅰ:an analysis of regional analogs[J]. Mon Wea Rev,126:437-455.
SHABBAR A,KNOX J L,1986. Monthly prediction by the analogue method[J]. Long-Range Forecasting Research Report Series,2(6):672-681.
VAN DEN DOOL H M,1987. A bias in skill in forecasts based on analogues and antilogues[J]. J Clim Appl Meteor,26:1278-1281.
VAN DEN DOOL H M,1989. A new look at weather forecasting through analogues[J]. Mon Wea Rev,117:2230-2247.
VAN DEN DOOL H M,1991. Mirror images of atmosphere flow[J]. Mon Wea Rev,119:2095-2106.

上海市无缝隙天气预报技术

储海 陈雷 戴建华 王海宾 李佰平 张欣 孙敏 刘梦娟

(上海中心气象台,上海,200030)

摘要:介绍了支持上海市气象局无缝隙预报业务的几类客观预报方法和业务系统。上海市气象局无缝隙格点预报产品涵盖0～45 d的五类不同分辨率、不同要素产品。从0～6 h与实况相衔接的逐10 min定量降水预报,24～240 h内的要素最优集成订正预报,到延伸期45 d的逐日趋势预报。依赖各类客观预报方法生成的格点预报背景产品,结合格点预报制作和预报检验系统,使得预报员在此基础上有效地制作发布格点预报,在满足现代化格点预报要求精确度的同时,最大限度地帮助预报员减轻了人工操作的负担。

关键词:无缝隙预报;客观订正;集成预报;非常规检验

Aspects on Seamless Weather Forecast Technologies of Shanghai City

CHU Hai CHEN Lei DAI Jianhua WANG Haibing LI Baiping
ZHANG Xin SUN Min LIU Mengjuan

(Shanghai Central Meteorological Observatory, Shanghai, 200030)

Abstract: Shanghai meteorology bureau has built its seamless grid forecast products with different forecast elements and resolution for term range from 0 to 45 days. In this article, the objective forecast technologies and operation systems used to form the grid forecast were briefly introduced. A background grid-forecast was made without manual intervention for the requirements on (1) 0-6 h blended quantitative precipitation forecast consisted with current observation; (2) 24-240 h consensus forecast of calibrated operation NWPs; (3) 11-45 d extended-range forecast. Finally, forecasters would give edit and publication on to the grid forecast with the help of our integrated forecast system and forecast evaluation system. By means of these objective tools, the grid forecast was able to get demanded precision without adding more burdens to forecasters.

Key words: seamless forecast; objective correction; consensus forecast; unconventional evaluation

0 引言

当前世界气象以精细化和自动化为发展趋势,美国、英国、澳大利亚等均已建立了格点精细化的自动

---

① 本文发表于《气象科技进展》2017年第6期。
第一作者:储海(1985—)。E-mail:chhaui@163.com。

预报系统,其产品相对于统预报准确率有较大提高,已成为预报员业务工作中的首选指导产品,同时产品也能够直接服务于用户。从实际使用效果看,它在提供更多预报产品的同时,并未让预报员投入更多精力。上海及华东区域是中国经济发展中心,实行精细化预报是未来上海气象局的工作重点,是中国气象局深化气象预报预测业务改革要求,也是社会经济发展和人民生活水平提高的保障。无缝隙预报最先由世界气候研究计划(World Climate Research Programme,WCRP)提出。由于天气到气候的变化本身是一个连续的过程,而实际预报本身是一个从天气到气候的无缝隙的服务,因而具有不同时间分辨率的各类天气气候模式应在具有普遍一致性的前提下,描述同一大气系统的不同方面特征[1]。尽管当前尚缺乏具备无缝隙预报能力的数值模式,但自 2012 年起,上海市气象局加快推进气象现代化的各项工作,不断提升气象预警预报及气候预测技术能力。2016 年,上海推出了覆盖华东地区 0~45 d、时空分辨率最高达到 10 min/3 km 的格点客观预报产品(表 1),并在此基础上推出"上下班时段天气预报""3 h 天气预报""未来 10 d 逐日天气预报"等一系列精细化预报产品,建立从临近预报到年度预测的无缝隙预报服务体系,预报质量稳步提升,晴雨、气温等天气预报准确率在全国处于领先水平。2015—2016 年 24 h 晴雨预报准确率达 89.75%,最高气温预报准确率达到 91.4%,强对流天气预警能力不断提高,预警时效延长到 144.3 min。

本文主要介绍上海 10 d 以内无缝隙天气预报格点指导产品的相关制作技术及特点。其中,0~6 h 逐 10 min 降水采用雷达实况外推与数值预报融合的 Blending 预报方法,分辨率为 3 km,0~24 h 逐 1 h、0~72 h 逐 3 h、1~10 d 逐 12 h 要素产品采用多模式综合订正集成的预报方法生成。

表 1 上海市气象局格点预报产品业务运行情况

预报时效	时间分辨率	空间分辨率	预报范围	更新频次	预报要素
0~6 d	10 min	3 km×3 km	长三角地区	每 30 min	降水量
0~24 d	1 h	3 km×3 km	华东区域	每天 4 次	天气现象、温度、降水量、云量、风向风力、相对湿度
0~72 h	3 h	5 km×5 km	上海	每天 2 次	天气现象、温度、降水量、云量、风向风力、相对湿度
1~10 d	12 h	5 km×5 km	华东区域	每天 2 次	天气现象、温度、降水量、云量、风向风力、相对湿度
11~45 d	24 h	9 km×9 km	华东区域	每天 1 次	天气现象、温度、降水量、云量、风向风力、相对湿度

1 预报方法介绍

1.1 0~6 h 降水融合 Blending 方法

对于短时定量降水预报,主流观点认为,将常规短时临近预报技术(主要是雷达外推技术)与数值预报融合是将预报时效提高至 2 h 以上的根本途径[2]。如英国的 NIMROD 系统通过给予外推与 NWP 预报不同时效的不同权重来进行短临强降水预报[3];美国国家大气研究中心(NCAR)的 NIWOT 系统则通过 NWP 预报调整外推回波范围[4];香港天文台的 SWIRLS 系统[5],通过位相修正和强度修正技术订正 NWP 预报,并由一个双曲函数确定外推与 NWP 的融合权重因子,有效提高了 0~6 h 的降水预报效果。由于 SWIRLS 系统相对较为成熟,并有较好的开放性,国内如北京市气象局、浙江省气象局等也都与其合作开展了相关工作。上海中心气象台在前期丰富的雷达应用经验基础上,利用长三角地区 10 部天气雷达的反射率因子观测实时外推和区域高分辨率数值模式模拟的雷达反射率因子预报,经过尺度分解、实时目标订正及权重融合、分类 Z-R 关系转换等技术,建立了 0~6 h 短时临近精细化客观预报定量降水融合产品。

1.1.1 雷达实况外推

采用陈雷等[6]改进的COTREC外推预报方法,首先对选取的长三角地区雷达资料进行实时质量控制,去除超折射和地物回波等非气象回波。非气象回波在低仰角资料中比较集中,通过高低仰角资料对比分析去除只存在于低仰角资料中的杂波。然后采用交叉相关法(Tracking Radar Echoes by Correlation,TREC)求出TREC风场,对求出的风场先进行平滑处理,去掉明显失真的风(风速极大或为0)。本文采用的是九点平滑方法。TREC法反演的风场只在有回波的区域有值,而没有回波的区域缺值(为0),这样经过水平无辐散处理后得到的COTREC风场会受到一定程度的削弱,特别是孤立的块状回波或线状回波受到的削弱更加明显。采用引入数值预报平均风场作为TREC风场引导流场的方法,较好地解决了TREC风场经过水平无辐散处理所受到的削弱问题。应用GFS模式3 h间隔预报风场,将数值预报各层次风场取矢量平均,并插值到TREC风场格点上,代替TREC风场中的缺值点。经过这一步骤后得到新TREC风场$(nu^0(i,j), nv^0(i,j))$,对新的风场进行水平无辐散限制求出COTREC风场$(u(i,j), v(i,j))$。最后,用COTREC风场将t_2时刻的回波外推到$t_2+\Delta t$时刻,完成回波的外推。如图1所示。

图1 COTREC法外推预报具体流程

1.1.2 区域模式预报反射率订正

为与雷达外推产品进行有效融合,以提高对短时降水、对流回波的预报效果,必然需要选用包含雷达反射率等资料同化的区域高分辨率数值模式结果。然而,由于模式预报的时效性相比外推预报仍有1~2 h左右的延迟,并且与实况观测相比,模式降水系统仍不可避免地存在强度及位置上的误差,因此在与外推结果进行融合之前必须先对模式的反射率预报进行订正。

首先需要对模式预报反射率的总体强度进行误差修正,采用程从兰等[7]的方法,用韦布尔(Weibull)函数对当前实况雷达反射率因子和模式对当前时次的反射率因子预报分别进行拟合,通过实况与预报的反射率因子概率分布对比,计算表征模拟与实况的Weibull函数分布态特征量,通过调整预报数据的量级、周期等,使其拟合的特征值接近实况,达到强度修正的目的。

经过强度误差修正之后,预报反射率因子达到与实况相似的强度频谱分布特征,之后才可以进行模式反射率因子位相误差修正。实况雷达回波与高分辨率模式预报的原始图像包含的信息往往过于复杂,不利于此后的目标识别及匹配,因此对于降水系统的位相误差,本方法重点关注模式对中尺度对流系统、天气尺度雨带这类目标的预报效果。在进行位相修正之前,首先使用快速傅里叶变换对预报与实况回波进行滤波,去除小尺度杂波和局地分散性降水的影响。然后使用上海中心气象台研发的自动目标识别检验方法,对预报与当前实况回波的目标进行匹配,计算二者的位置偏差,将其用于未来模式预报目标反射率的调整,以达到位相调整的目的。

1.1.3 权重融合及实时Z-R关系转化

采用一个双曲函数[8]计算权重系数,对实况COTREC外推及订正后的数值预报结果进行分时效加权平均,使得融合结果在前期与COTREC实况保持一致,而当时效延长外推预报失去效果时,平滑过渡至数值预报。对 2017 年 9—10 月影响上海地区的 4 次降水过程(9 月 20 日低槽切变线降水过程、9 月 25 日副热带高压边缘强对流降水过程、10 月 2 日系统性降水过程、10 月 15 日台风倒槽暴雨过程)进行 20 dBZ 以上组合反射率 TS 评分预报检验。结果显示,相比高分辨率数值模式,融合方法对 0~3 h 的降水预报有明显提高,而对 3 h 以后降水融合方法在提高外推时效的同时也对数值预报有一定的订正技巧(图 2)。

最后,应用实时 $Z\text{-}R$ 关系转换,将反射率融合结果转化为降水率。通过上海中心气象台前期的分类 $Z\text{-}R$ 关系统计临近时段内周边区域站点 1 h 降水量与雷达反射率因子数据,选取最佳的 $Z\text{-}R$ 关系结果用于当前最新预报,最终得到 10 min 分辨率的融合定量降水预报结果(图 3)。

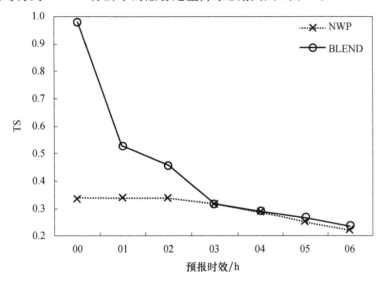

图 2 2017 年秋季 4 次降水过程 0~6 h 反射率预报 TS 评分检验
(阈值:20 dBZ,实线为融合方法评分,虚线为数值模式评分)

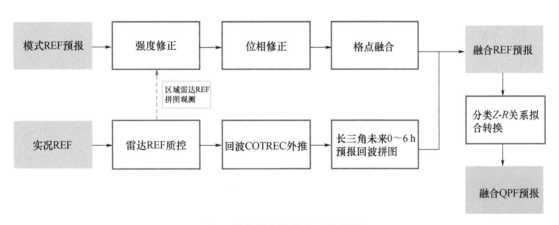

图 3 短临降水融合方法流程图

1.2 多模式最优集成方法

1.2.1 集成方法

对于中短期要素预报,当前的客观预报主要以数值预报为主要核心,而数值预报的迅速发展和提高使得以长期样本统计为基础的传统释用方法(如 MOS 法[9,10])难以取得稳定及可信的统计基础。同时,业务中存在诸多不稳定因素,如资料缺失、时空分辨率不一致等,也制约了单一模式的实际业务化应用效果。多种模式或多种预报方法的综合集成是当前国际上的发展趋势。多模式或多预报方法的集成,既能综合各预报结果的优势,又能减少单个预报成员预报偏差对最终结果的影响,增强了预报的稳定性。上海中心气象台使用数值模式作为指导产品,基于多模式最优集成技术(Optimal Consensus Forecast,OCF)进行中短期要素的订正及格点预报产品制作[11]。选取的业务模式资料,包括欧洲中期天气预报中心(European Centre for Medium-Range Weather Forecasts)全球高分辨模式(EC 模式)、日本细网格模式、T639 全球模式、全球预报系统(global forecast system,GFS)模式以及上海 SMB-WARMS 区域模式,实况资料则来自上海的乡镇自动站资料库(147 个站)及国家信息中心通过卫星下发的常规地面观测资料(531 个站)。集成方法分为两步:

首先,对各集成成员的预报结果进行预报偏差校正:①计算出各集成成员在过去 30 d 中的平均预报相对误差。②根据平均预报相对误差,将各集成成员的预报结果进行系统偏差校正。

然后,对各集成成员的预报结果进行绝对误差权重平均。

(1)定义预报误差(b):

$$b_{i,k,r,f} = F_{i,j,k,r,f} - A_{i,j,r+f} \tag{1}$$

式中,F 代表预报值;A 代表观测值;i 代表站点位置;k 代表集成成员;r 代表起报时间;f 代表预报时效。

计算各集成成员在过去一段时间内的平均预报绝对误差:

$$\hat{b}_{i,k,f} = \frac{Q_1 + 2Q_2 + Q_3}{4} \tag{2}$$

式中,Q_1,Q_2,Q_3 分别代表各成员过去一段时间内预报误差序列 b 的第一、第二、第三个四分位数。平均预报绝对误差是对各成员过去时间内预报系统误差的一个近似估计,取四分位数的平均估计系统误差能够去除极端成员的影响,从而得到预报值的无偏估计:

$$FB_{i,k,r,f} = F_{i,j,k,r,f} - \hat{b}_{i,j,k,f} \tag{3}$$

(2)由预报值的无偏估计,得到过去集成成员的预报平均绝对误差:

$$MAE_{i,k,f} = \frac{1}{N} \sum_r |FB_{i,k,r,f} - A_{i,r+f}| \tag{4}$$

根据平均预报绝对误差的大小,计算相应的权重系数(w),对各集成成员进行加权平均。平均预报绝对误差越大的成员,权重系数越小。

$$w_{i,k,f} = MAE_{i,k,f}^{-1} \left(\sum_k MAE_{i,k,f}^{-1} \right)^{-1} \tag{5}$$

最终,使用计算所得 w 对最新时次各集成成员的预报值无偏估计进行加权平均。得到 OCF 预报结果:

$$OCF_{i,k,r,f} = \sum_k w_{i,k,f} FB_{i,k,r,f} \tag{6}$$

针对实际应用中出现的一些问题,对误差订正方案进行更改,采用类递减平均(Decaying Average)的方法改进相对误差计算。在计算 b 时,增加最近时次预报误差权重,降低较早时次的误差权重,同时剔除预报误差特别大的预报样本。同时,考虑到转折性天气情况,在进行最新时次预报误差订正时,对预报要素明显偏离前期平均态的,则降低其订正效果。

对于风的集成采用略不同的方案。由于风向变化的不连续性和随机性,上述"偏差校正+权重平均"的集成方法不适合风向,实际操作中也无法取得满意的结果,因此采用"择优集成方法"。具体计算时,集成结果是选择过去 30 d 预报平均绝对误差最小的集成成员,定义为:

$$MAE = \frac{1}{N} \sum_{i=1}^{N} \sqrt{(Fu_i - Ou_i)^2 + (Fv_i - Ov_i)^2} \tag{7}$$

式中,Fu 表示预报风速分量;Ou 表示观测风速分量;Fv 表示预报风向分量;Ov 表示观测风向分量。

对于 OCF 降水预报来说,由于其为非连续变量,单纯的平均(或加权平均)往往会使预报降水范围相比单一模式的预报偏大,目前简单采用"晴雨+集成"的方法修正。首先计算过去 30 d 各家模式预报的降水晴雨率及雨量绝对误差,集成方案挑选前30 d中晴雨预报效果最佳的模式雨区预报结果作为OCF的降雨区预报,雨区降水量采用根据绝对误差计算的加权平均值得到。

释用方法于2013年开始试运行,期间经过若干调整,到2016年稳定运行。从2016年上海地区11个标准站24、48和72 h的最高、最低温度的检验可见,该方法能够有效订正数值模式预报误差,相比中国气象局(CMA)的MOS客观预报指导产品也有一定的优势(图4)。

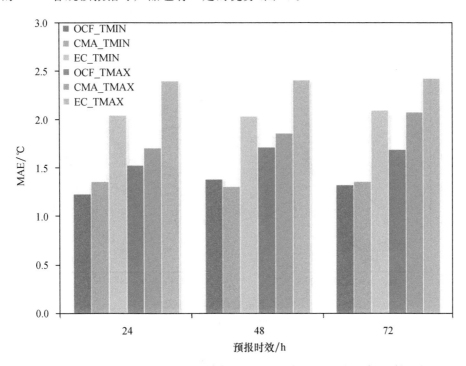

图4 2016年24、48和72 h的OCF释用方法、MOS、欧洲细网格对上海11个站点最高、最低温度预报的平均绝对误差

1.2.2 格点转换

以上结果是对站点的模式订正,要进一步得到格点精细化预报,还需对站点集成结果进行格点化订正及转换。目前精细化预报业务对格点化的分辨率要求12 h内达到3 km,72 h内达到5 km,而集成成员的分辨率最高为9 km,无法达到要求。因此,一方面,对于上海地区,通过147个加密自动站的误差订正能够一定程度上做出高分辨率的订正;另一方面,在进行格点转化时,综合考虑地形、下垫面分布因素进行修正,以提高产品精细化程度。

对站点预报结果进行格点化插值,采用多重cressman插值方案。尤其在进行2 m气温的格点化时,考虑格点地形高度差异进行插值订正。具体订正方法为:利用高分辨率地形高程数据(Shuttle Radar Topography Mission,SRTM;分辨率为90 m)插值得到的格点地形高度,采用经验温度递减率$\gamma=0.65$ ℃/(100 m),在cressman方法中,将格点周围站点温度从其原来不同的站点地形高度订正至同一标准高度(z_0)上,然后进行格点插值,得到该格点在z_0高度的插值温度T_0,然后根据SRTM资料插值的格点地形高度,将T_0再次用γ订正回到格点高度得到最终值(T)(图5)。

对2 m温度预报,在项目释用前期,由于部分模式预报资料在120 h之后的分辨率较粗,有的在12 h,因而在计算最高、最低温度时需要进行时间插值:

$$T(t_i)=T(t_1)+[Tc(h_i)-Tc(h_1)]+\frac{t_i-t_1}{t_2-t_1}[(T(t_2)-T(t_1))-(Tc(h_2)-Tc(h_1))] \quad (8)$$

式中,$T(t_i)$为插值时间温度;$Tc(h_i)$为历史统计t_i时刻温度;$T(t_1)$、$T(t_2)$为离$T(t_i)$最近的相邻两个时次OCF预报温度。

然而，后期由于各家模式预报资料分辨率的提升，在120 h后均可达到6 h，因而当前预报中可以直接使用模式预报场而不需进行插值。

对于不同下垫面的处理，目前仅对海陆格点进行区分。在进行格点插值时，对于陆面格点，仅使用陆面站点 OCF 订正进行插值；对于海面站点，由于缺乏海上观测资料，暂时只由模式直接预报值处理，因此格点温度场上具有很明显的海陆差异（图6）。

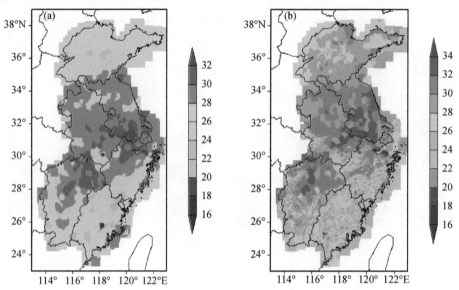

图5 2 m温度预报试验
(a)高程订正前；(b)高程订正后

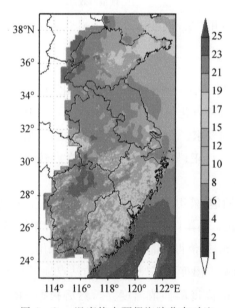

图6 2 m温度格点预报海陆分布对比

2 无缝隙预报支持

2.1 一体化预报制作

上海市气象局针对上海大城市天气预报服务和精细化预报业务需求，借鉴美国的 GFE 和中国气象局的MICAPS系统,于2013年研究建成基于WebGIS技术的一体化天气预报制作系统。该系统以大城市精细化预报服务需求为引领,学习国际先进技术发展,重点攻关模式客观释用关键技术、交互式格点预报编辑平台、智能工具库、基于格点预报的自动生成技术等关键技术,建立较为完善的精细化气象格点预报技术体系支撑。

目前,上海数值模式指导产品包括区域中尺度模式(STI-WARMS)、EC高分辨模式、日本高分辨模式(JAPAN)、OCF等,预报时效最长达240 h。其中,对于0~72 h短时临近/短期预报,采用OCF作为默认指导产品,在业务化使用中,每天两次起报,实现了华东区域分辨率5 km,上海地区分辨率3 km,240 h内逐6 h、96 h内逐3 h、24 h内逐1 h的中短期无缝隙客观精细化预报。其中直接集成的预报要素包括2 m温度、10 m风、定量降水、2 m相对湿度;诊断预报要素包括24 h最高/最低温度、云量、天气现象。同时,利用历史缺测查找补齐保证运行稳定性,目前使用情况良好。

图7为上海市气象局一体化格点预报制作系统示意图,虚线部分是精细化格点预报系统功能结构图,包括三个核心功能模块:数值模式指导产品、格点编辑工具库和预报产品生成器。格点预报系统是上海市气象局一体化预报制作业务的核心部分。业务系统界面如图8所示,预报员可以基于指导产品,通过格点编辑工具对其进行人工订正,最终经产品生成模块形成每个格点、城镇或者区域上的天气预报,生成各类产品。

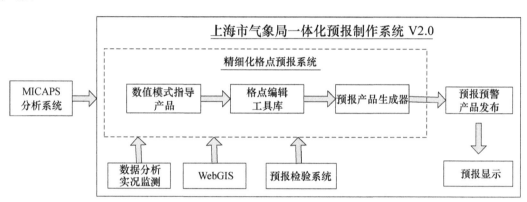

图7 上海市气象局一体化预报制作业务示意图

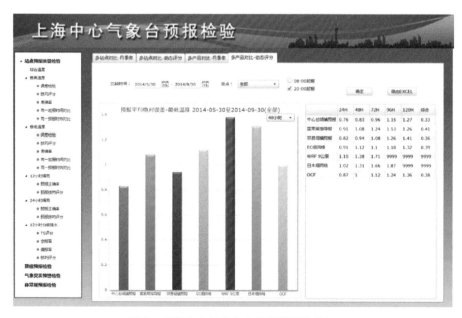

图8 上海中心气象台在线预报检验界面

格点编辑工具主要实现将天气概念模型应用到交互式图形化预报制作中区,并使格点数据订正结果在WebGIS界面中同步更新(图9)。上海精细化格点预报系统实现了平滑、过滤、增减、调整、插值、平移、区域操作等多种编辑工具。另外,系统实现了格点与站点的转换反演及时空的影响反演计算,建立了依赖基准站的曲线订正反演模型和站点预报影响模型,将模型中基准站点反演到面预报,面反演与时间序列反演结合,实现多时次预报快速订正。系统建立了时空、天气要素一致性调整基数,避免了天气现象、降水量、云量等不同气象要素可能出现的预报不一致现象[12]。

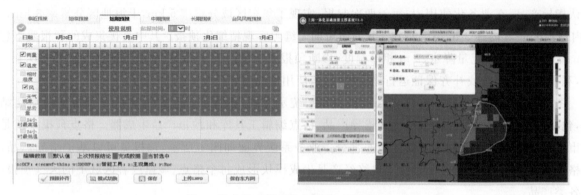

图9 格点管理器及格点编辑工具示意图

2.2 预报检验

对于上海地区常规要素检验,如温度绝对误差、降水分级TS评分等,预报员能够通过在线的检验平台(图8)直接进行自定义查询,对各类主客观预报做出定量评估判断,帮助进行预报决策。除此之外,针对本地影响较大的强对流天气,常规的误差、TS评分无法满足预报员对系统发展特征的误差判断,因此,特别进行了针对强对流天气的一些新的检验方法的开发,如目标对象法、尺度分离和模糊检验算法,与传统的检验方法一起,为局地灾害性天气检验预报业务提供技术支撑。

2.2.1 目标对象检验

目标对象检验法(图10)分为两部分——目标识别匹配和检验评价。该方法选取预报对象落区预报中的强度分布、区域面积、区域形态(长轴、短轴、轴向)、区域重心位置等诸多要素,与实况分布进行对比,并进行重心校正后的区域相似度分析(交叉相关),采用加权法综合各项检验指标对总体检验的集合贡献,从而最终获取强对流对象落区预报的检验结果。

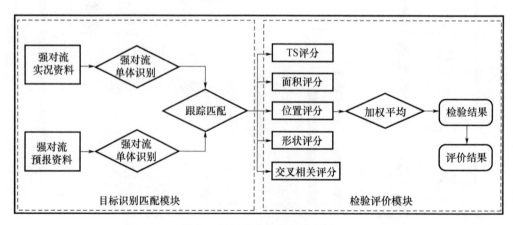

图10 目标对象检验法流程图

2.2.2 尺度分离和模糊检验

采用不同的时间、空间尺度的匹配方法,对选定的强对流预报对象进行评价和检验,即在对应预报格点周围一定的范围(尺度可变)内选取观测格点,将这些观测格点与预报格点进行匹配,再对该区域内预报值和观测值分别进行取平均、取阈值、分等级、计算概率分布等操作,然后进行对比检验与评价,采用TS、击中率(POD)等评分办法提供预报产品在不同尺度分布情况下的检验结果。主要的检验技术包括尺度模糊法、升尺度法、阈值对比法、最小覆盖法、基于概率分布的模糊逻辑法、多等级对比法等。

2.2.3 强对流预报检验系统

基于以上强对流检验新方法,开发基于网页版的强对流实时预报检验系统。最先主要针对灾害天气短时临近预报业务系统(SWAN)和NoCAWS的雷达回波外推和定量降水预报(Quantitative Precipitation Forecast,QPF)进行实时预报检验,并根据业务实际运行中出现的问题,不断对后台运行脚本和前端显示界面进行开发和完善。后期又增加高分辨率EC模式和上海市气象局区域模式的降水预报检验结果,开发对强对流天气落区展望预报的检验产品。针对部分检验产品增加检验评估部分,并针对不同用户(预报员、水文、航空)的需求特点,给出推荐的参数配置和产品显示。后期又将检验技术开发为"短临预报检验评估模块",集成至MICAPS/SWAN系统中(图11)。系统从2014年7月份开始,在上海中心气象台开展业务试验运行,并在江苏、安徽、湖北、江西等地试验运行并得以应用,为2014年南京青奥会气象服务提供检验技术支持。系统运行稳定、产品丰富,可以弥补原有检验方法和产品的不足,在各省级部门均实现业务运行。

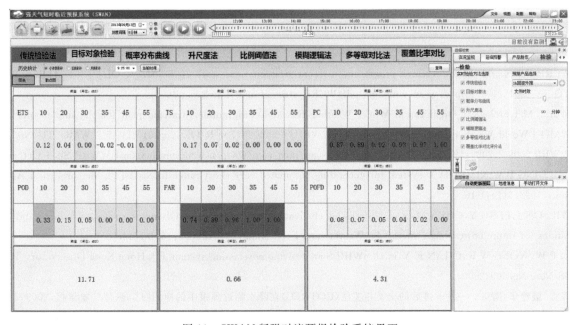

图11 SWAN版强对流预报检验系统界面

3 结论与展望

上海市气象局自2014年开展精细化格点预报订正业务,2016年一体化无缝隙预报业务正式投入,当前能够稳定提供0~240 h从短临到中短期,以及10 d以上的华东地区格点天气预报产品。预报要素包括降水率、天气现象、云量、温度、风、湿度、降水等,空间分辨率3~9 km,时间分辨率10 min~12 h。利用本文介绍的客观化技术生成的格点预报是上海市格点精细化业务的默认指导产品,为每日的格点精细化预报业务提供依据,并且所有预报结果在一体化平台上显示,结合实时检验平台,为预报工作提供参考。

(1)融合雷达实况外推与数值模式预报,制作0~6 h短时临近精细化客观预报定量降水融合产品。范围覆盖长三角地区,空间分辨率3 km,时间分辨率10 min。利用高分辨率雷达组网观测数据及区域数值模式预报产品,采用COTREC方法进行外推预报,通过强度误差修正及位相误差修正订正数值模式预报反射率,结合实时检验与历史统计方法进行产品订正及融合,进行分钟级的短时临近客观降水预测。

(2)集成全球及区域各主流数值模式预报,制作0~24 h逐1 h、0~72 h逐3 h、1~10 d逐12 h的中短期多模式集成要素预报产品。通过实时对各家模式前期预报误差进行历史统计来订正当前最新时次的模式预报,再通过误差权重进行集成,能够有效地校正模式系统误差。此外,采用地形高程、海陆分布资料对预报进行格点转换订正,使产品精细化程度进一步改进。

(3)结合上海中心气象台预报检验系统及强对流实时预报检验系统,上海市气象局一体化格点预报制作系统现已成为上海市气象局预报员进行日常预报业务的主要支撑平台。预报员在平台格点预报指导产品的基础上订正、编辑并制作各类短时临近、中短期精细化格点、站点预报产品,极大提高了无缝隙预报的工作效率,丰富了各类预报产品的发布形式。

(4)上海无缝隙预报取得了初步进展。然而,当前格点客观预报主要作为背景指导产品,需要在预报员对其进行订正、编辑之后才形成各类最终用户服务产品,离直接应用尚有差距。而在一些先进国家,如德国气象局的AutoWARN系统,能够利用客观系统完成事件观测识别、自动报警、自动预报、预报员互动决策、自动决策服务的一系列预报服务流程。未来若要进一步提高预报现代化水平,一方面要利用各类新技术如超级集合预报释用、机器学习技术等,减少预报员人工订正不确定性,提高客观预报水平;另一方面,需要面向服务对象,利用大数据分析及人工智能系统,实现传统人工天气预报向自动化影响天气预报的转变。

参考文献

[1] WCRP. The world climate research programme's recent strategic framework 2005-2015[R/OL]. [2015-11-15]. WMO/TD-No. 1291,2005. http://wcrp.wmo.int/pdf/WCRP_strategImple_LowRes.pdf.

[2] World Meteorological Organization World Weather Research Programme. Strategic plan for the implementation of WMO's World Weather Research Programme(WWRP)(2009-2017)[R/OL]. [2015-11-15]. WMO/TD No. 1505, WWRP 2009-2,2009:121. http://www.wmo.int/pages/prog/arep/wwrp/new/documents/final_WWRP_SP_6_Oct.pdf.

[3] GOLDING B W NIMROD. A system for generating automated very short range forecasts[J]. Meteorological Applications,1998,5(1):1-16.

[4] WILSON J,FENG Y,CHEN M,et al. Nowcasting challenges during the Beijing Olympics:successes,failures,and implications for future nowcasting systems[J]. Weather and Forecasting,2010,25(6):1691-1714.

[5] LI P W,WONG W K,CHAN K Y,et al. SWIRLS-an evolving nowcasting system[R]. Hong Kong Observatory Technical Note No. 100,2000.

[6] 陈雷,戴建华,陶岚. 一种改进后的交叉相关法(COTREC)在降水临近预报中的应用[J]. 热带气象学报,2009,25(1):117-122.

[7] 程从兰,陈明轩,王建捷,等. 基于雷达外推临近预报和中尺度数值预报融合技术的短时定量降水预报试验[J]. 气象学报,2013,71(3):397-415.

[8] 杨丹丹,申双和,邵玲玲,等. 雷达资料和数值模式产品融合技术研究[J]. 气象,2010,36(8):53-60.

[9] 刘还珠,赵声蓉,陆志善,等. 国家气象中心气象要素的客观预报——MOS系统[J]. 应用气象学报,2004,15(2):181-191.

[10] 赵翠光,赵声蓉. 华北及周边地区夏季分区降水客观预报[J]. 应用气象学报,2011,22(5):558-566.

[11] 漆梁波,曹晓岗,夏立,等. 上海区域要素客观预报方法效果检验[J]. 气象,2007,33(9):9-18.

[12] 王海宾,杨引明,范旭亮,等. 上海精细化格点预报业务进展与思考[J]. 气象科技进展,2016,6(4):18-23.

基于数值模式误差分析的气温预报方法

蔡凝昊[1,2]　俞剑蔚[1,2,*]

(1. 江苏省气象台,南京,210008；2. 南京大气科学联合研究中心,南京,210009)

摘　要：采用欧洲中期天气预报中心(ECMWF)全球确定性预报模式地面气温和国家地面站点观测资料,对模式初值场误差、历史误差以及卡尔曼滤波预测误差与实况误差之间的相关性进行分析,设计了4种回归方案订正日最高、最低气温预报偏差,并与ECMWF、中央气象台和全国城镇的预报产品进行检验对比。结果表明：采用模式近1～3 d 最高(最低)气温和模式最高(最低)气温历史平均误差、初值场误差以及卡尔曼滤波反演误差作为预报因子的改进方案效果最优,经对其2017年日最高和最低气温的预报检验,预报准确率较ECMWF原始模式预报有较明显提高,也明显优于中央气象台指导预报。在空间分布方面,其对地形较为复杂地区的改进效果更好。同时,与当前业务中质量最好的全国城镇预报相比,最高气温预报平均绝对偏差(Mean Absolute Error,MAE)较全国城镇预报低 8.24%～13.97%,预报准确率提高 1.24%～3.57%,日最低气温平均绝对偏差较城镇预报低 9.43%～17.69%,预报准确率提高 1.77%～2.72%。在 3 d 的预报中,对 24 h 时效内预报相对于 48 h 和 72 h 的改进幅度更大,订正效果更加明显。

关键词：偏差订正；线性回归；初值场误差；卡尔曼滤波

Temperature Forecasting Method Based on Numerical Model Bias Analysis

CAI Ninghao[1,2]　YU Jianwei[1,2]

(1. Weather Observatory of Jiangsu Provence, Nanjing, 210008；
2. Nanjing Joint Center for Atmospheric Research(NJCAR), Nanjing, 210008)

Abstract：In this study, based on the European Centre for Medium-Range Weather Forecasts(ECMWF) of 2 m surface air temperature and automatic weather observatory data of China, the correlation among the initial field bias, historical bias, Kalman filter predicted bias the real-time bias is analyzed. Four daily maximum and minimum temperature forecast regression schemes are designed. A comparison among ECMWF, CMA, and provincial forecasting is investigated. The results show that the improved scheme's predicted temperature, historical bias, initial field bias and Kalman filter inversion bias as the predictor are all optimal. In addition, the scheme's forecast quality for the maximum and minimum temperatures in 2017 is significantly better than that of both ECMWF and CMA, particularly in areas featuring com-

---

① 本文发表于《大气科学学报》2019 年第 6 期。
资助项目：北极阁开放研究基金——南京大气科学联合研究中心基金资助项目(NJCAR2016ZD04)；华东区域气象科技协同创新基金合作项目(QYHZ201602)。
通讯作者：俞剑蔚. E-mail:radargroup@foxmail.com。

plex terrain. Compared with the best-performing provincial forecasting, the maximum temperature MAE is 8.24%～13.97% lower than the provincial forecast, while the forecast accuracy is increased by 1.24%～3.57%, and the daily minimum temperature MAE is 9.43%～17.69% lower than the provincial ones. The forecast accuracy rate increased by 1.77%～2.72%. Additionally, it shows the greatest improvement within 1 day, thus indicating that the correction effect decreases considerably with the forecast lead time.

Key words: bias correction; linear regression; initial field bias; Kalman filter

  近十年来,数值天气预报技术发展迅猛,模式预报要素不断丰富,时间和空间分辨率不断提升,伴随集合预报技术的发展,有效预报时效不断延伸,各要素预报准确率也有了大幅度的提升。不过,相对于当前社会对要素预报精细化及精准度的需求,数值模式要素预报与实况观测的差距仍然较大(穆穆 等,2011)。为了提高预报的准确率,除了不断改进数值预报模式本身,针对模式产品进行后处理偏差订正是能明显改善预报准确率的一个重要途径,也是长期以来数值模式研究领域里的一个重要研究方向,在过去已取得了大量的研究成果。气温预报是天气预报的一项主要内容,特别是地面日最高、最低气温预报一直都是公众以及一些专业用户关注度最高的预报项目之一。目前国内外科学家针对确定性预报模式和集合预报气温预报方法开展了很多订正方法方面的研究,应用统计及物理方法减小日最高、最低气温预报的偏差是数值预报后处理的重要研究方向之一(林春泽 等,2009;智协飞 等,2010,2013;Krishnamurti et al.,2016;张颖超 等,2017)。赵声蓉(2006)采用神经网络方法建立了全国温度集成预报系统,为预报员提供预报参考。智协飞等(2009,2013)和Zhi等(2012)利用ECMWF、JMA、NCEP和UKMO四个中心的资料开展超级集合预报实验,发现其对24～168 h预报效果有明显提升。李佰平等(2012)对ECMWF气温预报的四种误差订正方法进行了比较研究,发现在短期预报中仅考虑最新预报结果的一元线性回归方法优于多元集成预报订正方法。研究表明,多模式集成预报在中期或延伸期上具有较好的性能(Arribas et al.,2005;Bowler et al.,2008;吉璐莹 等,2017;智协飞 等,2018a)。张庆奎等(2008)指出,卡尔曼滤波方法对温度的季节变化和数值模式变化等的适应性较好,但存在滞后时间长的缺点。翟宇梅等(2014)通过在模型中增加遗忘因子的方法解决建模数据时间过长导致的"数据饱和"现象,提高了模型的预报准确率。针对同样的问题,预报员在业务中经常将实况误差纳入预报考虑,也是有效的解决方案之一。综合以上研究,在气温的短期预报方面,线性回归方法较多元集成预报订正方法具有更多优势,并且卡尔曼滤波方法和初值场误差可以有效反映数值模式短期和当前起报时刻的误差变化情况。

  研究发现,确定性数值模式和集合预报直接输出的气温预报有时存在时间跳跃性和不一致性较强等问题(郭换换 等,2016a,2016b;智协飞 等,2018b),从而导致预报产生较大的偏差。例如,欧洲中期天气预报中心(ECMWF)全球确定性预报模式是业务台站进行气温预报时参考的重要模式,是众多数值模式中性能较为稳定、预报准确率较高的模式,但在某一段时间某个区域内会存在预报偏差持续较大的现象。2018年1月29日20时的24 h预报在安徽中部和江苏西南部地区预报了－16 ℃以下的负值中心,与实况不符,比实况偏低了平均6～8 ℃,最大误差值在10 ℃左右,远低于平时的模式预报水平。而且当时次预报的2 m气温初值场也在同一区域出现了偏差6～9 ℃的问题,该问题在随后数日预报中持续出现。

  针对模式气温预报在某一段时间里偏差较大和准确率低的问题,本文基于欧洲中期天气预报中心确定性模式预报产品,采用卡尔曼滤波和多元线性回归方法对日最高、最低气温预报进行订正,考虑引入模式邻近预报误差、模式历史误差、卡尔曼滤波预测误差以及模式初值场误差等新预报因子,通过各种误差与预报误差的相关性进行分析,分别建立四种订正模型方案进行试报效果对比。结果表明,新方法较为明显地提升了日最高、最低气温预报的准确率,并且明显优于现有主要业务预报产品。

1 资料和方法

1.1 资料

数值预报资料采用中国气象局下发的 ECMWF 高分辨率数值预报全国范围的 2 m 气温初值场、2 m 最低气温和 2 m 最高气温逐 6 h 预报等预报要素。资料时间段为 2015 年 1 月至 2017 年 12 月,每日 2 次预报,起报时刻分别为每日 08:00 和 20:00(北京时间,下同),空间水平分辨率为 0.125°×0.125°,时间分辨率为 6 h,并采用 cressman 客观插值方法将格点资料插值到检验站点。

选取全国 2436 个国家地面气象站的近 3 a 的观测资料作为训练和检验数据。除此以外,由于 ECMWF 细网格资料推送到业务处理系统的时间与起报时刻存在 7 h 左右的延迟,因此将 ECMWF 模式 12~84 h 的最高、最低温度预报应用于 1~3 d 的预报订正。

1.2 方法

1.2.1 初值场误差定义

将 ECMWF 的 2 m 气温初值站点插值结果与其起报时刻的站点实况观测气温相减,得到该时次预报的 2 m 气温初值场站点误差,具体公式如下:

$$B_{t0} = F_{t0} - O_{t0} \tag{1}$$

式中:F_{t0} 为 ECMWF 在 t_0 时刻的 2 m 气温初值;O_{t0} 为在 t_0 时刻的 2 m 气温实况观测值;B_{t0} 即为所求得的初值场误差。

1.2.2 多元线性回归方法

在多元线性回归建模过程中,以 2015 年 1 月至 2016 年 12 月的数据作为(离线 off-line)基础训练集,2017 年 1 月至同年 12 月的数据为(在线 on-line)测试集进行逐日建模,并将测试时刻之前的 2017 年测试集数据也补充至训练集中,逐日滚动建立各检验站点上某一预报时刻的日最高气温和最低气温误差的多元线性回归方程:

$$B_i = b_0 + b_1 x_1 + b_2 x_2 + \cdots + b_k x_k + e \tag{2}$$

式中:B_i 为 i 时效的日最高、最低气温误差订正值;b_0 为常数项;$b_1, b_2, \cdots, b_k$ 为预报因子 $x_1, x_2, \cdots, x_k$ 的回归系数;e 则为误差项。预报因子选取根据不同回归方案有所增减,在第 4 节中具体讨论。

1.2.3 卡尔曼滤波递减平均方法

卡尔曼滤波递减平均方法对订正时刻的各站点递减平均偏差进行估计,从而对该时刻的日最高、最低气温进行订正。其递减平均误差计算方法如下:

$$B_{\text{KLM},t0} = (1-w) B_{\text{KLM},t0-1} + w(f-o) \tag{3}$$

式中:$B_{\text{KLM},t0-1}$ 为上一订正时刻的递减平均误差;$(f-o)$ 为最近时刻预报与实况的偏差;w 为权重系数。本文进行了权重系数 w 取值的优选实验,在 0.01~0.4 范围内以 0.01 为步长,并根据其不同取值检验近 30 d 的 1~3 d 日最高、最低温度准确率,优选其中预报准确率最高时所对应的 w 值作为当前时刻的权重系数。

2 误差分析与回归因子选取

2.1 初值场误差分析

运用公式(1)计算得到 2015—2017 年 ECMWF 细网格 08 时和 20 时两个起报时刻的温度初值场误差,并对其与当次预报 1~3 d 最高、最低气温误差(本文中称为实况误差)的相关性特征进行研究(图 1)。从图 1a 可见,起报时刻为 08 时的 1~3 d 最高气温误差与其初值场误差相关性较小,均分布于 0 值略偏向正相关一侧,而且峰值频率值随预报时效延长有所提高;而 08 时初值场误差与最低气温误差的相关分布(图 1b)则在 1 d 时呈现明显的正相关关系,2~3 d 的区别不大,其峰值偏向正相关一端。对于起报时刻为 20 时的预报,初值场误差与 1 d 最高气温误差的相关系数分布(图 1c)峰值位于 0.25 左右,随预报时效的延长其相关系数也逐渐减小,并且分布区域也更为狭窄,峰值时的频率取值也更高;相较于最高气温误差的结果,1~3 d 初值场误差与最低气温误差相关分布(图 1d)随时效的变化较小,其相关系数分布峰值略小于最高气温误差的结果。本文采用 2015—2017 年 3 a 的数据,总体样本时间较长,最高、最低气温的出现时间总体符合气温平均日变化规律。因此,可以假设 1 d 内的最高气温出现在下午,而最低气温出现在清晨。08 时起报的数值预报其 1 d 预报时间为当日 20 时至次日 20 时,其最高气温应出现在次日下午,与起报时刻相隔 30 h 左右,而最低气温出现在次日清晨,相距起报时刻 20 h 以内。依此类推,20 时起报的 1 d 最高气温与起报时刻相距 20 h 以内,最低气温则相隔 30 h 左右。所以,当距离起报时刻更近

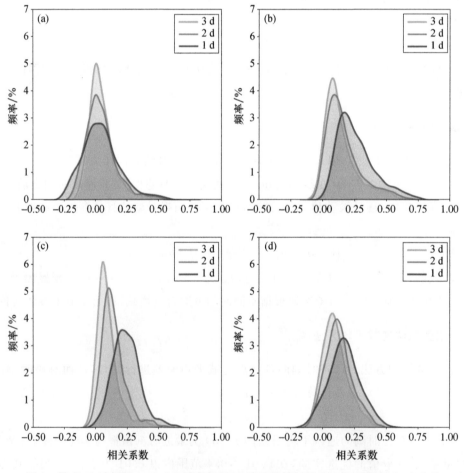

图 1 模式 08 时起报 1~3 d 日最高气温(a)及日最低气温(b),20 时起报 1~3 d 日最高气温(c)及最低气温(d)实况误差与初值场误差相关系数概率密度分布

时,初值场误差与最高、最低气温误差的相关性更高。同时,最高气温与初值场误差相关性随预报时效的衰减相较于最低温度更明显。

2.2 模式误差分析

不同起报时刻的数值模式对同一时刻实况进行多次预报。这里将当前模式起报时刻之前24 h、48 h和72 h起报数值模式对前一日日最高、最低气温的预报与观测实况之间的误差分别称为12～36 h、36～60 h和60～84 h历史误差。本节将分析1～3 d日最高、最低气温实况误差与历史误差之间的相关性。

通过公式(4)分别计算1～3 d日最高、最低气温与实况之间的误差,然后对模式前3 d起报的12～84 h最高、最低气温误差与当次误差求相关系数。

$$B_{t_{\max}/t_{\min}} = F_{t_{\max}/t_{\min}} - O_{t_{\max}/t_{\min}} \tag{4}$$

图2为1～3 d模式实况误差与历史误差相关系数概率密度分布。由于所对应实况均为同一日观测,12～36 h历史误差对应当前起报时刻前1 d的数值预报,而36～60 h和60～84 h时效预报,分别对应当前起报时刻之前第2日与第3日的数值预报。

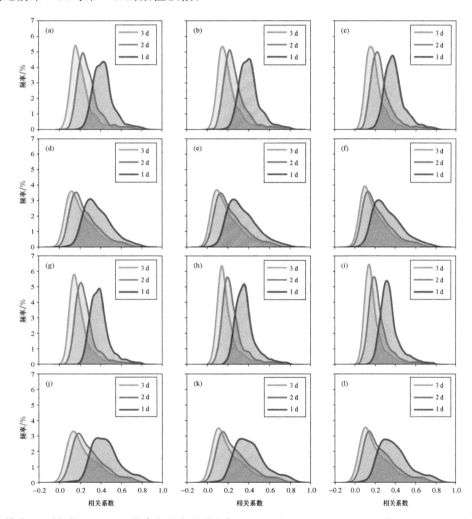

图2 模式08时起报1～3 d日最高气温实况误差与12～36 h(a)、36～60 h(b)以及60～84 h(c)历史误差,08时起报1～3 d日最低气温实况误差与12～36 h(d)、36～60 h(e)以及60～84 h(f)历史误差,20时起报1～3 d日最高气温实况误差与12～36 h(g)、36～60 h(h)以及60～84 h(i)历史误差,20时起报1～3 d日最低气温实况误差与12～36 h(j)、36～60 h(k)以及60～84 h(l)历史误差的相关系数概率密度分布

从图中可以看出,1~3 d 相关系数均明显偏向正相关一侧,并且其概率分布峰值随预报时效增加递减。对比 08 时和 20 时起报的日最高、最低气温误差相关概率密度分布情况,发现 08 时起报最高气温(图 2a、b、c)正相关略大于 20 时起报最高气温(图 2g、h、i),而 08 时起报最低气温正相关(图 2d、e、f)则与 20 时预报(图 2j、k、l)大体相当。对比最高、最低气温实况与历史误差相关系数概率密度分布形态发现,最高气温误差的分布较之最低气温误差明显偏窄,频率峰值也更大。与此同时,比较相同起报时刻的不同时效预报可以发现,随着当次预报时效的延长,其与下一时次最高、最低气温预报的误差相关性都有所减弱。

2.3 卡尔曼滤波递减平均方法误差分析

利用公式(3)计算得到 2015—2017 年逐日卡尔曼滤波递减平均误差,并对其与模式当次预报 1~3 d 最高、最低气温实况误差的相关性特征进行分析(图 3)。发现 08 时和 20 时起报预报的相关系数概率密度分布较为相似,最高气温(图 3a,3 c)1 d 预报的频率峰值均在 0.4 附近,最低气温(图 3b,3 d)1 d 预报的频率峰值均在 0.3 左右,并都呈现随时间相关性递减的情况,而且最高气温正相关性略大于最低温度,其分布宽度也较最低气温更窄。

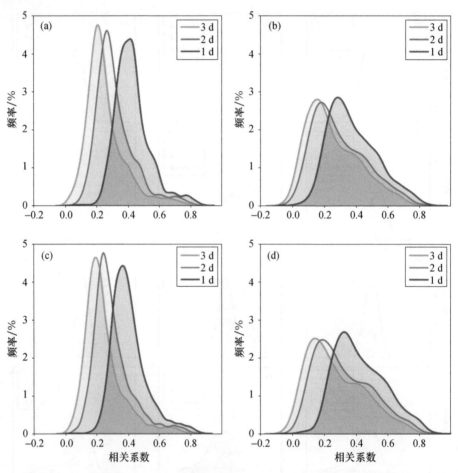

图 3 模式 08 时起报 1~3 d 日最高气温(a)、08 时起报 1~3 d 日最低气温(b)、20 时起报 1~3 d 日最高气温(c)以及 20 时起报 1~3 d 日最低气温(d)实况误差与卡尔曼滤波预测误差相关系数概率密度分布

综上所述,卡尔曼滤波预测误差和模式历史误差与模式实况误差的相关性在数值、衰减速度方面表现相当。12~36 h 相关系数在 0.4 左右,36~60 h 相关系数减少到 0.2~0.3,60~84 h 相关系数维持在 0.2 左右,因此将其作为预报因子。值得注意的是,由于起报时刻距离预报时刻较远,20 时起报日最高气温和 08 时起报日最低气温的相关性弱于 08 时起报日最高气温和 20 时起报日最低气温。而初值场误差

与模式实况误差的相关性明显低于前两者,仅与其他因子落后 48 h 的相关性相当,但是考虑到只有初值场误差包含了当次数值模式的误差信息,并且本文主要解决模式初值场出现较大偏差的问题,因此将其选为预报因子。与此同时,当次模式预报的 1～3 d 最高、最低气温预报值显然也会影响其预报误差,故也选为预报因子。

3 回归方案对比和改进

3.1 方案设计

根据第 2 节各个预报因子的误差相关性分析,将模式预报 1～3 d 最高、最低气温的初值场误差,卡尔曼滤波反演误差以及模式历史误差作为回归方案的备选因子进行回归建模。为了检验初值场误差和卡尔曼滤波反演误差对 1～3 d 日最高最低气温预报建模效果的影响,本文设计了以下 4 种统计回归的预报因子方案:

初始方案:采用模式预报 1～3 d 最高、最低气温和模式最高、最低气温历史误差作为预报因子进行统计建模。

改进方案一:在初始方案的基础上增加初值场误差作为预报因子进行统计建模。

改进方案二:在改进方案一的基础上增加卡尔曼滤波反演误差作为预报因子进行统计建模。

改进方案三:在初始方案的基础上增加初值场误差和卡尔曼滤波反演误差共同作为预报因子进行统计建模。

3.2 检验与分析

为了评估不同回归方案的预报性能,本文除了预报平均绝对误差 E_{MA} (Mean Absolute Error, MAE) 和均方根误差 E_{RMS} (Root Mean Squared Error, RMSE) 外,还应用气温预报小于等于 2 ℃ 准确率 (F_a) 进行检验,具体公式如下:

$$E_{MA} = \frac{1}{n} \sum_{i=1}^{n} | x_{obs,i} - x_{model,i} | \tag{5}$$

$$E_{RMS} = \sqrt{\frac{1}{n} \sum_{i=1}^{n} (x_{obs,i} - x_{model,i})^2} \tag{6}$$

$$F_a = \frac{N_r}{N_t} \times 100\% \tag{7}$$

式中:x_{obs} 为实况观测值;x_{model} 为模型预报值;F_a 为最高、最低温度预报绝对偏差不超过 2 ℃ 的百分比;N_r 为预报最高、最低温度与实况最高、最低温度之间不超过 ±2 ℃ 的次数;N_t 为预报的总次数。

表 1 列出上述 4 种方案 1～3 d 日最高气温和最低气温预报均方根误差及气温预报准确率的情况,可以看出,除 20 时起报的 3 d 日最高气温准确率改进方案二略优于改进方案三以外,其他各起报时刻的不同预报时效最高、最低气温 RMSE 和预报准确率均呈现 3 种改进方案优于初始方案,并且改进方案三最优,改进方案二次优的检验结果。综合对比 08 时起报和 20 时起报的最高气温预报检验结果发现,20 时起报的 1～2 d 预报的 RMSE 和预报准确率均高于 08 时起报结果,而 08 时 3 d 预报的 RMSE 和预报准确率表现则不一致,其中 RMSE 略优于 20 时,预报准确率则略逊于 20 时。对比不同起报时刻最低气温检验结果发现,20 时起报 1～3 d 预报的 RMSE 和预报准确率均优于 08 时。

表1 2017年初始方案与改进方案1~3 d日最高气温和最低气温预报检验

起报时刻	时效/d	日最高气温								日最低气温							
		均方根误差/℃				预报准确率/%				均方根误差/℃				预报准确率/%			
		初始	改进一	改进二	改进三	初始	改进一	改进二	改进三	初始	改进一	改进二	改进三	初始	改进一	改进二	改进三
08时	1	1.57	1.56	1.55	1.54	80.86	81.16	81.34	81.60	1.44	1.43	1.40	1.40	83.26	83.44	84.14	84.22
	2	1.85	1.84	1.83	1.82	73.81	73.94	74.41	74.54	1.62	1.61	1.57	1.57	78.87	79.02	79.87	79.93
	3	2.03	2.02	2.02	2.01	69.54	69.66	70.06	70.11	1.73	1.72	1.69	1.68	76.25	76.30	77.06	77.07
20时	1	1.53	1.52	1.51	1.50	82.23	82.59	82.76	83.10	1.34	1.33	1.32	1.31	85.79	85.98	86.50	86.57
	2	1.84	1.84	1.82	1.81	74.45	74.57	75.14	75.23	1.56	1.55	1.52	1.51	80.23	80.45	81.26	81.35
	3	2.06	2.05	2.04	2.04	69.65	69.66	70.25	70.20	1.67	1.66	1.63	1.63	77.48	77.64	78.39	78.42

总体而言,改进方案三最优,改进方案二次优,3个改进方案均较初始方案有所改善,其改进效果在起报时刻、预报时效以及最高、最低气温要素上均表现得较为稳定,仅在20时起报的日最高气温预报准确率检验中出现了改进方案二优于改进方案三的情况。

3.3 订正结果对比分析

3.3.1 与卡尔曼滤波方法比较

卡尔曼滤波递减平均方法是进行日最高、最低气温客观订正的常用方法。为了评价上述4种回归方案的改进效果,计算卡尔曼滤波递减平均方法与各方案RMSE之差,其正值越大表示改进效果越好。其结果如图4所示,对于日最高气温(图4a和c),改进方案一至三的改进效果依次递增,并且随着预报时效的增加其改进效果更为显著,在改进最大的3 d预报达到0.12 ℃以上。08时起报的日最低气温(图4b)的改进幅度最小,也呈现随预报时效改进效果逐渐递增的趋势,其中初始方案和改进方案一的改进幅度明显低于改进方案二与改进方案三。20时起报日最低气温(图4d)在1~3 d预报的平均改进幅度最大,并且其1 d和3 d改进幅度较2 d稍大。

总体看来,对两个起报时刻的1~3 d日最高、最低气温预报,初始方案和改进方案一至三均有明显改进,大体呈现随时效增加效果递增的趋势,1 d的RMSE改进0.05 ℃左右,2 d改进0.09 ℃左右,3 d则改进0.13 ℃左右,并且改进方案三在08和20时起报日最高、最低气温改进效果中均最大,为最优方案。

本文将最优方案(即改进方案三)与卡尔曼递减平均方法日最高、最低气温预报的RMSE进行对比分析,得到空间分布(图5)。分析对比可以看出,对于08和20时起报1~3 d日最高、最低温度,改进方案三在全国绝大部分地区均优于卡尔曼递减平均方法。随着预报时效的增加,08时起报日最高、最低温度以及20时起报日最高温度均呈现改进效果逐步提高的趋势,而这一现象在20时起报日最低气温预报中表现并不明显。值得注意的是,在西藏、青海、四川交界,云南、贵州、四川交界,天山,秦岭,皖南、浙西等地形变化剧烈、地形较为复杂的地区,改进方案三的改进效果更好。

3.3.2 与主要业务预报产品比较

检验中得到的最优方案即改进方案三与ECMWF确定性预报、中央气象台指导预报和各省城镇预报平均的日最高、最低气温平均绝对偏差与预报准确率见表2。为方便起见,这里将08和20时起报的MAE和预报准确率平均值作为最后结果。日最高气温和日最低气温1~3 d预报中,未订正的ECMWF预报MAE最大,预报准确率最低。中央气象台指导预报优于ECMWF预报,全国城镇预报优于中央气象台预报,而改进方案三的MAE最小,预报准确率最高。总体而言,各预报日最低气温预报均优于日最高气温预报。改进方案三的日最高气温MAE较全国城镇预报低8.24%~13.97%,预报准确率提高

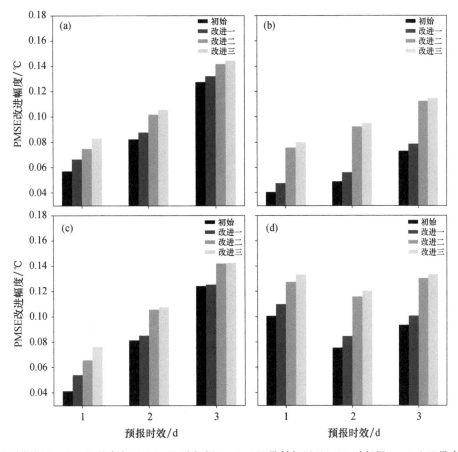

图4 模式08时起报1~3 d日最高气温(a)、08时起报1~3 d日最低气温(b)、20时起报1~3 d日最高气温(c)以及20时起报1~3 d日最低气温(d)初始及改进方案较卡尔曼滤波递减平均方法的RMSE改进效果

1.24%~3.57%,日最低气温MAE较城镇预报低9.43%~17.69%,预报准确率提高1.77%~2.72%。其1 d预报的改进幅度最大,并随预报时效延长呈减少的趋势,即预报时效越短订正效果越好。

表2 改进方案三与ECMWF、中央气象台、全国城镇预报2017年1~3 d日最高气温和最低气温预报检验对比

要素	时效/d	平均绝对偏差/℃				预报准确率/%			
		ECMWF	中央气象台	全国城镇预报	改进方案三	ECMWF	中央气象台	全国城镇预报	改进方案三
日最高气温	1	2.26	1.70	1.36	1.17	55.32	69.49	78.78	82.35
	2	2.35	1.82	1.54	1.41	52.76	65.98	73.23	74.89
	3	2.42	1.99	1.70	1.56	51.07	61.45	68.92	70.16
日最低气温	1	1.71	1.47	1.21	1.05	67.77	75.58	82.68	85.40
	2	1.76	1.51	1.34	1.21	66.20	74.43	78.83	80.64
	3	1.82	1.59	1.44	1.30	64.81	71.89	75.98	77.75

4 结论和讨论

通过对模式初值场误差、历史误差以及卡尔曼滤波预测误差与实况误差之间的相关性分析,设计4种日最高、最低气温预报回归方案,并与ECMWF、中央气象台、全国城镇预报进行对比检验,得到如下结论:

(1)卡尔曼滤波反演误差、模式历史误差在与模式实况误差相关性方面性能相当。初值场误差与之

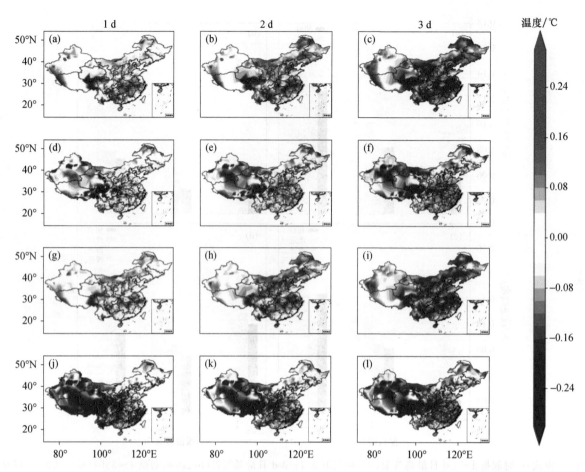

图5 模式08时起报1～3 d日最高气温(a、b、c)、08时起报1～3 d日最低气温(d、e、f)、20时起报1～3 d日最高气温(g、h、i)以及20时起报1～3 d日最低气温(j、k、l)改进方案三较卡尔曼滤波递减平均方法的RMSE差值
（台湾省资料暂缺）

的相关性明显低于前两者,与其他误差落后48 h后的相关性相当,并且各误差与模式实况误差之间都存在随预报时效增加相关性逐渐减小的趋势。

(2)将模式预报1～3 d最高/最低气温和模式最高/最低气温历史误差、初值场误差及卡尔曼滤波反演误差作为预报因子的改进方案三为4种预报方案中的最优方案。其改进效果在起报时刻、预报时效以及最高、最低气温要素上均表现得较为稳定,仅在20时起报日最高气温预报准确率检验中略逊于改进方案二。在空间分布上,订正效果在复杂地形区域更好,可能是由于ECMWF数值模式的空间分辨率不足、下垫面参数化方案适用性、地形数据不够精细等因素导致其在复杂地形区域温度预报质量稍差所引起的。

(3)应用改进方案三对2017年日最高、最低气温进行预报,结果显示,预报准确率明显优于ECMWF和中央气象台指导预报。与效果最好的全国城镇预报相比,日最高气温MAE较全国城镇预报低8.24%～13.97%,预报准确率提高1.24%～3.57%,日最低气温MAE较全国城镇预报低9.43%～17.69%,预报准确率提高1.77%～2.72%,并且在1 d时效内改进幅度最大,即预报时效越短订正效果越好。

参考文献

郭换换,段明铿,智协飞,等,2016a.基于TIGGE资料的预报跳跃性特征[J].应用气象学报,36(2):220-229.

郭换换,智协飞,段明铿,等,2016b. 数值天气预报中的不一致性问题综述[J]. 气象科学,36(1):134-140.
吉璐莹,智协飞,朱寿鹏,2017. 东亚地区冬季地面气温延伸期概率预报研究[J]. 大气科学学报,40(3):346-355.
李佰平,智协飞,2012. ECMWF模式地面气温预报的四种误差订正方法的比较研究[J]. 气象,38(8):897-902.
林春泽,智协飞,韩艳,等,2009. 基于TIGGE资料的地面气温多模式超级集合预报[J]. 应用气象学报,20(6):706-712.
穆穆,陈博宇,周菲凡,等,2011. 气象预报的方法与不确定性[J]. 气象,37(1):1-13.
翟宇梅,赵瑞星,高建春,等,2014. 遗忘因子自适应最小二乘法及其在气温预报中的应用[J]. 气象,40(7):881-885.
张庆奎,寿绍文,陆汉城,2008. 卡尔曼滤波方法在极端温度预报中的应用[J]. 科技信息(35):26-27.
张颖超,姚润进,熊雄,2017. 一种基于PSO-MEF的地面气温观测资料的质量控制方法[J]. 热带气象学报,33(2):273-280.
赵声蓉,2006. 多模式温度集成预报[J]. 应用气象学报,17(1):52-58.
智协飞,林春泽,白永清,等,2009. 北半球中纬度地区地面气温的超级集合预报[J]. 气象科学,29(5):569-574.
智协飞,陈雯,2010. THORPEX国际科学研究新进展[J]. 大气科学学报,33(4):504-511.
智协飞,季晓东,张璟,等,2013. 基于TIGGE资料的地面气温和降水的多模式集成预报[J]. 大气科学学报,36(3):257-266.
智协飞,彭婷,王玉虹,2018a. 基于BMA方法的地面气温的10～15 d延伸期概率预报研究[J]. 大气科学学报,41(5):627-636.
智协飞,胡耀兴,俞剑蔚,等,2018b. 基于TIGGE资料的东亚地区地面气温预报的不一致性研究[J]. 大气科学学报,41(3):298-307.
ARRIBAS A,ROBERTSON K B,MYLNE K R,2005. Test of a poor man's ensemble prediction system for short-range probability forecasting[J]. Mon Wea Rev,133:1825-1839.
BOWLER N E,ARRIBAS A,MYLNE K R,2008. The benefits of multi-analysis and poor man's ensembles[J]. Mon Wea Rev,136:4113-4129.
KRISHNAMURTI T N,KUMAR V,SIMON A,et al,2016. A review of multimodel superensemble forecasting for weather,seasonal climate,and hurricanes[J]. Reviews of Geophysics,54(2):366-377.

基于数值预报和随机森林算法的强对流天气分类预报技术[①]

李文娟[1]　赵　放[1,*]　郦敏杰[2]　陈　列[1]　彭霞云[1]

(1. 浙江省气象台,杭州,310017；2. 杭州市气象台,杭州,310057)

摘　要：随机森林(RF)算法是当前得到较为广泛应用的机器学习方法之一,有着很高的预测精度,训练结果稳定,泛化能力强,解决多分类问题有明显优势。本文将 RF 算法应用于强对流的潜势预测和分类,分短时强降水、雷暴大风、冰雹和无强对流四种类别,基于 2005—2016 年 NCEP 1°×1°再分析资料计算的对流指数和物理量,开展强对流天气的分类训练、0～12 h 预报和检验。2015—2016 年独立测试样本检验表明,针对强对流发生站点的点对点检验,整体误判率为 21.9%,85 次强对流过程基本无漏报,模型尤其适用于较大范围强对流天气。随机森林算法筛选的因子物理意义较为明确,和主观预报经验基本相符,模型准确率高,可用于日常业务。

关键词：强对流分类；对流指数；物理量；随机森林

Forecasting and Classification of Severe Convective Weather Based on Numerical Forecast and Random Forest Algorithm

LI Wenjuan[1]　　ZHAO Fang[1]　　LI Minjie[2]　　CHEN Lie[1]　　PENG Xiayun[1]

(1. Zhejiang Meteorological Observatory, Hangzhou, 310017;
2. Hangzhou Meteorological Observatory, Hangzhou, 310057)

Abstract：The random forest algorithm is currently one of the more widely used machine learning methods, featuring high prediction accuracy, stable training results and generalization ability, and has obvious advantages in solving the problem of multi-classification. This paper applied the random forest algorithm to the prediction and classification of severe convective weather, which is divided into four categories: short-time heavy rainfall, thunderstorm gale, hail and no severe convection. Then, based on the data of convection index and physics calculated from the NCEP data of 2005-2016, the training, 0-12 h forecasting and testing of classified severe convection are carried out. The results show that the whole misjudgment rate is 21.9% that is calculated out of the independent data of 2015-2016. It has almost no omission in 85 examples of severe convective weather and the model is especially suitable for a larger range of

---

[①] 本文发表于《气象》2018 年第 12 期。
资助项目：国家气象中心关键技术项目[YBGJXM(2018)02-13],浙江省科技厅重点项目(2017C03035)。
第一作者：李文娟,高级工程师,主要从事强对流天气预报及其研究。E-mail：liwenjuan1998@163.com。
通讯作者：赵放,男,研究员,主要从事短时临近天气预报与雷达资料开发。E-mail：e-zhaofang@163.com。

severe convective weather. The physical meaning of the factors used in random forest algorithm is relatively clear, and basically consistent with the subjective forecasting experience. It can be used in daily forecasting operation.

Key words: severe convection classification; convective index; physical quantity parameter; Random Forest(RF)

引 言

气象学中,对流指的是大气中由浮力产生的垂直运动所导致的热力输送,强对流天气通常指的是由深厚湿对流(DMC)产生的包括冰雹、大风、龙卷、强降水等各种灾害性天气,具有突发性、生命史短、局地性强、易致灾等特点。强对流天气预报,尤其是分类强对流天气一直是业务天气预报的难点之一,热动力物理参数敏感性分析及利用"配料法"、统计分析方法以及高分辨率数值模式进行强对流客观预报方法的研究逐渐成为预报强天气潜势的基础(郑永光 等,2015,2017;田付友 等,2015;漆梁波,2015;雷蕾 等,2011)。Doswell(2001)、俞小鼎等(2012)、孙继松等(2014)系统总结了 DMC 和不同类型强对流天气(冰雹、雷暴大风、短时强降水和龙卷)发生发展的环境条件、中尺度结构和特征。这些条件和结构特征是目前进行强对流天气分类预报的物理基础。近几年,国内一些学者基于数值模式计算的对流参数利用"配料法"和模糊逻辑法开展了分类强对流潜势预报的业务化试验。曾明剑等(2015)基于中尺度数值模式预报的对流参数,综合历史频率分布和权重分配,构建了分类强对流天气预报概率,并以优势概率作为分类判据,做出强对流分类预报。雷蕾等(2012)将统计的强对流天气判别指标应用到数值模式(快速更新同化系统),计算模式格点上的强对流发生概率,并针对冰雹、雷暴大风和短时暴雨天气下不同物理量的阈值范围,实现了对强对流天气的分类概率预报。机器学习等人工智能的方法多应用在强对流临近识别和概率预报中,Mecikalski 等(2015)使用 Logistic 回归和人工智能随机森森(Random Forest,RF)等方法发展了基于卫星资料和数值模式资料的对流初生(CI)临近概率预报技术。李国翠等(2014)和张秉祥等(2014)基于雷达三维组网数据利用模糊逻辑方法分别开发了雷暴大风和冰雹的自动识别算法;周康辉等(2017)将模糊逻辑算法用于雷暴大风的监测识别,实现了雷暴大风和非雷暴大风的有效区分;修媛媛等(2016)用机器学习中有监督学习模型支持向量机(SVM)来进行强对流天气的识别和预报。

RF 算法在近几年实际应用中得到了广泛关注,已经成为数据挖掘、模式识别等领域的研究热点,在生态学、水文学、经济学、医学等领域得到了广泛应用(张雷 等,2014;李欣海,2013;石玉立 等,2015;侯俊雄 等,2017;Mariana et al.,2016;Chen et al.,2017)。RF 是一种基于分类回归树的数据挖掘方法,是由 Breiman 和 Cutler 在 2001 年提出的一种较新的机器学习技术(方匡南 等,2011)。RF 算法通过聚集大量分类树来提高模型预测精度,与决策树一样,可用来解决分类和回归问题,预测精度很高,在异常值和噪声方面有很高的容忍度,且不易出现过度拟合现象(Breiman,2001)。国内外学者将 RF 算法与传统的神经网络、支持向量机(SVM)、Logistic 等机器学习方法做了对比,黄衍等(2012)证明 RF 泛化能力在多分类问题上优于 SVM;梁慧玲等(2016)在基于气象因子的塔河地区林火发生预测模型研究中得出,RF 模型的预测准确率高于传统 Logistic 模型 10% 左右;余胜男等(2016)的研究表明,RF 模型预测精度较高、稳定性好、泛化能力强,能有效预测年、月降水量,与 BP 神经网络模型和支持向量机模型相比,RF 模型效率更高、性能更优,尤其适用于大样本的逐月降水量预测;白琳等(2017)和 Zhang 等(2017)的研究均证明 RF 算法比传统的多元线性回归的结果更为理想,处理非线性和分级关系更具优势;Naghibi 等(2017)应用 RF、RFGA(Random Forest Genetic Algorithm)、SVM 三种模型评估地下水资料的潜势发现,RF 和 RFGA 比 SVM 更高效且更准确;Jan 等(2007)基于 RF 和 Logistic 模型建立了生态水文分布模型,对比得出 RF 的预测误差小于 Logistic 模型;Kampichler 等(2010)通过 5 种机器学习方法对比发现,RF 明显优于神经网络、支持向量机等方法。由此可见,大量的研究表明,RF算法在不同领域已取得较好的应用效果。

RF算法应运而生,给解决很多问题带来了新的方向,但将RF应用于强对流天气的分类预测相关研究为数不多。传统的"配料法"等通过挑选对不同类型强对流天气具有指示意义的物理量,根据历史个例的统计结果挑选预报因子,预测结果完全取决于天气学要素和物理量对强对流天气发生发展物理条件的代表性,而人工智能等机器学习算法可以建立在大数据集的应用基础上,通过智能化的筛选、组合多种因子进行预测分类,尤其在多分类预测方面有一定的优势,能够处理很高维度的数据,训练完后能够给出特征量的重要性排序,可以很好地预测多达几千个解释变量的作用。因此,本文将RF算法尝试性地应用于分类强对流的潜势预测,构建反映强对流天气发生发展环境条件的大数据集,通过训练学习达到预测分类的目的。

1 RF算法

1.1 RF算法原理

RF是由美国加利福尼亚州大学伯克利分校统计系教授Breiman(2001)提出来的一种统计学习理论。RF的基本组成单元是决策树,又称为分类回归树。基本思想是二分递归分割,在计算过程中充分利用二叉树,在一定的分割规则下将当前样本集分割为两个子样本集,使得生成的决策树的每个非叶节点都有两个分枝,这个过程又在子样本集上重复进行,直至不可再分成叶节点为止。由于单棵决策树模型往往精度不高,且容易出现过拟合问题,为此需要聚集多个模型来提高预测精度,RF中采用的是Bagging方法来组合决策树,其核心是重抽样自举。第一步,对样本量为N的原始样本集S进行有放回的随机抽样,得到一个容量为N的随机样本S_1(称自举样本);第二步,将自举样本视为训练样本,建立分类树T_1,重复上述两步M次,最终得到M个自举样本$S_1,S_2,\cdots,S_M$以及M个预测模型$T_1,T_2,\cdots,T_M$。然后组合M个决策树的预测模型,通过投票得出最终预测结果(图1)。RF的思路就是训练出在某一个方面有决策能力的决策树,这个决策树几乎不存在过度复杂和过分拟合数据的问题,相对而言它是一个弱决策树,但是多个方面的弱分类器集成能够形成一个强大的分类器。

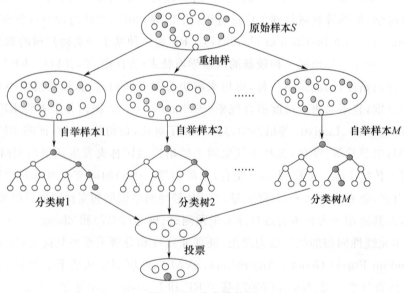

图1 RF分类结构图

1.2 泛化误差与重要性因子评价原理

RF 用 Bagging 方法生成训练集。样本容量为 N 的总训练集 S 中每个样本未被抽取的概率为 $(1-1/N)^N$，当 N 足够大时，$(1-1/N)^N \to 1/e = 0.368$，这表明原始样本集中约 37% 的样本不会出现在训练集中，这些数据称为袋外（Out-Of-Bag，OOB）数据。使用这些数据来估计模型的性能称为 OOB 估计。OOB 数据可以用来估计决策树的泛化误差，或用来计算单个特征的重要性。泛化误差是指分类器对训练集之外数据的误分率，泛化误差越小表示分类器性能越好，相反则表明分类器性能较差。每一棵树都可以得到一个 OOB 误差估计，将森林中所有树的 OOB 误差估计取平均，即可得到 RF 的泛化误差估计。Breiman 通过实验已经证明，OOB 误差是无偏估计，并且相对于交叉验证，OOB 估计是高效的，且其结果近似于交叉验证的结果（杨柳 等，2015）。

RF 测度输入变量重要性的基本思路是：对于解释变量重要性，一个直观的评价标准是，该变量越重要，其对预报结果的影响也越大。RF 算法的解释变量重要性评价采用类似标准：对所有检验样本，随机打乱某一解释变量取值，采用原 RF 算法对检验样本进行再次预报，袋外拟合误差增加愈多，该解释变量愈重要，表现为各类别的预测置信度变化明显，总体预测精度变化明显，袋外拟合误差增加量可用于定量评价解释变量重要性。因此，本文对输入变量重要性的评判指标采用预测精度的平均下降量测度输入变量对输出变量的重要性。

2 模型建立过程

本文将 RF 算法应用于强对流的环境场分类，基于 NCEP 1°×1° 08 时的分析场资料计算的若干对流指数和物理量指标作为输入变量，输出变量为短时强降水、雷暴大风和冰雹三种类别的强对流天气和无上述强对流天气。从理论和经验角度来看，不同的环境场有利于不同灾种的强对流发生，因此，采用多种物理量全面描述强对流发生的环境场，再应用机器学习算法对强对流天气进行预测及分类。预报模型的建立过程如图 2 所示，具体步骤如下。

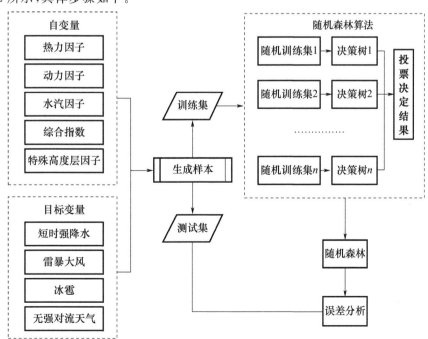

图 2 RF 预报模型的建立过程

2.1 选取预报因子

资料选用 2005—2016 年的强对流监测资料,基于 08 时 NCEP 1°×1°资料计算若干对流指数和 500、700、850、925 hPa 各层物理量场。这些要素涵盖强对流天气的构成要素,包括静力稳定度、水汽、能量及垂直风切变等。代表要素见表 1,表中物理量要素根据其计算公式和物理意义简单归为五类,其分类方法和条件参考刘建文等(2005)的研究。文后附录中列出了四种综合指数和条件-对流稳定度指数的计算公式和物理意义。因此,基于数值模式分析场资料计算的若干对流指数和物理量场共 68 类,组成预报因子数据集,构建强对流分类预报模型。

表 1 应用于 RF 算法的主要预报因子类型和要素

类型	要素
水汽因子	整层可降水量(P_w)
	比湿(q)
	水汽通量(Q_{flux})
	相对湿度(R_h)
	水汽通量散度(Q_{fdiv})
	温度露点差($T-T_d$)
	925 hPa 露点温度(T_{d925})
动力因子	散度(D_{iv})
	涡度(V_{or})
	垂直速度(ω)
	垂直风切变(S_{hr})
特殊高度层因子	0 ℃层高度(Z_{ht})
	−10 ℃层高度(M_{ht})
	−20 ℃层高度(F_{ht})
热力因子	假相当位温(θ_{se})
	K 指数(K_i)
	沙氏指数(S_i)
	最佳抬升指数(B_{li})
	最佳不稳定能量(B_{CAPE})
	不稳定能量(CAPE)
	总指数(TT)
	850 hPa 与 500 hPa 假相当位温差($\theta_{se500\sim850}$)
	850 hPa 温度(T_{850})
	条件-对流稳定度指数(I_{lc})
	下沉有效位能(D_{CAPE})
	温度差($T_{850\sim500\ hPa}$)
综合指数	强对流天气威胁指数(S_{weat})
	瑞士雷暴指数(S_{wiss00})
	修正深对流指数(M_{dci})
	风暴强度指数(S_{si})

2.2 选取目标变量

目标变量分为四类,分别是短时强降水、雷暴大风和冰雹等强对流天气以及无强对流天气。以

2005—2016年浙江省69个基准站点的监测实况为标准,实况选取时间段为00—20时,任一站点观测到冰雹记为一次过程,共监测到75次冰雹过程;为了区分强降水拖曳产生的局地性大风,雷暴大风样本选取影响范围相对较大的过程,至少3个站点出现8级或以上雷暴大风记为一次过程;短时强降水过程历史样本量较多,考虑到数据平衡性问题,仅选取全省11个地市代表站点的短时强降水过程,任一站点出现20 mm/h以上降水,记为一次过程。三种强对流天气往往是相伴产生的,在选取的强对流样本中,同时观测到短时强降水、雷暴大风或冰雹的有13次,观测到冰雹和雷暴大风相伴产生的有10次,因此,在分类过程中遵循一定的原则,根据灾害影响程度对强天气进行侧重分类,对于雷暴大风和短时强降水均出现的情况,一般记为雷暴大风过程,而雨强在50 mm/h及以上的极端降水则同时记为短时强降水和雷暴大风过程;对于冰雹和短时强降水均出现的情况,一般记为一次冰雹过程;对于雷暴大风和冰雹均出现的情况,则记为冰雹过程;对于无强对流天气样本,以2010—2015年全省69个基准站无雷暴日和10 mm/h以下的弱降水样本为主。因此,模型训练期为2005—2015年共1025个样本,模型验证期为2015—2016年共406个独立测试样本(表2)。

表2 模型训练和测试样本集

强对流分类	短时强降水	雷暴大风	冰雹	无强对流天气
训练样本集	255	181	73	516
测试样本集	163	82	2	159

2.3 预报模型构建过程

设RF包括M颗分类树,在第i颗决策树建立的过程中,首先通过随机方式选取k个输入变量构成候选变量子集X_i,依据变量子集X_i建立一棵充分生长的决策树,且无须剪枝。确定k的依据是:第一,决策树对袋外观测的预测精度,也称决策树的强度;第二,各决策树间的相互依赖程度,也称决策树的相关性。森林中包含的众多决策树形成一个组合预测模型,利用投票原则确定最后的预测结果。

本研究基于R语言的RF程序包进行强对流分类预报研究。RF算法包含2个参数,即M棵决策树和每棵树的输入变量k,M越大,RF算法过拟合效应越小;k越大,子预报模型间差异性越小。对于分类树,变量子集的大小k默认为$\sqrt{P}$,P为预报因子个数。M取值为500,k取值为8,以选取的68个预报因子作为解释变量(自变量),1025个分类强对流作为目标变量,构建RF模型,对解释变量进行重要性评价。

2.4 误差分析

利用袋外数据估计模型的泛化误差,为了更好地检验模型的预报性能,再利用检验期独立数据集进行验证,采取泛化误差和独立样本测试两种方式可以更全面地说明模型的预报效果。基于RF算法对全部观测做预测,计算混淆矩阵和整体的误判率。整体误判率=分类错误的样本数/总的样本数。

3 模型训练与预报结果分析

3.1 评判精度分析

3.1.1 独立样本测试

根据建立的RF模型,对2015—2016年检验期的406次独立数据进行预测。由于冰雹样本较少,选取影响浙江省较为严重的3次过程进行预测检验。结果见表3,检验期的独立测试样本均是点对点的验证,即针对站点监测到的短时强降水、雷暴大风实况进行预测,整体误判率为21.9%。由于2016年基准站点没有观测到冰雹,不能准确判断冰雹过程,因此预报出现冰雹必然会增加一定的错误率。3次冰雹过程中有1次判断为雷暴大风过程,实际情况既出现了冰雹又伴随大范围的雷暴大风,另外2次过程是2014年3月19日和2015年4月5日均发生的影响较严重的大冰雹天气,模型均准确判断出;无强对流天气过程判断准确率高。由于短时强降水和雷暴大风的实况很难明确客观地分类,导致两者的误判率相对较高;短时强降水和雷暴大风站点预测存在部分漏报的情况,但是从预报过程的检验来看,共有85次局地强降水过程和较大范围雷暴大风过程基本无漏报,预报落区偏差是导致站点漏报的主要原因。对于5个基准站点及以上出现强对流天气的较大范围过程,共有12次,仅1次雷暴大风过程误判为短时强降水,其余都预报正确;40次无强对流天气过程,4次为空报。总体来说,基于RF的预报分类模型效果比较理想,强对流过程基本能准确预报,尤其适用于较大范围的强对流天气。但是,由于强对流天气观测的原因,尽管我们采用了11年的观测数据,但还是存在样本不足的问题,使得RF模型存在一些缺陷,主要是存在训练不充分的情况,且模型的训练期样本在分类的过程中会出现混淆的情况,因此导致一定错误率的增加。

表3 2015—2016年模型检验期预测误差表

实况	预报				整体误判率/%
	无强对流天气	冰雹	短时强降水	雷暴大风	
无强对流天气	148		8	3	
冰雹		2		1	21.9
短时强降水	18	2	133	10	
雷暴大风	19	4	25	34	

图3列出了2016年的两次预测个例,可以简单说明模型的预报效果。2016年有2次过程均出现了较大范围的雷暴大风和短时强降水过程,据了解5月5日在浙南出现了局地小冰雹,6月1日在浙南出现了较大范围的雷暴大风。从预测效果来看,预报模型对出现的灾害性天气都有所反映,包括冰雹和雷暴大风的落区,不足的是,预报落区比实况范围大,落区也存在一定的偏差。

3.1.2 泛化误差

建立模型后需对训练模型与测试结果进行评估,其评估精度满足要求后模型才能被应用。以2005—2015年训练期的强对流样本基于RF预测模型构建的OOB误差见表4。由表4可见,RF对全部观测进行预测,预测误差很小,仅为0.39%,而单棵树的预测误差约为20%,说明由RF构建的预报分类模型效果比较理想。

表4 2005—2015年模型训练期OOB预测误差表

实况	预报				整体误判率/%
	无强对流天气	冰雹	短时强降水	雷暴大风	
无强对流天气	516				
冰雹		73			0.39
短时强降水			255		
雷暴大风		1	2	178	

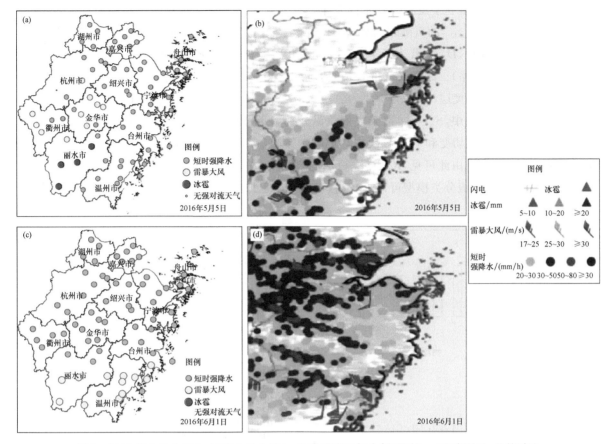

图 3　2016 年 5 月 5 日(a,b)和 6 月 1 日(c,d)强对流天气个例预测(a,c)和实况(b,d)的对比

3.2　强对流分类要素的重要性分析

RF 在计算过程中能根据预测精度的平均下降量计算各指标重要度。图 4 是 RF 算法对影响强对流分类要素的重要性排序，值越大表示越重要。从中可见，沙氏指数（S_i）在分类过程中的重要性高于其他要素，表明其对强对流天气分类的贡献程度最大。预测精度平均下降量筛选的前几位要素分别是 850 hPa 和 500 hPa 的温度差（$T_{850\sim500\,hPa}$）、低层相对湿度（R_{h850} 和 R_{h925}）、整层可降水量（P_w）、总指数（TT）、0～6 km 风垂直切变（$S_{hr\,0\sim6\,km}$）、最佳抬升指数（B_{li}）、强对流天气威胁指数（S_{weat}）、低层假相当位温（θ_{se850}）及风暴强度指数（S_{si}）。

从输入变量对分类强对流的重要性要素排序（图 5）来区分有无强对流天气的要素，能量、水汽条件是发生强对流天气的必要条件。其中，稳定度要素贡献较为显著，S_i 和 B_{li} 等稳定度要素表现较好，这两个要素是强对流天气主观预报的优选要素。其次，综合指数有较好的表现，如 S_{weat}、S_{si}、M_{dci} 可以综合反映中低层热力稳定度特性及适宜风暴发生动力环境对风暴发生所产生的共同作用。从短时强降水的重要性要素排序可见，短时强降水更倾向于表征水汽条件的要素，如 P_w、低层相对湿度、比湿、各层 θ_{se} 表征整层高温高湿的

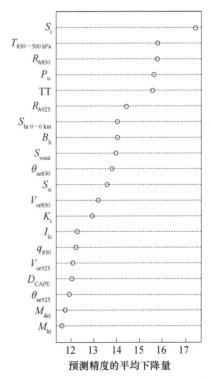

图 4　RF 对强对流天气分类前 20 项要素的重要性排序

环境场,即深厚的湿对流有利于短时强降水的发生。雷暴大风的环境场特征除了层结不稳定,代表环境温度直减率的要素 $T_{850\sim500\ hPa}$ 的贡献较为显著,环境大气有较大的温度递减率,既有利于强上升气流,也有利于强下沉气流。此外,D_{CAPE} 代表干下沉气流的作用以及低层相对湿度条件(R_{h850} 和 R_{h925}),对雷暴大风的贡献也较显著(图略)。冰雹天气的环境场,贡献最为显著的是不同高度层的风垂直切变,尤其是 $S_{hr\,0\sim6\ km}$,其次,$-20\ ℃$高度层(F_{ht})、$-10\ ℃$高度层(M_{ht})、$0\ ℃$高度层(Z_{ht})等特性层的高度对冰雹的形成起重要作用。综合指数中,S_{si} 对冰雹天气的贡献较为显著。S_{si} 的计算方法和垂直风切密切相关,由 $0\sim3600\ m$ 的环境风垂直切变和 CAPE 决定(刘建文 等,2005)。此外,温度直减率 $T_{850\sim500\ hPa}$ 较大同样有利于冰雹天气的发生。由此可见,RF 算法筛选的要素的物理意义较为明确,和主观预报经验基本相符,因此,RF 建立的强对流分类模型可信度较高,可以应用于日常业务。

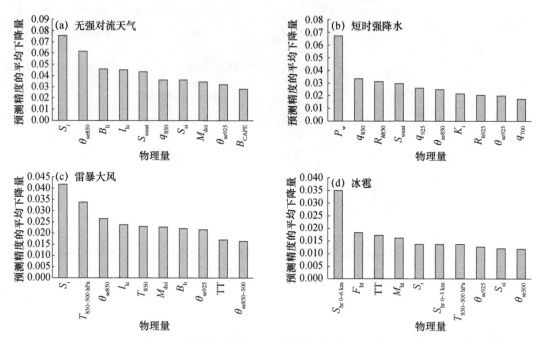

图 5 RF 对各类别前 10 项要素的重要性测度排序

3.3 预报因子特征分析

针对 RF 算法筛选的重要物理量绘制核密度估计图(图 6)。为了更加精确刻画变量的分布特点,可在变量的频率分布图上添加核密度估计曲线,将频率转化为概率密度,可直观对比不同组数值的分布形状以及不同组之间的重叠程度(李文娟 等,2017)。从不同灾害性天气的对流指数分布可以直观地看出,有无强对流天气表现在环境场的物理量指标存在较明显的差别,尤其是稳定度指标 S_i、K_i、B_{li} 和 S_{weat},黑色曲线(无强对流天气)和其他三条曲线分离度较高,峰值处于不同的阈值区间。例如,B_{li} 发生强对流天气的峰值在 $-5\ ℃$ 左右,无强对流天气时一般在 $0\ ℃$ 以上;S_{weat} 发生强对流天气的峰值一般集中在 $250\sim300$,而这区间对应无强对流天气的低概率密度区。S_{si} 在 $50\sim70$ 易发生强对流天气,冰雹天气在 $60\sim70$ 具有高概率密度,而短时强降水集中在 50 左右,雷暴大风的分布较不集中,说明 S_{si} 对冰雹天气的指示性较好。$T_{850\sim500\ hPa}$ 可以较好地区分短时强降水和风雹类强对流天气,$25\sim27\ ℃$ 是风雹类强对流天气的集中区,而短时强降水分布在 $22\sim24\ ℃$。和水汽条件密切相关的要素如 P_w、Rh_{925}、K_i,可以较好地指示短时强降水的发生条件,如短时强降水 P_w 峰值在 $60\ mm$,而风雹类强对流在 $50\ mm$,此外,雷暴大风的 R_{h925} 峰值在 $60\%\sim80\%$,明显低于短时强降水 90% 的相对湿度;D_{CAPE} 可以较好地表征雷暴大风类强对流天气,和短时强降水的曲线存在一定的分离度。同时发现,冰雹的环境指标有明显的双峰特征,如

P_w、R_{h925}、D_{CAPE}。分析环境指标不同季节的演变特征可以解释双峰特征,春季是浙江省冰雹天气的高发季,指标和夏季相比有明显的季节性特征,这里不做详细讨论。因此,RF算法可以自动筛选物理量的重要性,再结合核密度估计分布直观地反映物理量在分类过程中的作用及阈值区间,为主观预报提供参考。

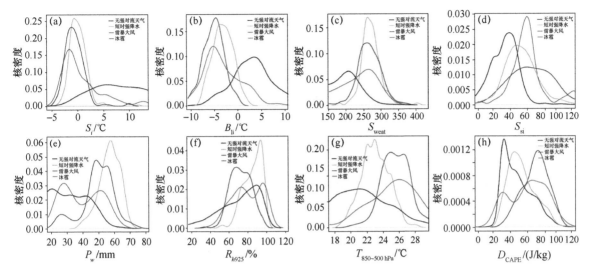

图6 物理量核密度估计分布

(a)沙氏指数(S_i);(b)最大抬升指数(B_{li});(c)强对流天气威胁指数(S_{weat});
(d)风暴强度指数(S_{si});(e)整层可降水量(P_w);(f)925 hPa相对湿度(R_{h925});
(g)850~500 hPa温度差($T_{850\sim500\ hPa}$);(h)下沉有效位能(D_{CAPE})

4 结论与讨论

随着大数据时代的到来,计算机辅助预测的方法日益丰富。一般来说,机器学习算法的性能会随着数据量的增多而提高,但是随着数据量的增大模型也容易出现过拟合的现象,从而影响模型性能。比如支持向量机、人工神经网络等机器学习模型都有着类似的特点,而RF算法具备训练结果稳定、泛化能力强的特点。因此,本文将RF算法运用于浙江省强对流天气的潜势分类,针对2005—2015年NCEP再分析资料计算的对流指数和物理量进行分类训练,建立模型预测强对流的潜势和类别。

误差分析结果表明,RF算法建立的模型准确率高,基于袋外观测的泛化误差仅为0.39%,基于2015—2016年独立测试样本的整体误判率为21.9%,85次强对流过程基本无漏报,模型尤其适用于较大范围的强对流天气,但是预报落区和范围仍存在一定的偏差。强对流分类要素的重要性分析表明,S_i、$T_{850\sim500\ hPa}$、R_{h850}、R_{h925}、P_w、TT、$S_{hr\ 0\sim6\ km}$、B_{li}、S_{weat}、θ_{se850} 及 S_{si} 等对RF强对流分类模型的贡献较为显著;$S_{hr\ 0\sim6\ km}$、$T_{850\sim500\ hPa}$ 及-20℃层高度对冰雹等强对流天气贡献显著。根据核密度估计分析,强对流天气 S_{weat} 集中在250~300;S_{si} 对冰雹的指示性较强,在60~70具有高概率密度。$T_{850\sim500\ hPa}$ 可以较好地区分短时强降水和风雹类强对流,25~27 ℃是风雹类强对流的集中区,而短时强降水分布在22~24 ℃。短时强降水易发生在整层高温高湿的环境场,短时强降水 P_w 的高概率密度区在60 mm左右;此外,雷暴大风的 R_{h925} 峰值在60%~80%,明显低于短时强降水90%的相对湿度;D_{CAPE} 也可以较好地表征雷暴大风类强对流天气。

该模型也存在不足之处,受历史强对流天气样本的数量限制,训练不够充分,在业务应用的过程中,需要不断动态训练模型,加入新的训练样本以及综合更多要素才能更好地发挥强对流分类模型的作用。

参考文献

白琳,徐永明,何苗,等,2017. 基于 RF 算法的近地表气温遥感反演研究[J]. 地球信息科学学报,19(3):390-397.
方匡南,吴见彬,朱建平,等,2011. RF 方法研究综述[J]. 统计与信息论坛,26(3):32-38.
侯俊雄,李琦,朱亚杰,等,2017. 基于 RF 的 $PM_{2.5}$ 实时预报系统[J]. 测绘科学,42(1):1-6.
黄衍,查伟雄,2012. RF 与支持向量机分类性能比较[J]. 软件,33(6):107-110.
雷蕾,孙继松,魏东,2011. 利用探空资料判别北京地区夏季强对流的天气类别[J]. 气象,37(2):136-141.
雷蕾,孙继松,王国荣,等,2012. 基于中尺度数值模式快速循环系统的强对流天气分类概率预报试验[J]. 气象学报,70(4):752-765.
李国翠,刘黎平,连志鸾,等,2014. 利用雷达回波三维拼图资料识别雷暴大风统计研究[J]. 气象学报,72(1):168-181.
李文娟,赵放,赵璐,等,2017. 基于杭州探空站资料的不同强度短时强降水预报指标研究[J]. 暴雨灾害,36(2):1-6.
李欣海,2013. RF 模型在分类与回归分析中的应用[J]. 应用昆虫学报,50(4):1190-1197.
梁慧玲,林玉蕊,杨光,2016. 基于气象因子的 RF 算法在塔河地区林火预测中的应用[J]. 林业科学,52(1):89-98.
刘建文,郭虎,李耀东,等,2005. 天气分析预报物理量计算基础[M]. 北京:气象出版社.
漆梁波,2015. 高分辨率数值模式在强对流天气预警中的业务应用进展[J]. 气象,41(6):661-673.
石玉立,宋蕾,2015. 1998—2012 年青藏高原 TRMM 3B43 降水数据的校准[J]. 干旱区地理,38(4):900-910.
孙继松,戴建华,何立富,等,2014. 强对流天气预报的基本原理和技术方法[M]. 北京:气象出版社.
田付友,郑永光,张涛,等,2015. 短时强降水诊断物理量敏感性的点对面检验[J]. 应用气象学报,26(4):385-396.
修媛媛,韩雷,冯海磊,2016. 基于机器学习方法的强对流天气识别研究[J]. 电子设计工程,24(9):4-11.
杨柳,王钰,2015. 泛化误差的各种交叉验证估计方法综述[J]. 计算机应用研究,32(5):1287-1290,1297.
余胜男,陈元芳,顾圣华,等,2016. RF 在降水量长期预报中的应用[J]. 南水北调与水利科技,14(1):78-83.
俞小鼎,周小刚,王秀明,2012. 雷暴与强对流临近天气预报技术进展[J]. 气象学报,70(3):311-337.
曾明剑,王桂臣,吴海英,等,2015. 基于中尺度数值模式的分类强对流天气预报方法研究[J]. 气象学报,73(5):868-882.
张秉祥,李国翠,刘黎平,等,2014. 基于模糊逻辑的冰雹天气雷达识别算法[J]. 应用气象学报,25(4):414-426.
张雷,王琳琳,张旭东,等,2014. RF 算法基本思想及其在生态学中的应用——以云南松分布模拟为例[J]. 生态学报,34(3):650-659.
郑永光,周康辉,盛杰,等,2015. 强对流天气监测预报预警技术进展[J]. 应用气象学报,26(6):641-657.
郑永光,陶祖钰,俞小鼎,2017. 强对流天气预报的一些基本问题[J]. 气象,43(6):641-652.
周康辉,郑永光,王婷波,等,2017. 基于模糊逻辑的雷暴大风和非雷暴大风区分方法[J]. 气象,43(7):781-791.
BREIMAN L,2001. Random forests[J]. Machine Learning,45(1):5-32.
CHEN T,TRINDER J C,NIU R Q,2017. Object-Oriented landslide mapping using ZY-3 satellite imagery,random forest and mathematical morphology,for the Three-Gorges Reservoir,China[J]. Remote Sensing,9(4):333.
DOSWELL III C A,2001. Severe convective storms[M]. Meteor Monogr. Boston:Amer Meteor Soc.
JAN P,BERNARD D B,NIKO E C V,et al,2007. Random forests a tool for ecohydrological distribution modelling[J]. Ecological Modelling,207(2/4):304-318.
KAMPICHLER,C,WIELAND,R,CALME,S,et al,2010. Classification in conservation biology:a comparison of five machine-learning methods[J]. Ecological Informatics,5:441-450.
MARIANA B,LUCIAN Dr GU,2016. Random forest in remote sensing:a review of applications and future directions[J]. ISPRS Journal of Photogrammetry and Remote Sensing,114:24-31.
MECIKALSKI J R,WILLIAMS J K,JEWETT C P,et al,2015. Probabilistic 0-1 hour convective initiation nowcasts that combine geostationary satellite observations and numerical weather prediction model data[J]. J Appl Meteor Climatol,54:1039-1059.
NAGHIBI S A,KOUROSH A,ALIREZA D,2017. Application of support vector machine,random forest,and genetic algorithm optimized random forest models in groundwater potential mapping[J]. Water Resources Management,31(9):

2761-2775.

ZHANG H, WU P B, YIN A J, et al, 2017. Prediction of soil organic carbon in an intensively managed reclamation zone of eastern China: a comparison of multiple linear regressions and the random forest model[J]. Science of The Total Environment, 592: 704-713.

附 录

对流指数	计算公式	物理意义
条件-对流稳定度指数 (I_{lc})	$I_{lc}=(\theta^*_{se500}-\theta_{se0})+(\theta_{se500}-\theta_{se0})$ θ_{se0} 表示地面假相当位温;θ^*_{se} 表示饱和假相当位温	条件性稳定度是考虑一小块空气上升得到的,而对流性稳定度是考虑厚度相当大的某一层空气抬升得到的,常把 I_l 与 I_c 相加称为条件-对流稳定度指数
修正的深对流指数 (M_{dci})	$M_{dci}=(T_{850}+T_{d850})/2+(T_s+T_{ds})/2-li$ T_s 与 T_{ds} 分别表示地面温度和地面露点温度;li 表示抬升指数	综合反映低层(地面—850 hPa)温湿特性及中低层条件稳定度的参数
强对流天气威胁指数 (S_{weat})	$S_{weat}=12\,T_{d850}+(TT-49)+2\,f_{850}+f_{500}+125(S+0.2)$ $S=\sin(\alpha_{500}-\alpha_{850})$ α_{500} 与 α_{850} 分别代表 500 hPa 风向和 850 hPa 风向	综合反映中低层热力稳定度特性及适宜风暴发生动力环境对风暴发生所产生的共同作用
瑞士雷暴指数(S_{wiss00})	$S_{wiss00}=SI_{850}+0.4S_{hr3\sim6}+0.1(T-T_d)_{600}$ $S_{hr3\sim6}$ 为 3~6 km 垂直风切变	当 $S_{wiss00}<5.1$ 时预报有雷暴,否则无雷暴
风暴强度指数(S_{si})	$S_{si}=100[2+(0.276\ln(S_{hr}))+(2.011\times10^{-4}CAPE)]$ S_{hr} 代表 0~3600 m 的环境风垂直切变;CAPE 代表对流有效位能	一个由 CAPE 和 S_{hr} 组合的函数,可将强雷暴与非强雷暴分开

基于评分最优化的模式降水预报订正算法对比

吴启树[1]　韩美[1]　刘铭[1]　陈法敬[2]

(1. 福建省气象台,福州,350001;2. 国家气象中心,北京,100081)

摘　要:使用2013年1月1日—2016年1月7日全国气象站观测资料,应用准对称混合滑动训练期,不改变雨带预报位置和形态,基于模式降水预报订正结果的TS评分最优化及ETS评分最优化,分别设计最优TS评分订正法(OTS)和最优ETS评分订正法(OETS)来确定预报日各级降水订正系数,对2014—2015年数值模式降水预报进行分级订正,并与频率匹配法(FM)对比。结果表明:在24 h累积降水的多个预报时效订正中,无论是对欧洲中期天气预报中心、日本气象厅、美国国家环境预报中心和中国气象局的全球模式降水预报,还是对四个模式降水预报的简单多模式平均,OTS和OETS较FM在TS评分和ETS评分等传统降水检验指标上均更优秀,其中OTS在所有时效均能提高模式降水预报质量,为三者最优。在概率空间的稳定公平误差评分方面,OTS在各时效、各单模式及多模式平均等方面优势明显。在预报员对应参考时效上,OTS在24~168 h降水预报的24 h累积TS评分也优于主观预报。

关键词:最优TS评分法;最优ETS评分法;频率匹配法;降水分级订正;训练期

A Comparison of Optimal-score-based Correction Algorithms of Model Precipitation Prediction

WU Qishu[1]　HAN Mei[1]　LIU Ming[1]　CHEN Fajing[2]

(1. Meteorological Observatory of Fujian province, Fuzhou, 350001;
2. National Meteorological Centre, Beijing, 100081)

Abstract:Based on date from national meteorological stations,one year quasi-symmetrical mixed running training period(QSRTP),and precipitation prediction from CMA(T639),ECMWF,NCEP,JMA,both optimal threat score(OTS) method and optimal equitable threat score(OETS) method are designed to conduct a comparison experiment on correction algorithms for model precipitation with frequency matching (FM) method. Through classification correction, three methods are used merely to calibrate model precipitation amount with the predicted rain-belt location and shape kept unchanged. The OTS method figures out correction coefficients of different precipitation classes by optimizing threat score (TS) of corrected precipitation within training period. OETS is similar to OTS but achieved by optimizing ETS. Correction experiments are conducted twice a day with forecast time at 0000 UTC and 1200 UTC, respectively. To consider seasonal background, 20 days before the forecast day and 20 days after the same

---

[1]　本文发表于《应用气象学报》2017年第3期。
资助项目:气象预报业务关键技术发展专项(YBGJXM201703-06)福建省自然科学基金社会发展引导性(重点)项目(2017Y-008)。
第一作者:吴启树。E-mail:172475076@qq.com。

day in the previous year are adopted to constitute training period. For each national meteorological station, there are 80 samples in total. The correction experiment shows that for either precipitation products of ECMWF,JMA,NCEP,CMA, or their ensemble mean, both OTS and OETS show much better performance than FM in 24 h accumulated precipitation classification calibration with different lead time according to traditional verification methods like TS and ETS. In particular, OTS is the best and can improve precipitation prediction in all lead times. After correction, both OTS and OETS incline to forecast larger precipitation area than observation for most classes but less precipitation amounts. Compared to FM, both methods tend to produce a little higher false alarm rates in middle and low classes, which is much less than the reduced missing rate, thereby leading to a higher threat score. In terms of ECMWF correction, OTS and OETS have a relatively stable BIAS score of 1.1, although there are much fewer samples in high class. By contrast, FM produces an unstable BIAS score, especially in maximum class with score over 2.2, indicating an excessively high missing rate. As for stable equitable error in probability space (SEEPS), OTS has superiorities over all lead times, all single models and multi-model mean. Furthermore, TS of corrected ECMWF precipitation using OTS method in 2015 are also better than subjective forecast from all aspects, with national averaged threat score of 1 d rainstorm forecast reaching 0.194.

Key words: optimal threat score method; optimal equitable threat score method; frequency matching method; Precipitation classification calibration; training period

引 言

数值预报模式输出的气象要素预报存在不同程度的系统性误差,通过统计分析及应用适当算法进行订正,可以减少误差,提高预报准确率[1-11]。不同模式在降水预报的系统性误差方面表现各异,如欧洲中期天气预报中心(ECMWF)、日本气象厅(JMA)等机构的全球模式具有弱降水预报偏空而强降水预报偏漏的情况[12,13],区域模式如中国气象局武汉暴雨研究的中尺度模式(AREM)[14]总体较实况偏强,特别是中低量级的降水,这些反映了模式对不同降水强度预报的误差非一致性和降水量在概率上的偏态分布。未分级的"消除偏差集合平均法"[15,16]预报质量能优于单模式确定性预报,但该方法对各量级的降水订正值相同,具有小量级降水订正幅度过大而大量级降水订正过小的缺点。孙靖等[17]在训练期和分级降水订正方面做了改进,使训练期样本更加接近预报日前后的季节背景,并对不同量级降水做相应偏差订正,但对中雨以上量级的模式降水预报阈值设定较为固定,还有优化的空间。周迪等[18]应用"观测概率匹配订正法",对四川盆地降水量为50 mm的集合预报平均值进行订正,改善了T213模式的暴雨预报。陈博宇等[12]根据概率匹配和融合产品各自特点,设计出概率匹配-融合法和融合-概率匹配法,对大量级的降水预报较融合产品有一定提高,但因融合产品百分位过于固定,大量级降水预报产生明显空报。李俊等[14,19]利用实况降水频率结合概率匹配平均法,能订正系统偏差和集合平均光滑的副作用。因此,无论单模式还是多模式集成,要提高模式降水订正后预报质量,确定训练期实况与预报两者之间的关系至关重要。

频率匹配法(FM)在降水分级订正中以降水预报偏差[19]达到最优化为算法核心。当训练期样本数量足够大且比较接近于预报日前后的季节背景时,该算法利用观测降水频率作为参考频率来订正模式预报效果较好,理论上能使预报与实况平均误差的数学期望值为零,且空报率与漏报率大致相同。但该方法应用于业务时,因降水存在明显的季节变化,且当实况或预报在大量级降水上的样本数偏少时,订正系数容易出现异常。即使订正后预报降水与实况面积相当,因降水落区与实况总是存在一定程度的差异,预报效果未必能达到最优。对于某一量级降水,在预报偏差为1的情况下,若适当把降水预报值略提高(降低),该量级增加(减少)的预报站点带来 TS 评分[20]增幅(降幅)大于(小于)原有评分时,相对于频率匹配法,预报质量就可以得到改进,而订正是否达到最佳可以通过计算该量级的 TS 评分来确定。

因此,本文基于训练期模式降水量预报调整后的 TS 评分最优化及 ETS 评分最优化进行建模,以确定预报日降水分级订正系数。分别设计最优 TS 评分订正法和最优 ETS 评分订正法,并与频率匹配法进行对比,优选订正质量既能在 TS 评分项目上达到最优,同时又能在 ETS 评分等降水预报检验项目上也较优的算法。由于三个订正算法本质上是分别基于训练期模式降水订正后的 TS 评分、ETS 评分和 BAIS 评分[19]的最优化设计,因此,本文还采用概率空间的稳定公平误差(stable equitable error in probability space,SEEPS)评分[21-24]作为三个订正算法的第三方检验方法。SEEPS 方法于 2010 年由 Haiden 等人研究与应用[22],其把站点按降水气候概率为基础划分成"干""小雨""大雨",克服因地域降水概率差异带来检验的不公平性,并获得一个可以代表降水预报整体性能的单一评分,2011 年被世界气象组织的确定性数值预报检验领导中心(LC-DNV)应用于业务。目前该方法也应用于中国气象局数值预报中心的模式检验,具有很高的公平性。

1 资料

1.1 实况资料

为确保计算效率及实况资料的可靠性,本文降水实况数据为 2013 年 1 月 1 日—2016 年 1 月 7 日全国 2014 个气象站资料。本文个别降水检验方法需要应用各站点气候概率,则使用 1984—2013 年共 30 年的历史资料。

1.2 数值预报产品

产品包括中国气象局下发并在气象系统日常预报业务使用率较高的欧洲中期天气预报中心高分辨率模式(ECWMF)、日本气象厅高分辨率模式(JMA)、美国国家环境预报中心的全球模式(NCEP)和中国气象局数值预报中心 T639 模式(T639)的确定性预报降水产品。2013—2015 年每日两次,起报时间分别为 00:00 和 12:00(世界时,下同)。ECMWF 水平分辨率为 $0.25°×0.25°$;JMA 和 NCEP 水平分辨率为 $0.5°×0.5°$;T639 水平分辨率为 $1.125°×1.125°$。ECMWF 预报时效最长,为 0~240 h,其他 3 个模式为 84~216 h 不等。

模式的格点资料采用 Cressman 客观插值法进行站点插值[25]。

2 方法

2.1 检验方法

本文设 10 个降水量级(k)应用于分级订正,阈值为 O_k,10 个降水量级分别为 0.1、1、5、10、25、35、50、75、100、150 mm。

结合天气预报业务的站点检验方法,本文采用 TS 评分、ETS 评分(E)[17]、空报率、漏报率、预报偏差(B)[19,20]、SEEPS 技巧评分[21-24]和 HSS 技巧评分(H)[24]等检验方法。对于第 k 量级的累积降水,部分检验方法计算如下：

$$E = \frac{N_A - R_a}{N_A + N_B + N_C - R_a} \quad (1)$$

$$B = \frac{N_A + N_B}{N_A + N_C} \quad (2)$$

$$H = \frac{N_A + N_d - R_a - R_d}{N_A + N_B + N_C + N_D - R_a - R_d} \quad (3)$$

$$R_a = \frac{(N_A + N_B) \cdot (N_A + N_C)}{N_A + N_B + N_C + N_D} \quad (4)$$

$$R_d = \frac{(N_B + N_D) \cdot (N_C + N_D)}{N_A + N_B + N_C + N_D} \quad (5)$$

式中，N_A 为预报正确站（次）数；N_B 为空报站（次）数；N_C 为漏报站（次）数；N_D 为预报和实况均未达到阈值的正确站（次）数；R_a 和 R_d 分别为空报站数与漏报站数相当时的随机预报 N_A 和 N_D 的数学期望。

TS评分、ETS评分、空报率、漏报率、预报偏差为目前中国气象系统降水预报业务检验考核项目。其中，ETS评分是对TS评分的改进，能对空报或漏报进行惩罚，使评分相对后者更加公平[24]。SEEPS技巧评分计算方法相对复杂，计算方法参考文献[21-23]，并应用各站点30年降水气候概率。该技巧评分对预报误差的"惩罚"与降水气候概率相关，可自动适应不同气候区域或降水季节的站点降水概率，使得多个站点的评分融合更加合理。HSS技巧评分也能对空报或漏报进行"惩罚"，且给随机预报和常量预报的期望评分为0，其数学属性为线性的且具有渐进公平性，也是公平的降水检验之一[24]。

2.2 算法设计原理及实现

本文对比3种降水分级订正算法，其中前两个算法为本文设计。对于某一模式、某一时效，模式降水订正算法分别对应算法1、算法2、算法3。

算法1：最优TS评分订正法（Optimal Threat Score Method，OTS）

$$y = \begin{cases} 0 & x < F_1 \\ O_k + (O_{k+1} - O_k) \times \dfrac{x - F_k}{F_{k+1} - F_k} & F_k \leqslant x < F_{k+1} \quad (k \text{ 为 } 1 \sim 9) \\ x \times \left(\dfrac{O_{10}}{F_{10}}\right) & x \geqslant F_{10} \end{cases} \quad (6)$$

式中，x 和 y 分别为模式降水的预报值和订正值；O_k 为第 k 量级降水阈值（k 为 $1,2,\cdots,10$）；F_k 为预报降水量订正到 O_k 时该量级对应模式降水阈值。$x < F_1$ 时，y 为0，称为消空订正；y 与 x 比值为订正系数[19]，F_k 相对 O_k 越小（大）订正系数就越大（小）。F_k 从训练期调整订正系数使第 k 量级累积降水TS评分达到最高时求得。

算法2：最优ETS评分订正法（Optimal Equitable Threat Score Method，OETS）

该订正算法与OTS相似，但模式订正阈值 F_k 由第 k 量级的累积降水ETS评分达到最高时的订正系数来确定。

算法3：频率匹配法（Frequency Matching Method，FM）[14,19]

其中参考频率为训练期实况降水，预报频率为训练期模式预报降水。具体步骤与文献[12]相似。

3种算法均只改变预报降水强弱，不改变预报的雨带位置和形态。假设经过FM订正后，某一量级（如暴雨）预报站点数与实况降水站点数相同，但只部分重合（图1a），那么OTS对降水TS评分提高的可能主要有两种情况。一种是订正时适当将预报降水量调大（图1b），这有可能造成空报率和漏报率的减小，从而提高TS评分等检验结果；另一种是适当调小（图1c），可能带来漏报率增大及空报率的减小，当空报率的减小幅度大于前者时在一定程度上也能提高 TS 评分。

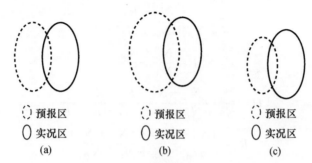

图 1　TS 评分或 ETS 评分提高的不同情况示意图
(a)预报站点数与实况相同；(b)预报站点数多于与实况；(c)预报站点数少于实况

实际计算中，以暴雨量级为例，F_k 求解示意见图 2。在调整预报降水订正阈值时，若 43.0 mm 为的 TS 评分达 0.272，为最高，则 43.0 mm 就为暴雨量级的 F 值。OETS 对预报质量的提高原理与 OTS 相似，3 种订正算法的优劣由预报检验进行评估。

为了控制因大量级降水样本偏少而引起订正系数异常偏大或偏小的情况，并考虑多数全球模式对大量级降水预报较实况偏弱的特点，对于任一算法，设 35 mm 及以上量级降水订正系数的上(下)限分别为 1.6(0.8)。

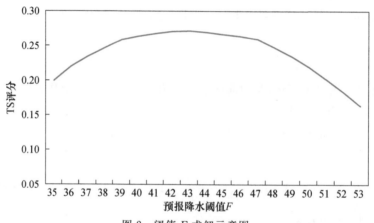

图 2　阈值 F 求解示意图

以最优 TS 评分订正法对 ECMWF 降水预报的订正为例，简写为"ECMWF_OTS"，另两个算法类似，分别简写为"ECMWF_OETS"和"ECMWF_FM"。

为了能更好地自适应预报日前后相似的季节背景，建模时应用 1 年期的准对称滑动训练期[25]，即滑动各取当年预报日之前和上一年预报日之后 20 d，每日两次共 80 次全国所有国家气象站样本，每次建模样本数量为 193120 站(次)，滑动计算出各量级降水订正阈值 F_k，统一应用于预报日所有站点降水的订正。

3　算法对比

对全国气象站 2014—2015 年每日两次预报进行检验。准对称混合滑动训练期需要应用 2013 年资料。

3.1　对 ECMWF 降水预报订正

3.1.1　24 h 预报时效检验

24 h 累积降水的 24 h 预报 TS 评分见图 3a，3 种算法对 ECMWF 降水预报有明显的订正能力，尤其对弱降水和强降水提高更为显著。3 种算法中,OTS 仅在 150 mm 量级上低于 OETS,其他量级均为最优。OETS 除了在 0.1 mm 和 5 mm 两个量级上略低于 FM,其他量级均优于后者。因此,在 TS 评分上,OTS 最优、OETS 次之。

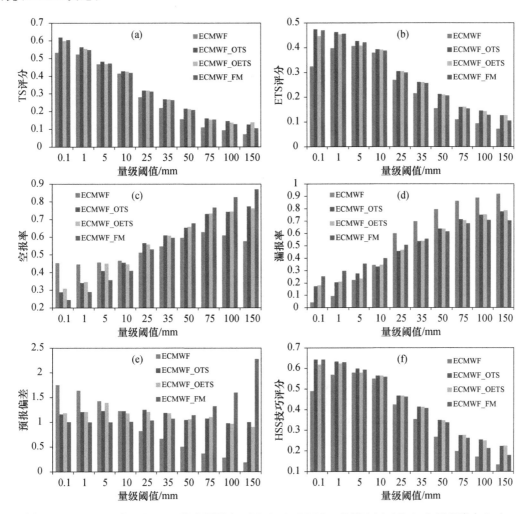

图 3　2014—2015 年 ECMWF 降水预报和 FM、OTS、OETS 3 种算法订正的 24 h 累积降水 24 h 预报的 TS 评分(a)、ETS 评分(b)、空报率(c)、漏报率(d)、预报偏差 BIAS(e)和 HSS 技巧评分(f)

ETS 评分中,因 ECMWF 降水预报在中量级(5 mm 和 10 mm)降水预报较好(图 3b),故 3 种算法在该量级评分提升幅度较小,但在其他所有量级均较模式有明显提升,特别是在弱降水和强降水方面提高更明显。3 种算法中,OTS 仅在 150 mm 量级上略低于 OETS,其他量级均为最优。OETS 在 0.1、1 和 5 mm 3 个量级上低于 FM,其他 7 个量级优于后者。因此,在 ETS 评分上,OTS 最优,OETS 次之。

ECMWF 降水预报存在低量级偏空报、大量级偏漏的情况(图 3 c～e),因此,3 种算法通过对最低量级的消空以及对其他低量级应用较低的订正系数减少空报和降低湿偏差[14],同时在大量级上应用较高的订正系数减少漏报。在样本数较多的中低量级(35 mm 及以下),FM 的空报率与漏报率相当(该算法的特点),且空报率在 3 种算法中最小,但在样本数较少的大量级(50 mm 及以上),FM 过度湿订正[14],BIAS 评分超过 1.3,且随量级增大而增加,最多超过 2.2(150 mm 量级),空报率为 3 种算法中最高。与 FM 相比,OTS 和 OETS 算法虽然在中低量级上(25 mm 及以下)湿偏差和空报略多,但湿偏差幅度较小(BIAS 评分为 1.06～1.40)且漏报较少,略增的湿偏差使漏报率降幅较空报率增幅更加显著,TS 评分更优。在较大量级上(35 mm 及以上),OTS 总体优于 OETS,两者不仅在漏报率方面与 FM 相当,且在空报率方面显著优于 FM。

HSS 技巧评分(图 3f)表明,三个算法对 ECMWF 降水预报有明显的订正能力,在弱降水和强降水两端提高更为显著。3 种算法中,OTS 仅在 75 和 150 mm 两个量级上略低于 OETS,其他量级为最优。OETS 在 0.1、1 和 5 mm 3 个量级上低于 FM,其他 7 个量级优于后者。因此,该检验 OTS 最优、OETS 次之。

上述分析表明,OTS 在 TS 评分、ETS 评分、BIAS 评分、HSS 技巧评分等方面均为 3 种算法中最优,OETS 次之。

3.1.2 多预报时效 TS 评分

图 4 列出四个代表量级(0.1、10、25、50 mm)24～240 h 预报时效 TS 评分。在所有量级各预报时效上,OTS 均有一定的订正能力且为 3 种算法中最优(图 4a～d)。在以 10 mm 为代表的中量级上,各算法提升幅度均最小,其中 FM 订正效果最差,甚至在 120 h 及以上预报时效出现反订正[17]。在 0.1 mm 量级上,FM 订正效果也最差,其在 192 h 及以上时预报时效为反订正,OETS 虽在 216 h 及以上时预报时效为反订正,但总体上优于前者。在 25 mm 量级和 50 mm 量级上,3 种算法在各时效时均优于模式自身,且 TS 评分均为 OTS 最优,OETS 次之。

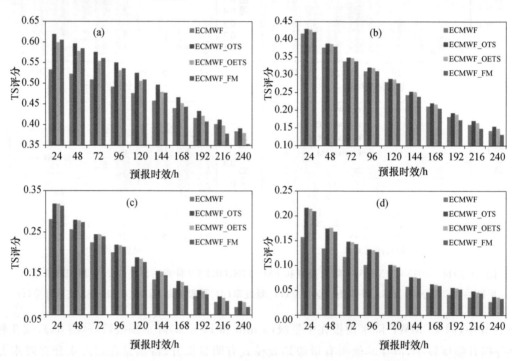

图 4 2014—2015 年 ECMWF 降水预报和 FM、OTS、OETS 3 种算法订正的
预报 0.1 mm(a)、10 mm(b)、25 mm 和 50 mm(d)24 h 累积降水 24～240 h 的 TS 评分

目前,天气预报业务上,降水预报以 TS 评分为主要考核项目,又因 24～240 h 预报时效的其他传统检验结果结论与 3.1.1 相近,不再详述。

3.1.3 SEEPS 技巧评分

除 HSS 技巧评分外,因 TS 评分、ETS 评分、空报比率、漏报率、BIAS 评分等传统的检验方法不是完全意义上的公平评分方法[24],预报员可以根据单一检验方法指标的不同特性,规避不利的预报;且 OTS、OETS 和 FM 3 种算法是分别基于 TS 评分、ETS 评分和 BIAS 评分的最优化设计;3.1.1 节的几个检验方法不是对所有量级的综合检验。因此,本文应用 SEEPS 检验方法作为 3 种算法的第 3 方检验。

SEEPS 技巧评分显示(图 5),各算法的技巧评分随预报时效增加而单调递减。FM 仅在 24 h 对

ECMWF降水预报略有正订正[17],其他时效均为反订正,OTS在所有时效、OETS在216 h以下时效对ECMWF降水预报均有正订正,OTS更为明显。因此,SEEPS技巧评分上,OTS为最优,OETS次之,FM最差。

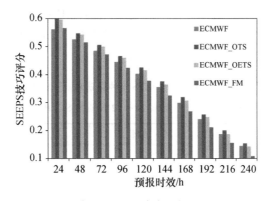

图5 2014—2015年ECMWF降水预报和FM、OTS、OETS 3种算法的24 h累积降水24～240 h预报的SEEPS技巧评分

3.2 对JMA、NCEP和T639降水预报订正

为进一步说明3种算法对比结果对模式的普适性,再对其他3种模式的降水预报进行订正并检验。JMA、NCEP和T639 3种模式预报时效不同,因不影响结果,本文仅计算三者均有的24 h累积降水的0～72 h降水预报。

3.2.1 TS评分检验

对于JMA、NCEP和T639降水预报的订正,因其他预报时效结论相近,本文仅列出24 h累积降水的24 h预报TS评分对比(图6)。可以看出,对于JMA和NCEP,3种算法在所有量级均有明显的订正能力,OTS仅在NCEP降水预报的150 mm量级上低于OETS,其他均为最优,FM表现最差。T639降水预报中,3种算法总体对模式有一定订正能力,但幅度小且集中在不大于25 mm的量级上;OTS占优的量级最多,OETS总体略好于FM。总体上,3种算法中OTS为最好,OETS次之。需要指出的是,对于T639的强降水预报,3种算法对模式提高幅度有限。这是由于3种方法虽能订正模式降水的相对强弱,但并不能改变降水落区,当预报落区与实况不太一致时(T639在50 mm及以上量级TS评分低于0.2),仅通过增大或减小降水强度,预报质量不一定能得到提升。但一般情况下,3种算法仍可对模式降水预报起到一定的正订正,且OTS最优。

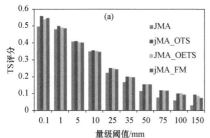 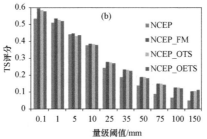 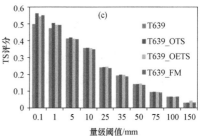

图6 2014—2015年JMA(a)、NCEP(b)和T639(c)降水预报及FM、OTS、OETS 3种算法订正的24 h累积降水24 h预报TS评分

3.2.2 SEEPS 技巧评分检验

从 SEEPS 技巧评分检验(表1)可以看出,OTS 和 OETS 两个算法对 JMA、NCEP 和 T639 降水预报的订正均为正技巧,以 OTS 算法为最优;而 FM 对 JMA 和 NCEP 降水预报在 48 h 及以下预报时效上有正技巧,72 h 预报时效上为负技巧;FM 对 T639 降水预报所有 3 个时效订正均为负技巧。因此,在三个模式降水预报的多个时效上,OTS 仍为最优,OETS 次之。

表1 2014—2015 年 FM、OTS 和 OETS 3 种算法对 JMA、NCEP 及 T639 降水预报订正的 24 h 累积降水 24～72 h 预报 SEEPS 技巧评分

模式	算法	24 h 预报	48 h 预报	72 h 预报
JMA	未订正	0.562	0.506	0.466
	FM	0.626	0.562	0.504
	OTS	0.640	0.572	0.513
	OETS	0.632	0.567	0.511
NCEP	未订正	0.525	0.493	0.474
	FM	0.572	0.533	0.495
	OTS	0.594	0.556	0.526
	OETS	0.588	0.55	0.519
T639	未订正	0.512	0.472	0.405
	FM	0.567	0.518	0.425
	OTS	0.602	0.546	0.453
	OETS	0.599	0.541	0.451

3.3 简单多模式降水预报平均

3.3.1 方法

对 ECMWF、JMA、NCEP 和 T639 4 个模式降水进行简单多模式平均[16](ensemble mean,EMN)形成预报场,再应用算法订正后检验对比。本文多模式平均计算公式为

$$X_{\text{EMN}} = \frac{1}{4}\sum_{i=1}^{4} x_i \tag{7}$$

式中,x_i 为第 i 个模式的降水预报值。

3.3.2 TS 评分检验

因结论相近,本节仅对 24 h 降水预报的 TS 评分进行对比,其他时效略。3 种算法对简单多模式降水平均有明显的订正能力(图7),特别是在降水量级两端(小量级和大量级)订正能力较强,这主要是由于简单多模式降水平均会产生小量级降水面积过多、大量级降水面积偏小和强度偏弱的缺点。OTS 仅在 25 和 75 mm 量级上略低于 OETS,其他量级均为 3 种算法中最优;而 FM 在所有量级上均为最差。

3.3.3 SEEPS 技巧评分检验

3 种算法对简单多模式降水预报平均的 SEEPS 技巧评分提高显著(图8),OTS 为最优,OETS 次之,FM 较差。简单多模式降水预报平均未订正的 SEEPS 技巧评分 24、48、72 h 分别为 0.492、0.467、0.431,

OTS算法则分别达0.601、0.556、0.516,提高值分别为0.109、0.089和0.085,幅度较大;其他两个算法也有不错的表现。

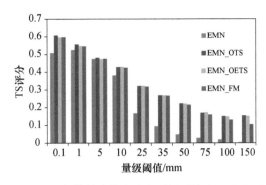

图7 2014—2015年简单多模式降水预报平均和FM、OTS、OETS
3种算法订正的24 h累积降水24 h预报TS评分

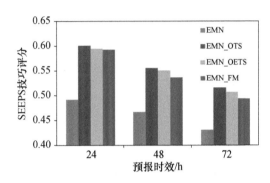

图8 2014—2015年简单多模式降水预报平均和FM、OTS、OETS
3种算法订正的24 h累积降水24～72 h预报的SEEPS技巧评分

4 业务应用对比

为了更好地说明3种算法在预报业务中的应用效果,本节对比2015年主观预报与3种算法中的最优算法预报,即ECMWF_OTS。因目前业务中预报员降水预报主要以量级为主,不做降水量预报,故检验以TS评分为主,未采用SEEPS技巧评分。

4.1 主观与客观降水预报

当前国家气象中心主观定量降水预报主要为每天00:00和12:00两次,核心任务及业务考核为24 h累积降水预报,00:00起报为24～168 h预报时效,12:00起报为24～72 h预报时效。基于预报员的可应用性,采用模式滞后12 h客观订正预报[25]、24 h累积降水的36 h时效订正预报与预报员的24 h主观预报对应,其他时效类推。主观预报的检验数据来自第3方检验机构,即国家气象中心预报系统开放实验室。

4.2 检验对比

表2是2015年国家气象中心预报员主观预报与ECMWF_OTS的24 h累积降水TS评分。在小雨到大暴雨共5级累积降水TS检验中(大暴雨数据略),所有起报时间、预报时效,ECMWF_OTS评分均优于预报员,其中00:00起报的24 h预报,预报员和ECMWF_OTS的暴雨TS评分分别为0.175、0.189,

后者较前者高 0.014；12：00 起报的 24 h 降水预报，预报员、ECMWF_OTS 的暴雨 TS 评分分别为 0.186、0.199，后者较前者高 0.013。表明采用 OTS 算法对 ECMWF 订正后，降水预报准确率高于同期预报员主观预报，具有较高的业务应用价值。

表 2 2015 年国家气象中心预报员与 ECMWF_OTS 24h 累积降水 TS 评分对比

起报时间	预报时效	小雨		中雨		大雨		暴雨	
		预报员	ECMWF_OTS	预报员	ECMWF_OTS	预报员	ECMWF_OTS	预报员	ECMWF_OTS
00：00	24 h	0.594	0.608	0.395	0.414	0.281	0.302	0.175	0.189
	48 h	0.572	0.590	0.359	0.376	0.250	0.263	0.147	0.153
	72 h	0.552	0.565	0.328	0.335	0.225	0.233	0.118	0.136
	96 h	0.522	0.538	0.298	0.305	0.198	0.206	0.094	0.118
	120 h	0.500	0.514	0.271	0.276	0.168	0.178	0.073	0.093
00：00	144 h	0.473	0.486	0.242	0.244	0.145	0.148	0.080	0.083
	168 h	0.442	0.448	0.208	0.214	0.124	0.133	0.050	0.062
12：00	24 h	0.590	0.609	0.392	0.414	0.285	0.304	0.186	0.199
	48 h	0.574	0.584	0.354	0.372	0.255	0.264	0.158	0.163
	72 h	0.55	0.559	0.323	0.337	0.222	0.236	0.125	0.150

5 结论和讨论

为了提高模式降水预报的订正效果，本文采用 ECMWF、JMA、NCEP 和 T639 的 4 个全球模式降水预报和全国气象站实况资料，设计了基于 TS 评分最优化订正的 OTS 及基于 ETS 评分最优化订正的 OETS 算法，并与基于 BIAS 评分最优化订正的 FM 算法对比，同时结合业务需求进行多时效、多种检验方法对比，结果表明：

(1) OTS、OETS 和 FM 3 种算法对 ECMWF、JMA、NCEP 和 T639 的 4 个模式降水预报及 4 个模式的简单多模式降水预报平均均有明显订正能力。从 TS 评分、ETS 评分等检验来看，前 2 个算法较第 3 个更优，尤其是在大量级上两者表现更突出。以降水预报业务质量考核更为常用的 TS 评分来看，无论在时效、量级及模式上，OTS 占优单项最多，为三者最优，OETS 次之。SEEPS 技巧评分对比中，OTS 在各时效、各模式中均明显更优。因此，OTS 的优异具有普适性。

(2) OTS 和 OETS 对模式降水预报进行分级订正后，大部分量级预报与实况保持一定的湿偏差，但湿偏差幅度小。与 FM 相比，前两者在中低量级上有略多的湿偏差，略增的湿偏差带来的漏报率减小幅度大于空报率增加的幅度，使 TS 评分更高。对 ECMWF 降水预报订正中，虽然大量级上样本较少，但 OTS 和 OETS 的 BIAS 评分仍保持在 1.1 左右，相对稳定，而 FM 不太稳定，其 BIAS 评分在最大量级上超过 2.2，过度湿订正。

(3) 应用 2015 年全国国家气象站资料，在预报员可参考的对应时效中，与国家气象中心定量降水预报对比，OTS 对 ECMWF 降水预报订正的各项检验均较前者更优，其 24 h 预报的全国暴雨 TS 评分平均达 0.194，表现出很好的可应用性。

本文分级订正算法采用将全国所有站点作为一个区进行统一订正的方案，并不一定是最优的分区订正方案。降水量在概率上的偏态分布，采用单站独立订正方案会产生模式降水订正阈值不稳定的情况，订正质量并不如前者；若对不同气候区进行更优的分区订正，质量可提升。本文订正算法对比的是站点预报，针对格点进行订正后、再插值到站点进行检验，结论与上述一致，目前算法也应用于业务的站点和网格预报。另外,因不影响结论,本文并未对多模式集成进行更深入研究。但若采用更优的权重对多模式集成后,再通过OTS算法进行降水分级订正,预报质量可能更优。

参考文献

[1] 周兵,赵翠光,赵声蓉.多模式集合预报技术及其分析与检验[J].应用气象学报,2006,17(S1):104-109.

[2] 赵声蓉.多模式温度集成预报[J].应用气象学报,2006,17(1):52-58.

[3] 林春泽,智协飞,韩艳,等.基于TIGGE资料的地面气温多模式超级集合预报[J].应用气象学报,2009,20(6):706-712.

[4] 范丽军,符淙斌,陈德亮.统计降尺度法对华北地区未来区域气温变化情景的预估[J].大气科学,2007,31(5):887-897.

[5] 赵声蓉,裴海瑛.客观定量预报中降水的预处理[J].应用气象学报,2007,18(1):21-28.

[6] 赵声蓉,赵翠光,赵瑞霞,等.中国精细化客观气象要素预报进展[J].气象科技进展,2012,2(5):12-21.

[7] 车钦,赵声蓉,范广洲.华北地区极端温度MOS预报的季节划分[J].应用气象学报,2011,22(4):429-436.

[8] 刘还珠,赵声蓉,陆志善,等.国家气象中心气象要素的客观预报——MOS系统[J].应用气象学报,2004,15(2):181-191.

[9] 马清,龚建东,李莉,等.超级集合预报的误差订正与集成研究[J].气象,2008,34(3):42-48.

[10] 王在文,郑祚芳,陈敏,等.支持向量机非线性回归方法的气象要素预报[J].应用气象学报,2012,23(5):562-570.

[11] 胡邦辉,刘善亮,席岩,等.一种Bayes降水概率预报的最优子集算法[J].应用气象学报,2015,26(2):185-192.

[12] 陈博宇,代刊,郭云谦.2013年汛期ECMWF集合统计量产品的降水预报检验与分析[J].暴雨灾害,2015,34(1):64-73.

[13] 张宏芳,潘留杰,杨新.ECMWF、日本高分辨率模式降水预报能力的对比分析[J].气象,2014,40(4):424-432.

[14] 李俊,杜钧,陈超君.降水偏差订正的频率(或面积)匹配方法介绍和分析[J].气象,2014,40(5):580-588.

[15] 王海霞,智协飞.基于TIGGE多模式降水量预报的统计降尺度研究[J].气象科学,2015,35(4):430-437.

[16] 智协飞,季晓东,张璟,等.基于TIGGE资料的地面气温和降水的多模式集成预报[J].大气科学学报,2013,36(3):257-266.

[17] 孙靖,程光光,张小玲,2015.一种改进的数值预报降水偏差订正方法及应用[J].应用气象学报,26(2):173-184.

[18] 周迪,陈静,陈朝平,等.暴雨集合预报-观测概率匹配订正法在四川盆地的应用研究[J].暴雨灾害,2015,34(2):97-104.

[19] 李俊,杜钧,陈超君."频率匹配法"在集合降水预报中的应用研究[J].气象,2015,41(6):674-684.

[20] 王雨,闫之辉.降水检验方案变化对降水检验评估效果的影响分析[J].气象,2007,33(12):53-61.

[21] RODWELL M J,RICHARDSON D S,HEWSON T D,et al. A new equitable score suitable for verifying precipitation in numerical weather prediction[J]. Quart J Roy Meteor Soc,2010,136:1344-1363.

[22] HAIDEN T M,RODWELL M J,RICHARDSON D S. Intercomparison of global model precipitation forecast skill in 2010/11 using the SEEPS score[J]. Mon Wea Rev,2012,140:2720-2733.

[23] 陈法敬,陈静."SEEPS"降水预报检验评分方法在中国降水预报中的应用试验[J].气象科技进展,2015(5):6-14.

[24] JOLLIFFE I T,STEPHENSON D B. 预报检验[M]. 李应林,等,译. 北京:气象出版社,2016.

[25] 吴启树,韩美,郭弘,等.MOS温度预报中最优训练期方案[J].应用气象学报,2016,27(4):426-434.

A Convection Nowcasting Method Based on Machine Learning[①]

SU Aifang[1,2]   LI Han[1,2]   CUI Liman[1,2]   CHEN Yungang[3]

(1. Key Laboratory of Agrometeorological Safeguard Application Technique, CMA, Zhengzhou, 450003;
2. Henan Provincial Meteorological Observatory, Zhengzhou, 450003;
3. Beijing Presky Technology Co., Ltd, Beijing, 100195)

Abstract: In this study, a convection nowcasting method based on machine learning was proposed. First, the historical data were back-calculated using the pyramid optical flow method. Next, the generated optical flow field information of each pixel and the Red-Green-Blue (RGB) image information were input into the Convolutional Long Short-Term Memory (ConvLSTM) algorithm for training purposes. During the extrapolation process, dynamic characteristics such as the rotation, convergence, and divergence in the optical flow field were also used as predictors to form an optimal nowcasting model. The test analysis demonstrated that the algorithm combined the image feature extraction ability of the Convolutional Neural Network (CNN) and the sequential learning ability of the Long Short-Term Memory Network (LSTM) model to establish an end-to-end deep learning network, which could deeply extract high-order features of radar echoes such as structural texture, spatial correlation and temporal evolution compared with the traditional algorithm. Based on learning through the above features, this algorithm can forecast the generation and dissipation trends of convective cells to some extent. The addition of the optical flow information can more accurately simulate nonlinear trends such as the rotation, or merging, or separation of radar echoes. The trajectories of radar echoes obtained through nowcasting are closer to their actual movements, which prolongs the valid forecasting period and improves forecast accuracy.

Key words: radar reflectivity; ConvLSTM; pyramid optical flow method; nowcasting

1 Introduction

Convection nowcasting refers to the short-term forecasting of the convective weather system and the catastrophic convective weather they may produce, up to 0-6 hours beyond the current observation time. At present, convection nowcasting based on radar data mainly involves thunderstorm

---

[①] Advances in Meteorology, 2020, Article ID 5124274, https://doi.org/10.1155/2020/5124274.
Corresponding author address: SU Aifang, Henan Provincial Meteorological Observatory, Zhengzhou, China, 450003. E-mail: afsu011@sohu.com.

identification, tracking, and automated extrapolation techniques (Yu et al., 2012), e. g., the centroid tracking method (Crane, 1979; Dixon et al., 1993; Johnson et al., 1998), the Tracking Radar Echo by Correlation (TREC) method (Rinehart et al., 1978; Wilson et al., 1998; Chen et al., 2007), and the optical flow method (Gibson, 1950; Germann et al., 2002; Yilmaz et al., 2006; Han et al., 2008; Sakaino, 2013; Cao et al., 2015).

The centroid tracking method can only be used for convective precipitation systems, while the TREC method can also track the layered cloud precipitation system. When faced with a fast-changing strong convective precipitation system, the TREC failure rate increased significantly (Cao et al., 2015). Optical flow is a dense field of displacement vectors which defines the transition of each pixel in a region. It is computed using the brightness constraint, which assumes brightness constancy of corresponding pixels in consecutive frames. Thus, optical flow is commonly used as a feature in motion-based segmentation and tracking applications (Yilmaz et al., 2006). These advantages can solve the shortcomings of both the centroid tracking and TREC methods to some extent. However, the success of optical flow based methods is limited because it does not consider the physical meaning of radar echo development. As such, it is challenging to predict short-term local convection with rapid generation or extinction (Shi et al., 2015; Wang et al., 2017).

The traditional optical flow method is only applicable to small movements of image features, although the pyramid delaminating technique can improve the calculation accuracy and convergence speed of this technique (Adelson et al., 1984; Bouguet, 1999; Jiang et al., 2007). According to Zhang et al. (2014), even though the pyramid optical flow method has advantages in forecasting the convective radar echoes, it can only track echo characteristics that already exist in an image and cannot predict either the generation or dissipation of echoes. In addition, since atmospheric motion often exhibits highly nonlinear and random disturbance behavior, the optical flow method is insufficient over a short valid forecasting period (0-6 h).

Traditional image recognition processing requires manually setting specific features such as the shape, length-width ratio, and area of a connected region in order to extract the desired information; occasionally all of the pixels in an image are used as the basic information for classifier training and classification. The former cannot guarantee the successful extraction of relevant or significant features, while the latter tends to introduce a great deal of redundant information.

Inspired by the cognitive mechanism of biological natural vision, Hubel et al. (1962) found that a unique network structure can effectively reduce the complexity of neural network feedback, and proposed the concept of the Convolutional Neural Network (CNN). The CNN effectively solved the forecasting problem of the traditional fully-connected network algorithm via local connectivity, parameter sharing, and downsampling. Specifically, with local connections, each neuron is not connected to every neuron in the upper layer, effectively reducing the amount of parameters during training. Each group of connections share the parameter of convolutional kernel, which reduces the amount of training required and accelerates the training speed. The pooling method used for downsampling greatly reduces the feature dimensions and amount of calculation, and avoids overfitting. Compared with other neural networks, the CNN shows superior performance in automatically extracting the salient features of an image (Lawrence et al., 1997; Dahl et al., 2010; Krizhevsky et al., 2012). Liu et al. (2016) used CNN technology to extract extreme weather events (tropical cyclones, atmospheric rivers, and frontal activities) from climate datasets and obtained meaningful results.

Another neural network method is the Recurrent Neural Network(RNN), which is applicable to data with time series features(Mikolov et al., 2010; Donahue et al., 2015; Karpathy et al., 2015). During the actual training process, gradient disappearance and explosion occurred sometimes. To solve these problems, a more complex RNN named Long Short-Term Memory Network(LSTM) was proposed by Hochreiter et al. (1997). Using LSTM, the medium and long-term data information of a time series can be well-preserved. It is advantageous to train and model meteorological data with time series characteristics and to conduct forecast research using the LSTM. Akram et al. (2016) used 15 years of hourly weather data to train the multilayer LSTM model and to forecast meteorological conditions out to 24 and 72 hours. They found that the LSTM can forecast general weather variables with a better accuracy.

The CNN and LSTM are suitable for extracting spatial features and processing time series, respectively. A technique for integrating the two is worth exploring, in order to learn and train datasets with spatial and temporal features more efficiently. Since short-term nowcasting based on radar echoes is a forecasting problem that involves sequences of both spatial and temporal features, Shi et al. (2015) proposed the ConvLSTM algorithm, based on both the CNN and LSTM. In the ConvLSTM, on the basis of the traditional LSTM, the convolutional structure was added to each LSTM unit. Compared with the traditional LSTM, this hybrid network has a greater ability to extract the spatial features of radar echoes. By using three-dimensional radar data, Kim et al. (2017) applied ConvLSTM to predict the rainfall amount, and the results show that the ConvLSTM is better than FC-LSTM and linear regression.

Although the ConvLSTM algorithm performs well in convection nowcasting based on the training and learning of evolutionary patterns of the radar echoes using historical data, it lacks of dynamic field data(such as U, V wind field) in the input predictors that may be one of the potential factors to the generation or dissipation of echoes. In this paper, we simulate the nearly-ideal background wind field(U, V) obtained from radar data using the pyramid optical flow method. Two components of background wind field(U, V) and three channels of Red-Blue-Green images of radar echoes are taken as 5 forecasting factors of ConvLSTM. As the results shown below, it will confirm that the new model performs better than optical flow method. The present paper is organized as follows. In section 2, a brief description of the machine learning system, and the materials and methods are provided. Results of two cases are presented in section 3, followed by a summary and conclusion in section 4.

2 Data and Methodology

In this study, Henan Province in China is selected as the area of interest, and the radar data from 2016 to 2017 are used. Based on the Red-Green-Blue(RGB) image information of the radar images, the improved machine learning algorithm was used to study the nowcasting technique of convective echoes, by which we provide a technical support for nowcasting and early warning of severe convective weather.

2.1 Data processing

The monitoring data collected by 8 radars(Fig. 1) located in Zhengzhou City, Luoyang City, Puyang City, Sanmenxia City, Pingdingshan City, Shangqiu City, Nanyang City, and Zhumadian City in Henan Province were selected. After non-meteorological clutter had been filtered, beam blockage compensated, and frequency attenuation corrected(Steiner et al. ,2002;Kessinger et al. ,2001;Fulton et al. ,1998),the nearest-neighbor and vertical linear interpolations were applied in the radial and azimuthal directions(Xiao et al. ,2006;Klazura et al. ,1993)in order to transform polar coordinates into grid points. Thereafter, the exponential weight function method was utilized to process the radar data from different scanning modes,different bands,and different generation times into data in a 3D Cartesian coordinate system of unified observation time and resolution,resulting in a CAPPI composite map consisting of 31 contour planes on the same base map. The vertical extension height is 18 km,and the vertical grid spacing(Δz)is 0.25 km below a height of 3 km,and stretches to 0.5 km up to 9 km,after which Δz remains constant at 1 km. The horizontal spacing and time resolution are respectively 0.01° and 6 min. The maximum of 31 levels of CAPPI is then mapped to the same layer to obtain the combined reflectivity,which is used as the input radar images to optical flow and ConvLSTM.

Fig. 1 Distribution of 8 radars in Henan Province

In order to make the model more generalized,data of non-severe and severe weather were both included in the training dataset to learn the evolution characteristics of echoes under various weather conditions. This experiment selected volumetric data of the radars composite from May 2016 to April 2017 for each two consecutive hours. A total of 50000 samples were used as data sets,80% of which were used as training sets and 20% as test sets. The first 5 radar images(30 minutes)of radar echo image at each initial time were used as input data,and the next 20 images(2 hours)as output. Due to the low proportion of strong convective weather in the sample set,the sample equalization(i. e. ,resample and data-augmentation technologies) was processed to make the proportion of samples in each batch more balanced. Then,about 5000 samples selected from May to August 2017 were used as test sets to evaluate the forecasting ability of the model,and 7 strong weather events(Table 1) were selected from the test sets as case studies. Only radar echoes greater than 10 dBZ were considered for the evaluation of echo prediction bias,while echoes greater than 35 dBZ were considered for prediction accuracy and positional deviation.

Table 1 Characteristics of the 7 cases

Date(DD/MM/YYYY)	Severe weather phenomena	Mesoscale features
19/05/2017	Heavy rainfall, maximum wind gust(20 m/s), maximum diameter of hail(1 cm)	Shear line
22/05/2017	Heavy rainfall, maximum wind gust(25 m/s)	Trough and shear line
12/06/2017	Heavy rainfall, maximum wind gust(21 m/s)	Trough in Northern China
20/06/2017-21/06/2017	Thundershower(light rainfall), maximum wind gust(23 m/s)	Cold air diffusing southward in northwesterly airflow
04/07/2017	Heavy rainfall, maximum wind gust(20 m/s)	Cyclonic vortex with shear line
14/07/2017	Heavy rainfall, maximum wind gust(20 m/s), maximum diameter of hail(1 cm)	Cyclonic vortex with shear line
01/08/2017	Continuous heavy rainfall with thunder and lightning	Typhoon trough

2.2 Nowcasting based on machine learning

In convection nowcasting, the pyramid delaminating technique(Jiang et al. ,2007) was used to classify and preprocess the mosaic radar data. During the process, the dense optical flow method was applied to calculate and track the evolution of each pixel in the network data in order to obtain the dense optical flow field. Feature points were selected from the strong convective cells, and the pyramid delaminating technique was utilized to reduce the dimensions in order to obtain the top-level wind field information satisfying the assumptions of the optical flow method. Through hierarchical iteration from the top to the bottom layers of the pyramid, the sparse optical flow field of the feature points of the severe convective cells in the network data was obtained and then the dense optical flow field was corrected. Finally, a corrected optical flow field, the background horizontal wind field(U, V), was generated.

The CNN structure in the ConvLSTM includes a convolutional layer, a downsampling layer, and a fully connected layer, among which the convolutional and downsampling layers may have a multilayer structure(deep convolutional). The convolutional and downsampling layers were not connected one by one. The next sampling layer could be connected after multiple convolutional layers in order to extract features of the output image in each dimension. With the output feature of the convolutional layer as the input, the fully connected layer acted as a classifier. Compared with the traditional recurrent neutral network(RNN), LSTM introduces three kinds of gate structures(forget gate layer, input gate layer, output gate layer) to protect and control information. The first thing that LSTM has to solve is to decide which information to pass through this neuron, which is done by the forget gate layer. The input gate layer determines how much information needs to be saved to the current state at the current moment. After the selective memory and update steps, the output gate layer determines the information passed to the next neuron. The convolution kernel only needed to establish full-connection sampling for the local area of the image to extract the underlying information, including local information such as edges and corner points. Through the weight-sharing mechanism, a convolution kernel that had been trained based on a particular characteristic could complete the resampling of similar characteristics for the entire image (Liu,2018). Following convolution training, the image features were used as the information to be input into the LSTM. This information was protected and controlled through point-by-point multiplication, as well as the sigmoid activation function(Hochreiter et al. ,1997) of gate structure of the LSTM, so as to finally obtain the extrapolation nowcasting results of the radar reflectivity factor within 0-2 h.

Fig. 2 shows the short-term nowcasting process based on pyramid optical flow and ConvLSTM machine learning. The main steps are summarized as follows:

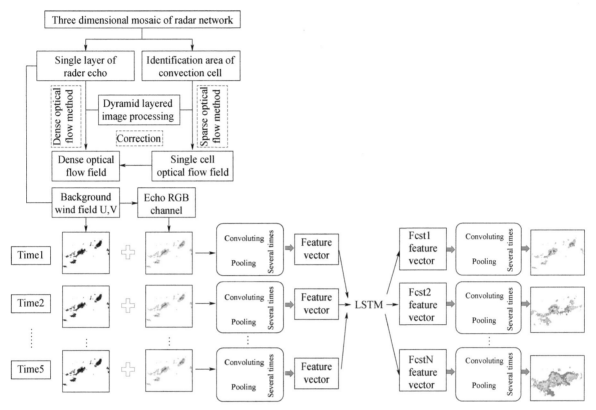

Fig. 2 Flowchart of the convective echo extrapolation algorithm based on ConvLSTM machine learning

Step 1: optical flow field computation. The sparse and dense optical flow field were obtained after preprocessing the mosaic radar data using pyramid optical flow method. The dense optical flow field was calculated and tracked from every pixel of the radar data, while the sparse optical flow field only obtained from pixels of the strong convective cells. Therefore, we correct the dense optical flow field with the sparse optical flow field to obtain a more accurate background wind field (U,V).

Step 2: model configuration. We developed a ConvLSTM that has 3 ConvLSTM2D layers with 40 units in each layer. The ConvLSTM2D layer is an extension of FC-LSTM layer, which replaced fully connected structures to convolutional structures in both the forget gate and input gate of LSTM.

Step 3: model training. The input data of ConvLSTM, which is extracted from 5 radar images in the past 30 minutes, contains 5 dimensions in each training sample: 2 components of background wind field and 3 channels of Red-Green-Blue(RGB) image information of radar echo. The output is 20 radar images in the next 2 hours.

2.3 Measures of model skill

In this study, the value of 35 dBZ is taken as the test threshold. To quantitatively describe the position, intensity and valid period of the forecasted radar echoes, the resulting echoes were compared to the observed grid point by grid point. Considering the impact of the wind field, each grid point of the forecast field and its adjacent 3 × 3 grid point area were compared in the actual assessment. Both magnitude, bias and forecast accuracy, were tested; ratio bias(BIAS), root-mean-square error(RMSE), and error of mean (EM), were used as magnitude bias indicators, and probability of detection(POD), false alarm ratio (FAR), critical success index(CSI) and correlation coefficient(CC) where used as forecast accuracy indicators. The specific equations are as follows:

$$POD = \frac{NA}{NA+NC}$$

$$FAR = \frac{NB}{NB+NC}$$

$$CSI = \frac{NA}{NA+NB+NC}$$

$$CC = \frac{\sum_{j}^{n}(g_{i,j}-\bar{g}_i)(r_{i,j}-\bar{r}_i)}{\sqrt{\sum_{j}^{n}(g_{i,j}-\bar{g}_i)^2(r_{i,j}-\bar{r}_i)^2}}$$

The explanations of verification index(NA, NB, NC) are listed in Table 2. If the actual grid point value and the forecasted grid point value were both greater than the threshold value, then the grid point was considered to be a successful forecast (NA). If the actual grid point value was greater than the threshold, while the forecasted grid point value was less than the threshold, then the point was considered to be a missed forecast(NC). If the actual grid point value was less than the threshold, while the forecasted grid point value was greater than the threshold, then the point was considered to be a false forecast(NB).

Table 2 Verifying classification

Actual value	Forecast value	
	Greater than the threshold	Less than the threshold
Greater than the threshold	NA	NC
Less than the threshold	NB	—

2.4 Computational platform

A C++ and Python mixed programming of optical flow method and ConvLSTM model was performed on a computer with 2 Intel Xeon E5-2650V3 CPUs(12 cores) and 2 NVIDIA GEFORCE GTX TITAN X GPUs, and the running efficiency of both algorithms is satisfied. For instance, the pre-processes of 8 radars' data cost approximately 45 seconds. The calculation time of optical flow for each single radar composite image every 6 minutes is within 10 seconds. Although Training the ConvLSTM model needs a relatively long time for about 30 hours, the extrapolation time for 2 hours took only 120 seconds after training. Thus, both two nowcasting algorithms in this paper can complete a single computation in 5 minutes.

3 Results

Seven severe convective weather events from May to August 2017 were selected for testing and analyzing nowcasting results based on the model in this study. The selected cases are listed in Table 2. A more in deep analysis was done with the events on July 14 and August 1.

3.1 Comprehensive assessment

Figs. 3 and 4 show the test results of optical flow and ConvLSTM for the 2 forecasts starting at 1100UTC July 14 2017 and 0400UTC August 1 2017, respectively.

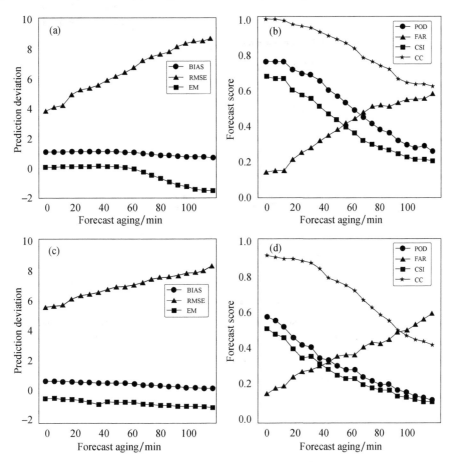

Fig. 3 Forecast bias and forecast score for the 2 hours starting at 1100UTC July 14 2017
(a)(b)the results of ConvLSTM method;(c)(d)the results of optical flow method

Fig. 3(a-c) show the comparison of average forecast bias of radar echoes greater than 10 dBZ within 2 h between optical flow and ConvLSTM. For ConvLSTM, the BIAS was maintained near 1, the absolute value of the EM was within 0.5, and the RMSE was kept within 7 dBZ, whereas the RMSE of optical flow was 1-2 dBZ higher than ConvLSTM on average. The similar results can also be seen from Fig. 4(a) and (c), indicating a more accuracy in the extrapolation of radar echo location and strength in ConvLSTM.

According to the average forecast bias of strong echoes greater than 35 dBZ in Fig. 3(b) and (d), the forecast accuracy of two algorithms both decreased as the valid forecasting period increased. The POD, CSI and CC of ConvLSTM show a change range of 0.73 to 0.28, 0.65 to 0.21 and 0.96 to 0.61, respectively. While the same indicators of optical flow show a range of 0.58 to 0.14, 0.52 to 0.13 and 0.91 to 0.43, respectively. From the comparison of Fig. 4(c) and (d), the result of optical flow method shows a faster descent curve of POD, CSI and CC than ConvLSTM as well. The results indicate that the average forecast bias of strong echoes of ConvLSTM was less than optical flow within the 2-h valid

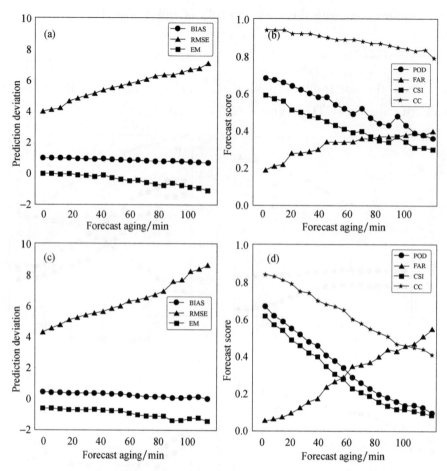

Fig. 4　Forecast bias and forecast score for the 2 hours starting at 0400UTC August 1 2017
(a)(b)the results of ConvLSTM method; (c)(d)the results of optical flow method

forecasting period.

　　Based on the 7 selected weather phenomena listed in Table 1, more than 20 forecast-initiation events were chosen for each process, resulting in a total of more than 150 forecast-initiation events. The TREC, optical flow, and ConvLSTM machine learning proposed in this study were used to conduct 0-2 h extrapolation forecasts on the selected forecast-initiation events. According to the average test results (Table 3), compared with the TREC extrapolation method, both the optical flow method and the ConvLSTM extrapolation method proposed in this study achieved qualitative improvements in forecast accuracy and correlation, as well as precision of forecast position and intensity. The average accuracy improved by more than 30%, illustrating the advantage of the nonlinear extrapolation method. The ConvLSTM method with pyramid optical flow field information further improved forecast accuracy compared with the optical flow method, reflecting the learning ability of the machine learning method.

Table 3　Results of the comprehensive assessment

Forecast method	Valid period of forecast/min	POD	CSI	CC	Forecast centroid deviation/km	RMSE/dBZ
TREC	60	0.291	0.254	0.489	18.1	21.5
Optical flow method		0.482	0.428	0.728	12.8	15.3
Machine learning method		0.544	0.465	0.821	11.3	12.9

续表

Forecast method	Valid period of forecast/min	POD	CSI	CC	Forecast centroid deviation/km	RMSE/dBZ
TREC	120	0.182	0.114	0.320	27.5	28.4
Optical flow method		0.216	0.195	0.466	17.8	20.3
Machine learning method		0.383	0.287	0.513	17.1	15.8

3.2 Case Analysis

3.2.1 Case 1

Influenced by cyclonic vortex shear, there were thunderstorms in the northern and central parts of Henan Province on July 14-15, 2017, accompanied by strong convective weather such as lightning, strong wind, and heavy rainfall. During this event, echoes developed rapidly, lasted for an extended period, and exhibited a wide range of influence. However, the convective system was scattered and new echoes were constantly being generated, making it difficult to conduct an early-warning forecast of short-term heavy rainfall and thunderstorms.

Fig. 5(a) and (c) show the convective echoes intensity and movement from 1100UTC-1300UTC. The convective echoes generally moved eastward, although the speed of echo movement varied from region to region. The echoes in the central and southern areas moved relatively faster, whereas the echoes in the northern region propagated more slowly. The convective clouds in the northern region were generated at 1100UTC and tended to decay by 1254UTC, while the clouds in the south area evolved from stratified mixed clouds into strong convective clouds during this period. It can be seen from Fig. 5(b) and (d) that based on the composited radar echo data of the 2.5 km height layer of the front and rear frames, the horizontal moving speed of the radar echo obtained using the pyramid optical flow technique has certain indication significance for the radar intensity change trend: The cloud system in the north had a low inversion wind speed and was divergent, whereas in the south, the northwest wind had a high speed and the wind field was characterized by obvious convergence, which was consistent with the variation pattern of the echoes, that is, moving quickly and being generated in the south while moving slowly and dissipating in the north.

The timing when numerous convective echoes initiated(i.e., 1100UTC July 14) was selected as the onset time for the 0-2 h extrapolation forecast. Fig. 6 shows the echo forecast based on optical flow and machine learning. The forecasting start time was 1100UTC July 14 2017, and the valid periods of the forecast were the next 30, 60, 90, and 120 minutes. Comparison of the distribution of ConvLSTM forecasted echoes and the corresponding real-time radar echoes shows that the forecasted position and shape of the echoes over the 2-hour forecast period were in good agreement with the actual weather conditions. The forecasted echoes in southern Henan Province developed faster to the east with the increase of deviation degree and the valid forecasting period; the maximum amount of deviation can be controlled to a range of 0.03-0.05°(about 3-5 km). Based on the echo intensity, the characteristics indicating the evolution of the stratiform cloud system into the convective cloud system failed to be predicted, although they were reflected in the dissipation process of convective cells in the north, and there were similar forecasts of the

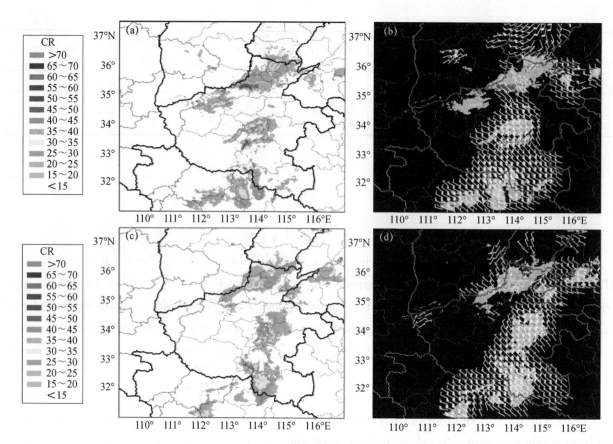

Fig. 5 Evolutionary characteristics of the composite reflectivity(a and c) and wind field (b and d) at 1100UTC and 1254UTC July 14 2017

evolution of echo formation in central Henan Province. For this case, the traditional optical flow method only predicts the strong echo in northern Henan Province while failed to predict its weakening of strength, and the evolution of radar echoes in southern and central Henan Province. During the development of this cloud system, in addition to the echo forms changing significantly, there was also the generation and dissipation of convective cells, occurrences that were reflected in the extrapolation algorithm proposed in this study. This also illustrates the improvement of this method compared to the traditional extrapolation method. With machine learning and the pyramid optical flow method, the convergence and divergence characteristics of the wind field can be accurately described, indirectly reflecting upward and downward air motions. Therefore, not only can this method accurately forecast the position and shape of radar echoes, but it can also forecast the generation and dissipation of local convection.

3.2.2 Case 2

Affected by the westward movement of the inverted trough of Typhoon Haitang, on August 1, a spiral-shaped strong convective system moving from east to west developed in Henan Province. Of strong intensity and long duration, this system brought a wide range of heavy rainfall to the eastern part of Henan Province, with a heavy rainstorm occurring in the northeastern area. As shown in Fig. 7, the spiral cloud band of the typhoon was moving from east to west at 0400UTC on August 1, affecting the eastern and southeastern areas of Henan Province. There was an arc-shaped strong convection zone in the northern section of the cloud cluster, with strong echoes constantly rotating into Henan Province. As

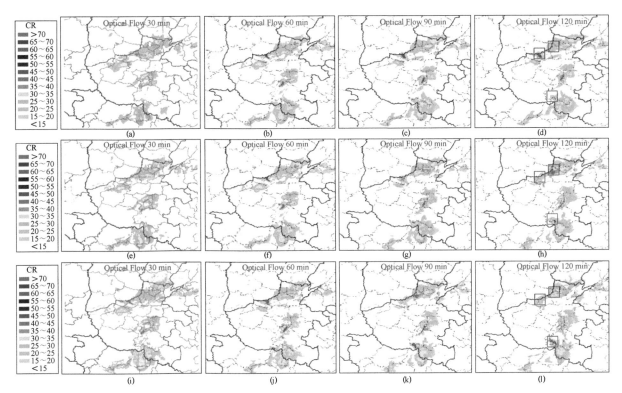

Fig. 6 Comparison between the composite reflectivity factor forecast results and the actual weather conditions for the 2 h starting at 1100UTC July 14 2017
(a)(b)(c)(d) the forecast results of optical flow method 30,60,90,120 min after the start time of 1100UTC; (e)(f)(g)(h) the forecast results of ConvLSTM method 30,60,90, 120 min after the start time of 1100UTC; (i)(j)(k)(l) the actual weather conditions 30, 60,90,120 min after the start time of 1100UTC; red square lines represents the enhancement area of observe radar echoes, and blue square lines for the weakened area

shown in Fig. 7(a) and (c), the strong precipitation echoes in the southern section of the cloud cluster were separated from the main rain band during its movement towards the west, and the southern part of the system became an independent precipitation cloud cluster within 2 h. In addition, it can be seen from Fig. 7(b) and (d) that there were obvious counterclockwise rotation characteristics in the wind field. The northerly wind speed was relatively high, but the wind speed in the south was low, and the wind was blowing westward, consistent with the separation process of the echoes in the southern part of the system.

At 0400UTC, when the convective system became mature, the 0-2 h extrapolation forecast started. Fig. 8 shows the distribution of the forecasted radar echoes and the corresponding actual radar echoes for the next 30,60,90, and 120 min, starting at 0400UTC August 1 2017. Due to the concentrated nature of the convective system during this period, the forecasted shape and position of the echoes of ConvLSTM method in the 2-h forecast window were highly consistent with the actual radar echoes. The shape and movement trends of the convective cloud clusters in the stratiform cloud system were also accurately captured. In particular, the strong echoes rotating into the eastern part of the system could be predicted. However, the traditional optical flow method failed to predict the weakening process of echoes in southern and central Henan Province, and the strength of forecasted strong radar echo in northern Henan Province was also weaker compared with observation. The above results indicate that the learning ability of the algorithm proposed in this paper has significant advantages over the traditional extrapolation

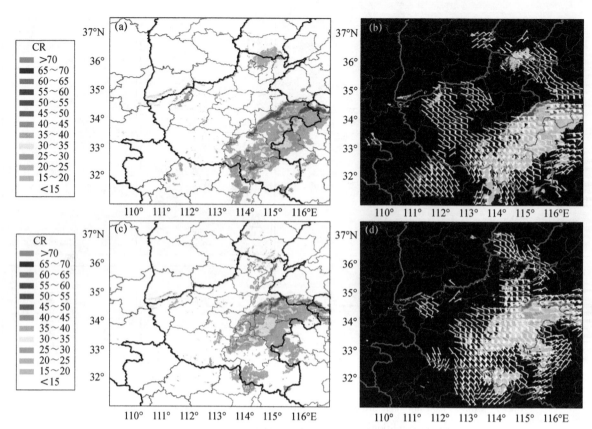

Fig. 7 Evolutionary features of the composite reflectivity(a,d) and the wind field(c,f) at 0400UTC and 0554UTC August 1 2017

algorithm. The central portion of the forecasted echoes gradually dissipated and the southern section became separated, which agreed with the evolution of the actual radar echoes. However, the dissipation and separation speeds of the forecasted echoes were slower than those of the observed radar elements. Therefore, even with the optical flow field information added to the machine learning training set, the ability of the proposed method to separate, combine, and forecast convective systems still has a room for significant improvement.

4 Conclusions

Our experimental results demonstrated that the global optical flow field generated by pyramid delaminating technique can better reflect the movements of radar echoes, including nonlinear motions such as rotation, convergence, and divergence of local radar elements. By adding the background horizontal wind field(U, V) and Red-Green-Blue(RGB) image information to the ConvLSTM extrapolation nowcasting model, the forecast accuracy of echo position and intensity was further improved, the generation, dissipation, and merging of convective cells were also better identified.

Compared with the TREC method in business application, the machine learning method achieved qualitative improvements in forecast accuracy and correlation, as well as precision of forecast position and intensity. Moreover, the machine learning method also effectively eliminate the false merging of convective cells and improve the accuracy of convective cell recognition. Compared with the traditional optical flow method, the extrapolation of machine learning method is closer to the actual trajectory of radar

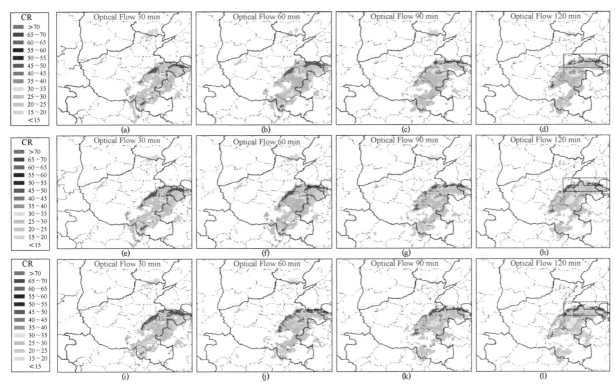

Fig. 8 Comparison between the composite reflectivity factor forecast results and the actual weather conditions for the 2 h starting at 0400UTC August 1 2017
(a)(b)(c)(d)the forecast results of the optical flow method 30,60,90,120 min after the start time of 1100UTC;(e)(f)(g)(h)the forecast results of the ConvLSTM method 30, 60,90,120 min after the start time of 1100UTC;(i)(j)(k)(l)the actual weather conditions 30,60,90,120 min after the start time of 1100UTC;red square lines represents the area where the predicted radar echoes of ConvLSTM are closer to the observation than the prediction of optical flow method

echoes,especially in the case of cyclonic systems,for which the rotation vectors and curve movement trajectories can be generated.

As for scattered convective weather,the valid forecasting period of ConvLSTM for fast-developing convective cells is relatively short, ranging from 30-60 min. When it comes to systematic convective weather covering a broad expanse, with long system duration and complete echo structure, the ConvLSTM machine learning method has a greater learning capacity and higher forecasting accuracy than the traditional method(Johnson,1998;Shi et al.,2017;Zhou et al.,2017).

Acknowledgments. This study was supported by the China Meteorological Administration 2017 Special Fund for the Development of Key Technologies in Weather Forecasting [YBGJXM(2017;2018) 02-08] and the National Key R&D Program of China(2017YFC1502000).

References

ADELSON E H,ANDEROSN C H,BERGEN J R,et al,1984. Pyramid methods in image processing[J]. Rca Engineer,29 (6):33-41.

AKRAM M,EL C,2016. Sequence to sequence weather forecasting with long short-term memory recurrent neural networks [J]. International Journal of Computer Applications,143(11):7-11.

BOUGUET J-Y,1999. Pyramidal implementation of the Lucas Kanade Feature tracker description of the algorithm[R]. USA:Microsoft Research Labs.

CAO C Y,CHEN Y Z,LIU D H,et al,2015. The optical flow method and its application to nowcasting[J]. Acta Meteor Sinica,73(3):471-480.

CHEN M X,WANG Y C,YU X D,2007. Improvement and application test of TRCE algorithm for convective storm nowcast[J]. Appl Meteor Sci,18(5):690-701. (in Chinese)

CRANE,R. K,1979. Automatic cell detection and tracking[J]. IEEE Trans Geosci Electron,17(4):250-262.

DAHL J V,KOCH K C,KLEINHANS E,et al,2010. Convolutional networks and applications in vision[C]// IEEE International Symposium on Circuits and Systems. IEEE:253-256.

DIXON M,WIENER G,1993. TITAN:thunderstorm identification,tracking,analysis,and nowcasting- a radar-based methodology[J]. J Atmos Oceanic Technol,10(6):785-797.

DONAHUE J,HENDRICKS L A,ROHRBACH M,et al,2015. Long-term recurrent convolutional networks for visual recognition and description[M]// AB initto calculation of the structures and properties of molecules. Elsevier:85-91.

FULTON R A,BREIDENBACH J P,SEO D J,et al,1998. The WSR-88D rainfall algorithm[J]. Weather & Forecasting,13(2):377-395.

GERMANN U,ZAWADZKI I,2002. Scale-dependence of the predictability of precipitation from continental radar images. Part Ⅰ:description of the methodology[J]. Monthly Weather Review,130(12):2859-2873.

GIBSON J J,1950. The Ecological approach to visual perception[M]. Boston:Houghton Mifflin.

HAN L,WANG H Q,LIN Y J,2008. Application of optical flow method to nowcasting convective weather[J]. Acta Scientiarum Naturalium Universitatis Pekinensis,44(5):751-755.

HOCHREITER S,SCHMIDHUBER J,1997. Long short-term memory[J]. Neural Computation,9(8):1735-1780.

HUBEL D H,WIESEL T N,1962. Receptive fields,binocular interaction and functional architecture in the cat's visual cortex[J]. Journal of Physiology,160(1):106-154.

JIANG Z J,YI H R,2007. An image pyramid based feature detection and tracking algorithm[J]. Geomatics and Information Science of Wuhan University,32(8):680-683.

JOHNSON J T,MACKEEN P L,WITT A,et al,1998. The storm cell identification and tracking algorithm:an enhanced WSR-88D algorithm[J]. Weather and Forecasting,13(2):263-276.

KARPATHY A,LI F F,2015. Deep visual-semantic alignments for generating image descriptions[C]//Computer Vision and Pattern Recognition. IEEE:3128-3137.

KESSINGER,C. , S. ELLIS, J. VANANDEL, 2001. NEXRAD data quality:the AP clutter mitigation scheme[C]//Preprints,30th Int Conf on Radar Meteorology,Munich,Germany,Amer Meteor Soc,707-709.

KIM S,HONG S,JOH M,et al,2017. DeepRain:ConvLSTM Network for precipitation prediction using multichannel radar data[R].

KLAZURA G E,IMY D A,1993. A description of the initial set of analysis products available from the NEXRAD WSR-88D System[J]. Bull Amer Meteor Soc,74(7):1293-1312.

KRIZHEVSKY A,SUTSKEVER I,HINTON G E,2012. ImageNet classification with deep convolutional neural networks[C]//International Conference on Neural Information Processing Systems. Curran Associates Inc:1097-1105.

LAWRENCE S,GILES C L,TSOI A C,et al,1997. Face recognition:a convolutional neural-network approach[J]. IEEE Transactions on Neural Networks,8(1):98-113.

LIU F P,2018. Application of neural network and deep learning[M]. Beijing:Publishing House of Electronics Industry.

LIU Y,RACAH E,PRABHAT,et al,2016. Application of deep convolutional neural networks for detecting extreme weather in climate datasets[R].

MIKOLOV T,KARAFIát M,BURGET L,et al,2010. Recurrent neural network based language model[C]//INTERSPEECH 2010, Conference of the International Speech Communication Association, Makuhari, Chiba, Japan, September. DBLP:1045-1048.

RINEHART R E,GARVEY E T,1978. Three-dimensional storm motion detection by conventional weather radar[J]. Nature,273(5660):287-289.

SAKAINO H,2013. Spatio-temporal image pattern prediction method based on a physical model with time-varying optical flow[J]. IEEE Transactions on Geoscience & Remote Sensing,51(5):3023-3036.

SHI X,CHEN Z,WANG H,et al,2015. Convolutional LSTM network:a machine learning approach for precipitation nowcasting[J]. Advances in Neural Information Processing Systems,9199:802-810.

SHI X,GAO Z,LAUSEN L,et al,2017. Deep learning for precipitation nowcasting:a benchmark and a new model[C]//31st Conference on Neural Information Processing Systems(NIPS 2017).

STEINER M,SMITH J A,2002. Use of three-dimensional reflectivity structure for automated detection and removal of non-precipitating echoes in radar data[J]. Journal of Atmospheric & Oceanic Technology,19(19):673-686.

WANG Z B,XIAO Y J,WU T,2017. Motion estimation for radar image based on improved optical flow method[J]. Comput Technol Dev(12):170-175.

WILSON J W,CROOK N A,MUELLER C K,et al,1998. Nowcasting thunderstorms:a status report[J]. Bull Amer Meteor Soc,79(10):2079-2099

XIAO Y J,LIU L P,2006. Study of methods for interpolating data from weather radar network to 3D grid and mosaics[J]. Acta Meteor Sinica,64(5):647-657. (in Chinese)

YILMAZ,A. ,JAVED,O. ,and SHAH,M,2006. Object tracking:a survey[J]. ACM Comput Surv,38(4):1-45.

YU X D,ZHOU X G,WANG X M,2012. The advance of the nowcasting techniques[J]. Acta Meteor Sinica,70(3):311-337. (in Chinese)

ZHANG L,WEI M,LI R,et al,2014. Improved optical flow method application to extrapolate radar echo. Science Technology and Engineering,14(32):133-137.

ZHOU Z H,FENG J,2017. Deep forest:towards an alternative to deep neural networks[C]//Proceedings of the 26th International Joint Conference on Artificial Intelligence.

日极端气温的主客观预报能力评估及多模式集成网格释用[①]

吴乃庚[1,2]　曾　沁[2]　刘段灵[2]　陈炳洪[2]　赵声蓉[3]　张红艳[2]

(1. 中国气象局广州热带海洋气象研究所/区域数值天气预报重点实验室,广州,510080；
2. 广东省气象台,广州,510080；3. 国家气象中心,北京,100081)

摘　要：精细格点天气预报是中国未来几年天气预报业务的发展重点。本文利用 ECMWF、GRAPES 等业务数值模式和广东站点观测资料,对日极端气温的主客观预报能力进行多角度综合评估。统计结果表明,T_{max}、T_{min} 的主客观预报误差均存在明显季节差异,在系统误差相对平稳的夏半年主观预报具有较明显订正能力,模式预报呈现一定流依赖特征,温度越高(低)的负(正)偏差越明显；主客观预报误差空间分布均受地形影响,随着时效延长误差总体增幅不大,主观订正能力也较稳定。根据以上评估特征和网格预报特点,研究开发了一套多模式动态集成网格释用技术方案(McGF)。结果表明,相比单个模式的预报和主观预报,McGF 较明显提升了 T_{max} 的预报技巧；T_{min} 的模式预报偏差总体较 T_{max} 偏小,McGF 提升幅度相对较小；网格释用后的广东区域预报能较合理反映气温空间和强度特征,较周边未经释用区域明显更优。

关键词：最高(低)气温；预报评估；多模式集成；网格释用

Evaluation on Subjective and Objective Diurnal Extreme Temperature Forecasts and Multi-model Consensus Gridded Forecast Application

WU Naigeng[1,2]　ZENG Qin[2]　LIU Duanling[2]　CHEN Binghong[2]
ZHAO Shengrong[3]　ZHANG Hongyan[2]

(1. Guangzhou Institute of Tropical and Marine Meteorology/Key Laboratory of Regional Numerical Weather Prediction, CMA, Guangzhou, 510080; 2. Guangzhou Meteorological Observatory, Guangzhou, 510080; 3. National Meteorological Center, Beijing, 100081)

Abstract: Digital gridded weather forecast is the developing trend of weather forecasting operation in China. Based on NWP model products, meteorological observation data and an evaluation on subjective and objective forecast, a diurnal extreme temperature Multi-model Consensus Gridded Forecast (McGF) system was developed. Statistic results show that there are significant seasonal forecast error differences in both subjective and objective forecasts, and forecasters have more forecasting skills in summer, when the

---

[①] 本文发表于《气象》2017年第5期。
资助项目：公益性行业(气象)科研专项(GYHY201406003)；中国气象局预报员专项(CMAYBY2015-052)；华南区域气象科技攻关重点项目(GRMC2014Z02)。
第一作者：吴乃庚,主要从事热带灾害性天气气候机理诊断及预测研究。E-mail:wunaigeng@hotmail.com。

NWP systemaitc errors are more stable. NWP model shows a flow dependent (conditional bias) charateristic. When the temperature is higher (lower), the nagative (positive) forecast error is bigger. Both subjective and objective forecasts are effected by topography and there are relatively significant forecast errors in the northern mountainous areas. As the lead time of forecast extends, the growth of forecast errors is not big and the subjective forecasting skills are stable relatively. Based on these results, McGF interpretation application system was devloped with four modules, including real-time verification, station-based interpretation, gridded application and performance-weighted averages. The results showed that T_{max} forecast of McGF are better than subjective and objective forecasts, with its mean absolute errors less than 2 ℃ within 72 h. Relatively speaking, T_{min} forecast errors are much lower and the enhancements of McGF are relatively smaller. The cases of extreme high/low temperature showed that McGF gridded forecasts in Guangdong Province can more reasonably reflect the spatial distribution and intensity feature.

Key words: maximum (minimum) temperature; forecast evaluation; multi-models consensus; gridded interpretation

引 言

随着过去几十年科学技术的不断发展积累,数值天气预报模式水平提高引起的"悄然革命"得到了各界高度关注[1],数值天气预报模式已逐渐成为各国气象预报业务机构最重要的技术支撑。应用研究表明,尽管近几年模式的预报水平快速提高,但相对于形势预报,其要素预报性能离实际需求仍有一定差距,业务中应用需对模式直接输出结果(Direct Model Output,DMO)进行偏差订正[2-6]。

关于模式的偏差订正,20世纪70年代初Glahn等[7]根据数值模式不断发展的特点提出的模式输出统计方法(Model Output Statistics,MOS)在气象业务应用最为广泛,也是美国国家天气局要素预报业务中至今一直在发展应用的重要方法。很多气象学者利用MOS方法结合不同地区和不同要素特点进行模式释用并取得了一定的效果[8-10]。由于传统MOS方法需要较长时间的模式历史预报资料,建立的方程更多地反映采样期间各变量和预报对象的平均关系,并不能根据最新变化及时更新,在现代数值模式快速更新发展、气象要素预报需求越发精细的背景下其弊端逐渐显现[2,11,12]。因此,在传统MOS方法基础上,一些学者运用只需较少模式资料即可滚动更新的卡尔曼滤波等方法,对一些较连续的要素预报进行改进。谢庄等[13]指出,利用卡尔曼滤波技术能一定程度解决MOS方程建模更新慢的问题,实现方程自适应更新。张庆奎[14]针对最高最低温度预报,利用卡尔曼波对基于MOS方法建立的多元回归方程进行动态调整,试验结果表明动态调整后的结果有一定改善,特别是在秋冬转换季节中效果更突出。即便如此,卡尔曼滤波也存在MOS方法类似的对预报因子选取比较敏感且预报有滞后的问题[15]。另外,业务和研究均表明,多模式综合集成预报既能发挥各模式预报结果的优势具有更好的预报技巧,且不会因某个模式性能变化导致综合结果的较大变动[16-19],近年来,模式释用订正也逐渐向多模式集成发展。Krishnamurti[16,17]最早提出了多模式超级集合预报的思想,并随后对850 hPa风场、降水和飓风路径强度等进行超级集合预报试验,发现其能有效减少误差。智协飞等[20-23]利用近年TIGGE科学计划的多模式集合预报资料,应用多模式集合预报技术对北半球中纬度气温、降水和西太平洋台风进行分析应用,亦取得了较单模式更优的结果。在业务应用方面,Woodcock等[19]利用澳大利亚多家业务数值模式和区域MOS产品,综合考虑不同模式近期预报表现基础上动态集成(Operational Consensus Forecasts,OCF)建立了全国客观站点指导预报,性能较模式DMO结果改善明显。漆梁波等[24]将OCF方法应用于上海区域模式及释用产品进行站点预报。释用结果表明,OCF方法较卡尔曼滤波和模式DMO均有一定正效果,其中冬、春季节气温预报水平与主观预报相当,夏、秋季节结果改善不明显。张秀年等[25]利用T213和ECMWF模式产品,对集成MOS方法在温度预报方面做研究,结果表明多模式集成MOS方法较传统MOS方法效果更好。刘还珠等[4,26]持续发展中国MOS客观气象要素预报系统,通过大量因子建模、多模式集成等方面改进,有效提高了城镇站点客观要素指导预报水平。同时,针对近年精细格点预报业务变革,Glahn等[27,28]在过去几十年一直发展的站点MOS系统基础上,把Bergthorssen、Cressman和Doos提出的逐步订正的插值分析方法(简称BCD法),拓展为增加考虑不同下垫面和地形高度影响的格点应用(简称BCDG法),将NCEP-GFS模式的站点MOS结果插值分析到2.5 km分辨率的格点,为美国格点预报业务提供支撑。但该方案也难以避免"插值分析"带来的不确定误差,并且需获取大量精细的下垫面信息,站点布局需有一定海拔落差,计算量也较大。

总体而言,在传统的MOS方法基础上,模式的释用方法逐渐呈现出从固定方程向自适应调整,从单模式向多模式集合转变,从站点释用向精细格点释用拓展的趋势,但目前来说,综合订正技术发展的三大趋势,特别是格点预报的集成应用技术研究仍较少见。未来几年,从传统的站点预报向格点预报转变将是中国天气预报业务改革发展重要方向。面对大量精细格点预报,有必要加强对高分辨率数值模式的格点释用订正技术研究,综合发挥客观技术和预报员主观经验作用。

本文结合模式偏差订正技术发展趋势及较早在国内开展格点预报探索的广东省气象台业务实际,针对日极端气温预报,在对主客观预报能力综合评估基础上,开发出一套以近期检验为基础动态更新、多模式集成的网格释用技术方案,以期为今后格点预报业务开展提供一些参考。

1 资 料

预报资料:2014年1月1日至12月31日,每日12UTC起报的欧洲中期天气预报中心(ECMWF)的全球模式温度预报数据,分辨率为0.25°;中国气象局/广东区域数值预报重点实验室区域模式(GRAPES)的0.12°和0.36°分辨率温度预报数据;广东省气象台(GDMO)5 km分辨率的格点温度预报数据。由于资料传输和存储故障部分数据有缺失,ECMWF、GRAPES和GDMO年内有效数据分别为357 d、298 d和365 d。

实况资料:2014年1月1日至12月31日,广东省86个国家级自动气象站日最高、最低气温观测数据。

地形资料:广东省1 km分辨率地理信息高程数据,包括高度、坡向、坡度等。

对于同一起报时间,实际预报业务制作时主要参考模式提前12~24 h起报的产品,但考虑到本文主要目的并非分析主观预报订正技巧,简便起见,下文提及"订正技巧"均指同起报时次预报。

2 日极端气温主客观预报能力评估

开展模式偏差订正,从主客观预报能力深入分析其在本地预报的系统偏差(可订正)特征十分重要。为直观对比,模式和主观预报分别选取国内外应用最广泛的ECMWF模式和全国格点预报业务试点省广东省气象台网格预报为代表,通过空间插值将网格预报产品插值到气象观测站进行评估。对温度预报的定量评估,已有研究[19,24-25,29]采用能反映整体预报偏差幅度的平均绝对误差(MAE)或均方根误差(RMSE),但两者均存在不能反映误差正负方向问题。平均误差(ME)虽出现正负相抵情况,难以反映整体偏差幅度,但也能提供重要的天气信息[30,31],ME表征的方向性适合用于定量预报订正,预报员思考应用更多的也是ME。因此,为综合衡量主客观预报能力和针对性地设计释用方案,下面综合应用MAE和ME进行评估。ME和MAE计算见公式(1)和(2),其中T_{fc}、T_{ob}分别为预报和观测值,n为需要空间平均或时间平均的样本数量，k 为第 k 个样本。

$$\mathrm{ME}=\overline{T_{\mathrm{fc}}-T_{\mathrm{ob}}}=\frac{1}{n}\sum_{k=1}^{n}\left[T_{\mathrm{fc}}(k)-T_{\mathrm{ob}}(k)\right] \tag{1}$$

$$\mathrm{MAE}=\overline{|T_{\mathrm{fc}}-T_{\mathrm{ob}}|}=\frac{1}{n}\sum_{k=1}^{n}|T_{\mathrm{fc}}(k)-T_{\mathrm{ob}}(k)| \tag{2}$$

2.1 $T_{\max}$ 和 $T_{\min}$ 预报偏差的季节差异

图1(a)为ECMWF模式在广东区域站点平均的24 h $T_{\max}$ 预报的逐日ME、|ME|、MAE变化序列，|ME|与MAE的值越接近越能反映预报大范围内一致性偏高(低)。从图中可见，ECMWF的 $T_{\max}$ 预报偏差季节差异显著，冬半年(11月—次年4月)MAE约2 ℃，夏半年(5—10月)则超过3 ℃。配合ME和|ME|曲线可知，夏半年的偏差为稳定大范围的预报偏低(ME持续为负偏差、|ME|与MAE绝大部分时间重合)。总体来看，ECMWF对 $T_{\max}$ 预报以偏低为主，夏半年 $T_{\max}$ 预报误差更大，但为大范围内稳定偏低，而春秋过渡季节绝对误差较小，但持续偏向性差一些。

GDMO与ECMWF的MAE曲线明显相反(图1b)，呈现为夏半年误差小(1.4 ℃)、冬半年(1.8 ℃)误差大的特征，其中在夏半年较ECMWF显著偏小。这表明尽管夏半年模式 $T_{\max}$ 预报误差较大，但预报员对持续稳定的误差有明显订正能力。

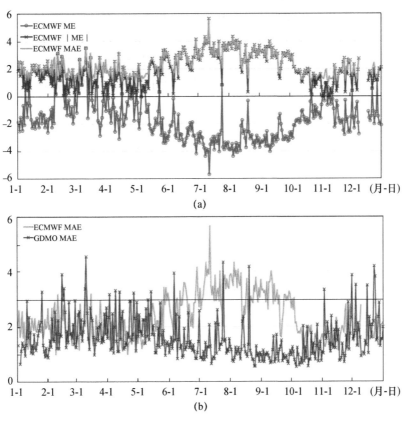

图1　2014年广东区域平均的 $T_{\max}$ 24 h预报误差序列图(单位：℃)
(a)ECMWF的ME、|ME|、MAE；(b)ECMWF和GDMO的MAE

对比 $T_{\max}$ 而言，$T_{\min}$ 预报MAE明显较小，且季节分布特征相反，呈现"冬半年大(约1.5 ℃)夏半年小(约1 ℃)"的特征。由主客观对比可知，在 $T_{\min}$ 相对平稳的夏半年，GDMO有一定正技巧，但在冬半年，特别是春秋过渡季节，平均正技巧并不明显(图2)。

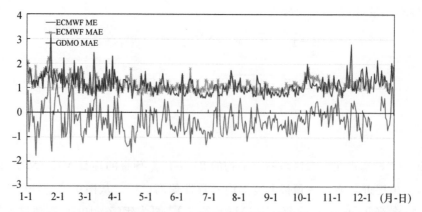

图2　2014年广东区域平均的 T_{min} 24 h 预报 ECMWF 的 ME、MAE 和 GDMO 的 MAE 序列图（单位：℃）

2.2　不同温度强度下 T_{max} 和 T_{min} 预报偏差的差异

为考察不同温度强度（一定程度上反映天气类型）下 T_{max}、T_{min} 预报偏差的差异，图3给出了 ECMWF 模式 T_{max}、T_{min} 的平均误差（$T_{fc}-T_{ob}$）及其气温预报（T_{fc}）的散点分布。已有研究表明[5,30,31]，如果 ME 独立于预报和围绕一个固定值变化，说明存在非条件偏差（Unconditional Bias）；而如果 ME 是流依赖（Flow Dependent），例如误差依赖于预报本身或其他参数，则存在条件性偏向（Conditional Bias）的系统性误差，该情况表明其预报误差与大尺度天气流型密切相关。图3（a）可见，T_{max} 越高（低）呈现出越大的负（正）偏差，表明模式的 T_{max} 预报呈现流依赖特征，存在明显条件性偏差，对较高（低）的气温模式预报更偏低（高），特别是30 ℃以上高温预报严重偏低。T_{min} 的预报亦呈现一些类似特征，但相对没那么明显（图3b）。

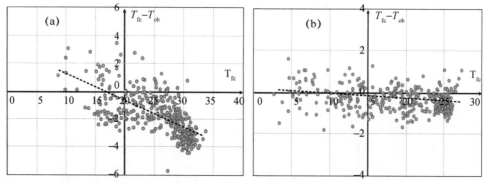

图3　2014年广东区域平均 ECMWF 模式 T_{max}（a）、T_{min}（b）24 h 预报平均误差及其预报值的散点分布图（单位：℃，虚线为线性趋势线）

2.3　T_{max} 和 T_{min} 预报偏差的地区差异

图4给出的是 T_{max}、T_{min} 预报的 MAE 空间分布。从 ECMWF 预报结果可见，其 T_{max}、T_{min} 预报 MAE 均呈现出"北部高、沿海低"分布特征，且南北差异十分明显（北部部分地区误差较沿海偏高超过 2 ℃），而 GDMO 预报有类似特征，但南北差异相对较小。由此可见，尽管广东总体海拔不算太高，相对南部沿海来说，北部南岭山脉地形影响仍使得预报产生更大的偏差。

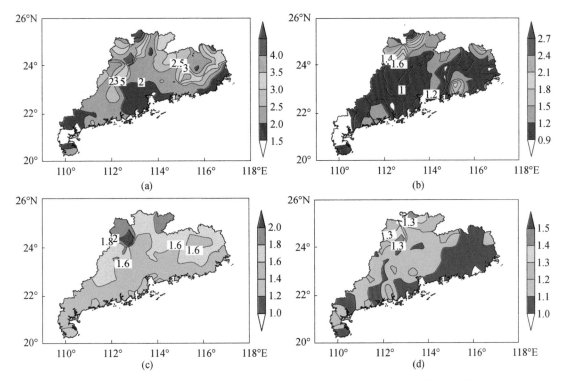

图 4 2014 年平均的最高(低)气温 24 h 预报平均绝对误差 MAE 空间分布(单位:℃)
(a)(b)ECMWF 的 T_{max} 和 T_{min} 误差;(c)(d)GDMO 的 T_{max} 和 T_{min} 误差

2.4 不同预报时效 T_{max} 和 T_{min} 预报偏差的差异

从不同预报时效看(图 5),T_{max} 和 T_{min} 均为时效越长误差越大,但总体误差增长不算太大,特别是 T_{min} 预报误差仅增加约 0.3 ℃。同时,GDMO 预报有类似变化特征,且不同预报时效主观订正能力也较稳定(T_{max} 约 0.7 ℃,T_{min} 约 0.1 ℃)。该结果一定程度上也反映了不同时效的主观预报订正思路和订正的系统误差可能具有一定相似性。

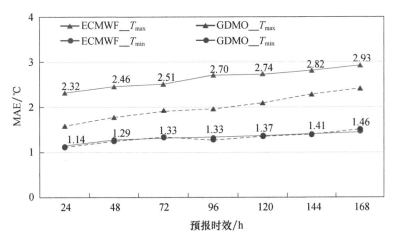

图 5 2014 年平均的 T_{max}/T_{min} 不同预报时效的预报 MAE 序列图(单位:℃)

3 日极端气温的多模式动态集成网格预报释用及效果

由前期主客观预报评估分析可知,在时间和强度方面,模式对日极端气温预报往往存在"高温偏低、低温偏高"特征并具有一定的时间持续性,而主观预报的订正能力也主要体现在持续性的系统偏差;在空间分布和预报时效方面,山区地形对预报偏差产生影响,不同时效主观预报订正能力差异并不大。由此可见,对广东网格预报的后处理释用中,时间持续性偏差和地形影响偏差值得重点考虑。

根据评估分析结果,结合多模式集成和格点释用技术发展趋势,开发一套多模式动态权重集成网格释用技术方案(Multi-model Consensus Gridded Forecast,McGF)。业务程序方案包括实时站点检验、单模式站点释用、格点应用订正和多模式动态权重集成四大部分(图6),程序实现模块化设计,模式数量、订正方案以及格点应用等可实现按用户需求定制。

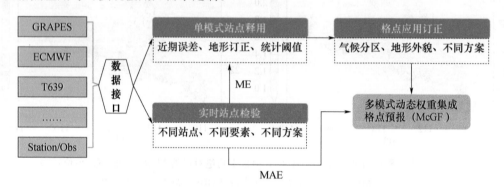

图6 多模式动态权重集成网格释用业务技术流程图

3.1 站点实时检验

实时站点检验包括两部分:①ME检验,考虑ME正负方向性适用于模式的定量预报订正,针对不同数值模式不同站点动态ME评估,用于各模式偏差订正。②MAE检验,考虑MAE更能稳定反映模式的平均性能,对各集成成员模式的预报结果进行动态MAE评估,用于各模式权重集成。另外,考虑到模式误差的流依赖特征以及天气系统的短期波动,本方案时间滑动训练期默认设置为7 d。

3.2 模式偏差订正

不同于传统站点应用,精细格点预报业务中具有大量数字网格,作为主观网格预报制作的初始场,客观订正方案应意义直观明确。基于前期模式性能本地评估特点,重点考虑时间上的持续系统性偏差、空间上的山区地形造成偏差两方面进行订正。

(1)时间持续偏差订正:针对模式预报常存在持续系统性偏差的情况,计算出各集成成员在不同观测站过去7 d的ME,据此分别进行订正。站点订正结果$T(t)$计算见公式(3),其中根据滑动训练期n取7,另考虑到转折性天气较大误差波动以及模式业务中偶尔出现部分时次资料丢失情况,定义滑动训练期内模式极端气温预报(T_{fc})与实况(T_{ob})偏差超过5 ℃则认为该训练日不具有参考价值,将其剔除。

$$T(t)=T_{fc}(t)+\mathrm{ME}(t)=T_{fc}(t)+\frac{1}{n}\sum_{k=1}^{n}\left[T_{fc}(t-k)-T_{ob}(t-k)\right] \quad (3)$$

(2)空间地形影响订正:由前期评估结果可知,尽管广东的国家观测站海拔不高(超过150 m的仅有6个),地形对预报误差仍有一定影响。同时广东丘陵地形多,模式网格地形与真实地形存在较大差异,对精细格点气温分布将有较大影响。因此,进行地形影响订正。格点订正结果T'的计算见公式(4),其中

H_{true}、H_{model} 分别为真实地形和模式地形高度，垂直温度递减率 γ^* 采用基于广东立体气候梯度观测站计算结果[32]。

$$T' = T_{\text{fc}} + \Delta T = T_{\text{fc}} + \frac{1}{100}(H_{\text{model}} - H_{\text{true}}) \times \gamma^* \quad (4)$$

3.3 站点向格点应用

有关格点预报释用，目前的应用研究仍较少。对于格点预报业务而言，简单的站点插值应用意义不大，而以区域站点组合发展代表方程的方式虽具有一定代表性，但准确率较单站点差且存在明显的边界不连续问题，给精细格点预报带来较大影响。Glahn 等[27]在原有 MOS 系统基础上开展的 GRIDDING MOS 方案主要考虑地形和下垫面影响，将站点 MOS 结果插值分析到格点，较好地解决了精度和边界问题，但其实施一方面引入了"插值"误差，另一方面也需大量精细下垫面信息，站点布局需有一定海拔落差（>130 m）且计算量较大。

基于业务实际，预报员关注更多的是站点实况结果，且下垫面对气温的影响深入研究目前而言并不足够精细合理，为减少引入类似"插值"带来的不确定误差，根据气候分区、地形特点和站点距离等影响，本文设计了一套站点向格点应用的方案。站点与格点关联原则如下（按先后顺序）：

（1）格点与相关站点应属同一气候分区（本文分区基于刘黎明等[33]研究结果）；

（2）格点与相关站点应地貌相似（根据 1 km 地理信息资料计算，坡向接近，角度在 90°以内）；

（3）与格点距离最近的站点（若出现多个满足条件站点，以最近的为准）。

根据站点与格点映射对应关系（图 7），将基于站点观测的 ME、MAE 等应用至相应格点，为模式偏差订正以及模式动态权重集成提供基础（该映射对应关系可提供台站的本地进一步动态修正）。

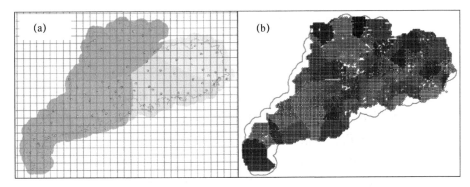

图 7 考虑气候分区、地形地貌的站点与格点关联示意图
(a)不同的气候分区，圆点代表的是观测站点；(b)不同站点映射关联的区域

3.4 动态权重集成方案

研究表明，根据模式最近一段时间的误差来确定权重（Performance-weighted Averages）往往比固定权重平均的合成效果好[19,34]。本文成员权重根据过去 7 d 的 MAE 检验结果动态计算，MAE 越大的成员，权重系数越小。系数计算见公式（5），其中 n 代表成员总数，i 表示某一成员，w 表示权重。

$$w_i = \text{MAE}_i^{-1} \left(\sum_{i=1}^{n} \text{MAE}_i^{-1} \right)^{-1} \quad (5)$$

集成释用结果（McGF）：根据前面各模式偏差订正后的结果（T'_i）进行动态权重集成。

$$\text{McGF} = \sum_{i=1}^{n} (w_i T'_i) \quad (6)$$

3.5 释用效果评估

为了解释效果,下面从业务常用的 MAE、准确(偏差)率分别进行评估,并结合实例给出格点释用空间分布效果。

(1) MAE 评分

对比主客观预报(图8)可见,对于 T_{max} 预报,McGF 的 MAE 评分从 24、28、72 h 分别为 1.30、1.57 和 1.81 ℃,均较模式和主观预报有较大提升,其中较 ECMWF、GRAPES 模式和 GDMO 预报分别提高约 0.7、0.4 和 0.3 ℃。T_{min} 主客观预报误差均较 T_{max} 偏低,McGF 较 GRAPES 模式提升约 0.4 ℃,但较 ECMWF 和 GDMO 提升幅度不大(<0.1 ℃)。

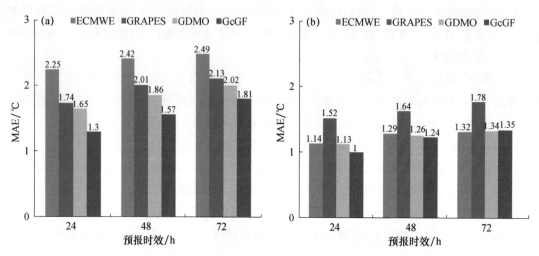

图 8 2014 年广东区域的平均主客观日极端气温预报 MAE(单位:℃)
(a)最高气温;(b)最低气温

(2) 预报准确(偏差)率

为进一步考察 McGF 的预报性能和稳定性,表1给出了准确率和偏差率对比(若对当天广东区域平均预报 MAE<1.5 ℃视为"准确",MAE>2.5 ℃则视为"偏差";准确(偏差)率=准确(偏差)天数/总天数×100%)。由表1可见,对于 T_{max} 预报,McGF 方案对 72 h 内预报准确率分别达 75%、62% 和 52%,明显较其他模式高;而偏差率也明显更小,特别是 24 h 预报偏差率,仅为 4%。

T_{min} 预报准确率明显较 T_{max} 高,McGF 的准确率较表现最好的成员(ECMWF)提高仅约 1%~2%,而偏差率甚至更高一些。这表明对于模式整体误差较小且无明显稳定性系统偏差的 T_{min} 预报,集成方案准确率能保持在较高水平,但异常偏差率较最好成员略有增多。

表 1 2014 年广东区域平均 T_{max}、T_{min} 预报准确率和偏差率

	预报时效	ECMWF (357 d)	GRAPES (298 d)	GDMO (365 d)	McGF (362 d)
T_{max} 预报 (MAE<1.5 ℃)	24 h	17%	43%	58%	75%
	48 h	12%	30%	50%	62%
	72 h	10%	26%	41%	52%
T_{max} 预报 (MAE>2.5 ℃)	24 h	38%	14%	11%	4%
	48 h	42%	23%	18%	11%
	72 h	43%	26%	21%	17%

续表

	预报时效	ECMWF (357 d)	GRAPES (298 d)	GDMO (365 d)	McGF (362 d)
T_{min}预报 (MAE<1.5 ℃)	24 h	89%	60%	85%	89%
	48 h	77%	54%	78%	79%
	72 h	73%	47%	72%	74%
T_{min}预报 (MAE>2.5 ℃)	24 h	0.0%	7%	1%	2%
	48 h	0.2%	10%	2%	6%
	72 h	1.1%	14%	5%	7%

(3) 格点预报的实例

McGF方案除了传统站点预报技巧评分外，精细格点预报分布也是其重要特色。为考察格点预报效果，图9分别给出了一个显著的高温和低温实例。McGF输出为5 km分辨率的矩形网格场，其中图7所示的广东省内采取偏差订正和权重集成，广东省外则为各模式加权平均。图中可见，对于T_{max}、T_{min}预报，McGF释用订正后的广东区域结果均较好地体现了气温的精细分布，特别是广东不多见的大范围38 ℃以上高温区域、大范围0 ℃以下低温区域均与实况有较好对应。同时，对比广东省和周边区域可见，经过偏差订正和权重集成后的广东区域预报效果明显较好，而周边地区偏差较大（T_{max}预报偏低，T_{min}预报偏高）。

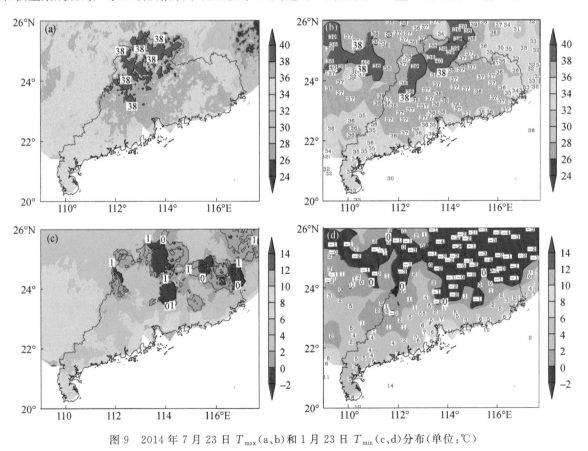

图9 2014年7月23日T_{max}（a、b）和1月23日T_{min}（c、d）分布（单位：℃）

4 小结与讨论

本文结合模式偏差订正技术发展趋势以及格点预报试点省业务实际，针对日极端气温预报，在对主客观预报能力综合评估基础上,开发了一套多模式动态集成的网格释用技术方案。

(1)季节差异方面,T_{max}主客观预报偏差季节差异显著,EC模式夏半年较冬半年明显偏大,并出现大范围持续性预报偏低,而主观预报偏差特征则相反(夏半年误差小、冬半年误差大),对夏半年模式相对持续稳定的误差有较强订正能力;T_{min}主客观预报偏差较T_{max}小,呈现夏(冬)半年误差小(大)的特征,在相对平稳的夏半年主观预报有一定正技巧。

(2)强度预报方面,模式预报呈现一定流依赖特征,当气温越高(低)呈现出越大的负(正)偏差,30℃以上T_{max}预报偏低严重。地区差异方面,主客观预报均受地形影响,北部山区误差相对较大。预报时效方面,主客观预报随预报时效延长总体误差增长不大,主观订正能力也较稳定。

(3)结合前期评估分析,建立了包括实时站点检验、单模式站点释用、格点应用订正和多模式动态权重集成的业务网格释用方案(McGF)。结果表明,相比单个模式的预报和主观预报,McGF较明显提升了T_{max}的预报技巧;T_{min}的模式预报偏差总体较T_{max}偏小,McGF提升幅度相对较小。McGF订正后的广东区域预报较好地体现了气温较精细的空间分布和强度特征,比周边未经网格释用区域明显较好。

值得注意的是,数值模式具备较完善动力热力过程,其产品后处理统计释用可以订正模式的偏差,也可能会引入不少不稳定误差。因此,业务客观释用方案重点在于对模式部分较稳定的系统性偏差进行订正,便于有效发挥预报员和模式的综合作用。目前而言,McGF工作仍较初步,释用结果对模式系统偏差订正有一定能力,但冬春季节总体订正能力不高,转折性天气过程存在着调整滞后、偏差扩大等问题需主观预报进一步订正。另外,结合本地精细观测对主客观预报能力多维度动态评估、精细气候分区、模式过去误差的时间窗选择和下垫面对气温影响规律等也需更进一步开展工作。

参考文献

[1] BAUER P, THORPE A, BRUNET G. The quiet revolution of numerical weather prediction[J]. Nature, 2015, 525(7567):47-55.

[2] MONACHE D, NIPEN T, LIU Y B, et al. Kalman filter and analog schemes to post-process numerical weather predictions[J]. Mon Wea Rev, 2011, 139(11):3554-3570.

[3] 陈法敬,矫梅燕,陈静. 一种温度集合预报产品释用方法的初步研究[J]. 气象,2011,37(1):14-20.

[4] 赵声蓉,赵翠光,赵瑞霞,等. 中国精细化客观气象要素预报进展[J]. 气象科技进展,2012,2(5):12-21.

[5] ANDERSSON E. User guide to ECMWF forecast products: Version 1.1[Z]. ECMWF,2013.

[6] 陈良吕,陈静,陈德辉,等. 基于T213集合预报的延伸期产品释用方法及初步试验[J]. 气象,2014,40(11):1293-1301.

[7] GLAHN H R, LOWRY D A. The use of Model Output Statistics (MOS) in objective weather forecasting[J]. J Appl Meteor, 1972,11(8):1203-1211.

[8] LOWRY D A, GLAHN H R. An operational model for forecasting probability of precipitation—PEATMOS PoP[J]. Mon Wea Rev, 1976,104(3):221-232.

[9] 丁士晟. 中国MOS预报的进展[J]. 气象学报,1985,43(3):332-338.

[10] GLAHN B. Determining an optimal decay factor for Bias-correcting MOS temperature and Dew point forecasts[J]. Weather & Forecasting, 2014,29(4):1076-1090.

[11] JACKS E, BOWER J B, DAGOSTARO V J, et al. New NGM-based MOS guidance for maxima and minima temperature, probability of precipitation, cloud amount, and surface wind[J]. Wea & Forecasting, 1990,5(5):128-138.

[12] VISLOCKY R L, FRISCH J M. Generalized additive models versus linear regression in generating probabilistic MOS forecasts of aviation weather parameters[J]. Wea Forecasting, 1995,10(4):669-680.

[13] 谢庄,黄嘉佑. 卡尔曼滤波在MOS方程自适应更新中的应用介绍[J]. 北京气象,1993(2):16-17.

[14] 张庆奎. 基于MOS方法的客观温度预报模型的研究与应用[D]. 南京:南京信息工程大学,2009.

[15] 任宏利,丑纪范. 统计-动力相结合的相似误差定制法[J]. 气象学报,2005,63(6):988-992.

[16] KRISHNAMURTI T N, KISHTAWAL C M, LAROW T E, et al. Improved weather and seasonal climate forecasts from multimodel superensemble[J]. Science,1999,285(5433):1548-1550.

[17] KRISHNAMURTI T N, KISHTAWAL C M, ZHANG Z, et al. Multimodel ensemble forecasts for weather and seasonal climate [J]. Journal of Climate,2000,13(23):4196-4216.

[18] ELIZABETH E EBERT. Ability of a poor man's ensemble to predict the probability and distribution of precipitation [J]. Mon Wea Rev,2001,129(10):2461-2480.

[19] WOODCOCK F, ENGEL C. Operational consensus forecasts[J]. Weather & Forecasting,2005,20(1):101-111.

[20] 智协飞,林春泽,白永清,等. 北半球中纬度地区地面气温的超级集合预报[J]. 气象科学,2009,32(5):569-574.

[21] ZHI X F, QI H X, BAI Y Q, et al. A comparison of three kinds of multimodel ensemble forecast techniques based on the TIGGE data[J]. Acta Meteorologica Sinica,2012,26(1):41-51.

[22] 张涵斌,智协飞,王亚男,等. 基于TIGGE资料的西太平洋热带气旋多模式集成预报方法比较[J]. 气象,2015,41(9):1058-1067.

[23] 李佰平,智协飞. ECMWF模式地面气温预报的四种误差订正方法的比较研究[J]. 气象,2012,38(8):897-902.

[24] 漆梁波,曹晓岗,夏立,等. 上海区域要素客观预报方法效果检验[J]. 气象,2007,33(9):9-18.

[25] 张秀年,曹杰,杨素雨,等. 多模式集成MOS方法在精细化温度预报中的应用[J]. 云南大学学报:自然科学版,2011,33(1):67-71.

[26] 刘还珠,赵声蓉,陆志善,等. 国家气象中心气象要素的客观预报——MOS系统[J]. 应用气象学报,2004,15(2):181-182.

[27] GLAHN B, GILBERT K, COSGROVE R, et al. The gridding of MOS[J]. Weather & Forecasting,2009,24(2):520-529.

[28] RUTH D P, GLAHN B, DAGOSTARO V, et al. The performance of MOS in the digital age[J]. Weather & Forecasting,2009,24(2):504-519.

[29] 周兵,赵翠光,赵声蓉. 多模式集合预报技术及其分析与检验[J]. 应用气象学报,2006,17(S1):104-109.

[30] NURMI P. Recommendations on the verification of local weather forecasts[Z]. ECMWF Tech Mem,2003.

[31] WILKS D. Statistical methods in the atmospheric sciences[M]. London:Academic Press,2006.

[32] 刘蔚,王春林,陈新光,等. 基于立体气候观测的粤北山区热量资源特征多[J]. 应用生态学报,2013,24(9):2571-2580.

[33] 刘黎明,陈创买. 广东的气候分区[J]. 热带气象学报,1998(1):47-53.

[34] YOUNG G. Combining forecasts for superior prediction preprints[C]//16th Conf on Probability and Statistics in the Atmospheric Sciences. Orlando, Amer Meteor Soc,2002.

基于SWAN雷达拼图产品在暴雨过程中的对流云降水识别及效果检验[①]

张勇　吴胜刚　张亚萍　刘伯骏　龙美希　邹倩

（重庆市气象台，重庆，401147）

摘　要：本文基于SWAN雷达拼图产品，选取组合反射率因子、组合反射率因子水平梯度、回波顶高及垂直累积液态水含量作为识别参数，采用模糊逻辑法对暴雨过程中的对流云与层状云降水进行分类试验，对发生在重庆的12次区域性暴雨天气过程分类结果进行验证；并以ADTD地闪资料作为对流云降水的实况观测数据，分别采用4种不同半径的空间匹配与4种不同时间匹配方式对识别出的对流云降水产品进行定量检验。检验结果显示：随着空间匹配半径的增大，正确率明显提高，而6 min地闪相对于6 min拼图产品提前6、3、0 min及滞后3 min的4种时间匹配方式，其正确率变化很小。对12次暴雨过程的总体评分较高，检验方法具有清楚的物理意义，在不同的时空匹配方式下的评分结果符合实际情况，同时说明对流云与层状云降水分类效果较好，也对对流云降水识别定量检验的一次探索。

关键词：SWAN拼图产品；暴雨；降水分类；地闪；检验

Identification and Effect Verification of Convective Cloud Precipitation in Rainstorm Processes Based on SWAN Mosaic Products

ZHANG Yong　WU Shenggang　ZHANG Yaping　LIU Bojun　LONG Meixi　ZOU Qian

(Chongqing Meteorological Observatory, Chongqing, 401147)

Abstract: Based on SWAN(Severe Weather Automatic Nowcast System) radar mosaic products, we selected the composite reflectivity factor and its horizontal gradient, echo top height and vertically integrated liquid water content as identification parameters, and conducted a classification test for convective cloud and stratiform cloud precipitation in rainstorm processes using the fuzzy logic method. The results were verified by 12 regional rainstorm processes that occurred in Chongqing. Taking the Advanced TOA and Direction System lightning data as objective observation data of convective cloud precipitation, we tested the related products quantitatively by four different radiuses of spatial matching and four different time matching method, respectively. The verification results showed that with the increase of space matching radius, the correct rate improves significantly. However the correct rate of 6 min cloud-to-ground lightning flashes relative to 6 min mosaic products ahead of 6 min, 3 min, 0 min and lagging

---

[①]　本文发表于《气象》2019年第2期。

资助项目：中国气象局预报员专项(CMAYBY2016-059)；重庆市气象部门业务技术攻关项目(YWJSGG-201903)。

第一作者：张勇，主要从事天气预报工作及雷达资料应用研究。E-mail: zangy110@126.com。

3 min four time matching method, changes slightly. The overall score of the rainstorm is high and the test method has clear physical meaning. The score results under different temporal and spatial matching patterns are consistent with the actual situation, which means the classification outcome of the convective and stratiform rainfall is good. This is an exploration of convective cloud precipitation quantitative test as well.

Key words: SWAN(Severe Weather Automatic Nowcast System) mosaic products; rainstorm; precipitation classification; cloud-to-ground lightning; verification

引 言

大范围内的暴雨天气过程,常常是层状云与对流云降水的混合,在大片的降水中嵌入对流降水系统。因为对流云与层状云降水具有不同的特点,可以区分出对流云与层状云降水,这对进一步研究对流云与层状云降水机理、定量降水估测、强对流天气监测预警、人工影响天气作业指挥、航空航天及数值预报中的资料同化以及对流活动的气候统计分析等方面起到重要作用(仲凌志 等,2007;Rao et al.,2011;Chen M Z et al.,2012;Chen Z C et al.,2014)。

目前,国内外对降水或降水云分类方面的研究较多,根据使用不同的资料可分为:天气雷达(Churchill et al.,1984;Steiner et al.,1995;Biggerstaff et al.,2000;王静 等,2007;肖艳姣 等,2007)、风廓线雷达(Williams et al.,1995;Rao et al.,2008a,2008b;黄钰 等,2015)、气象卫星(Adler et al.,1988;Baum et al.,1997;白慧卿 等,1998;师春香 等,2002)及雨量计(Houze,1973;Baldwin et al.,2005),也有的利用多种资料共同识别或相互验证。从时空分辨率及空间覆盖范围方面考虑,采用天气雷达资料来进行降水分类具有明显优势,国内外这方面的研究较多。在国外,Churchill等(1984)利用雷达反射率因子阈值确定对流中心,再固定一个对流中心影响的半径以确定对流云的面积。Steiner等(1995)在上述基础上将固定的影响半径改为反射率因子的函数,同时将反射率因子阈值改为局地平均背景反射率因子的函数,对识别效果有所改进。Biggerstaff等(2000)考虑0℃层亮带的影响及对流云与层状云的三维结构特征,采用雷达反射率因子来对其进行识别。在国内,仲凌志等(2007)在Steiner等的"峰值法"的基础上,根据中国暴雨的特点,对方法中的步骤与参数设置进行调整,实现本地移用,并分析降水分类在定量降水估测中的应用。王静等(2007)采用神经网络方法对雷达资料进行降水分类研究,取适当的参数时取得较好的试验效果。肖艳姣等(2007)采用模糊逻辑法对雷达回波进行对流云与层状云的分类,结果显示分类效果较好。方德贤等(2016)综合利用雷达、探空资料将风暴按强度进行分类,并应用于人工防雹。前面的研究均采用单雷达资料,一方面,单雷达本身观测具有一定的局限;另一方面,对降水分类结果的检验大多建立在主观识别的基础上或利用其他资料(如降水)验证,检验结果具有一定的主观性,检验范围也具有一定的局限性,一定程度上限制了实际业务应用。近几年,一些算法在多雷达拼图资料上得到扩展或改进(勾亚彬 等,2014;李国翠 等,2014;杨吉 等,2015),同时,闪电的应用研究也越来越广泛(郄秀书 等,2014)。本文利用多部雷达组网优势(肖艳姣 等,2006),并根据对流云与层状云表现出的不同特点,采用SWAN(Severe Weather Automatic Nowcast System)雷达拼图产品基于模糊逻辑法(Fuzzy Logic Method,FLM)(Zadeh,1968;肖艳姣 等,2007)来区分降水类型,并以ADTD(Advanced TOA and Direction system)闪电资料作为实况资料来定量检验对流云的识别效果。这里假定对流云产生雷电,并被ADTD探测到,由此得到定量的检验结果,利用闪电来检验对流云降水具有清楚的物理意义。

1 对流云与层状云降水分类

1.1 模糊逻辑法

模糊逻辑法(Fuzzy Logic Method,FLM)最早由 Zadeh(1968)提出,其最大特点是不需要识别量的具体值,仅需要识别量较宽松的分级区间,即可得到较为合适的结果,具有较强的扩充性和兼容性,在冰雹识别及云分类方面得到广泛应用(曹俊武 等,2005;刘黎平 等,2007;Baum et al.,1997)。传统的 FLM 包括四个过程:模糊化、规则推断、集成与退模糊。FLM 隶属函数采用梯形函数系的基本形式,隶属函数表达式为(肖艳姣 等,2007)

$$T(x,x_1,x_2) = \begin{cases} 1 & x \geqslant x_2 \\ (x-x_1)/(x_2-x_1) & x_1 < x < x_2 \\ 0 & x \leqslant x_1 \end{cases} \quad (1)$$

式中,x 表示识别量。根据对流云与层状云降水特点,选取 4 个识别量,分别是组合反射率因子(CR)、组合反射率因子水平梯度(GCR)、反射率因子为 18 dBZ 的回波顶高(ET)、垂直累积液态水含量(VIL),其中 CR、ET、VIL 直接使用 SWAN 输出产品,参与 SWAN 拼图的雷达包括重庆 4 部及周边的 8 部,雷达站点及型号见表 1。GCR 参考肖艳姣等(2007)的计算方法,其表达式为

$$\text{GCR} = \max\left(\left|\frac{\lg(|Z_{i+n}-Z_{i-n}|)}{2n}\right|, \left|\frac{\lg(|Z_{j+n}-Z_{j-n}|)}{2n}\right|\right) \quad (2)$$

式中,Z 表示 Z_{CR},单位为 mm^6/m^3;i 与 j 分别代表 x 方向第 i 个格点与 y 方向的第 j 个格点;n 表示格点数,并且当 $n=2$ 时,层状云与对流云的 GCR 差异最明显,GCR 的单位为 dB/km。对于识别参量门限值 x_1 与 x_2 的取值,严格来说,需要通过大量的样本统计来确定,但由于统计大量样本存在困难,且区分层状云与对流云降水的识别量也较主观,实际业务中一般结合经验给出门限值。这里通过暴雨个例中识别量的大致取值范围并结合经验及参考相关文献(肖艳姣 等,2007;Steiner et al.,1995;Chen et al.,2012),4 个识别量门限值分别为:对于 CR,$x_1=25$,$x_2=45$;对于 GCR,$x_1=0.4$,$x_2=1.0$;对于 ET,$x_1=6$,$x_2=12$;对于 VIL,$x_1=2$,$x_2=10$。确定识别量的门限值后,其模糊基函数也就确定了。在区分层状云与对流云降水时,首先将识别量通过模糊基函数模糊化;再通过识别参数模糊值按照各自的权重累加得到确定值,即计算 $T=k_1T(\text{CR})+k_2T(\text{GCR})+k_3T(\text{ET})+k_4T(\text{VIL})$,这里取权重系数 $k_1=k_2=k_3=k_4=0.25$;最后将 T 值与设置的识别阈值比较,这里设置识别阈值为 0.5,当 $T>0.5$ 时即识别为对流云降水,否则判断为层状云降水。

表 1 SWAN 拼图雷达站点及型号

序号	雷达站点	雷达型号
1	重庆	CINRAD/SA
2	万州	CINRAD/SB
3	永川	CINRAD/SA
4	黔江	CINRAD/CD
5	成都	CINRAD/SC
6	宜宾	CINRAD/SC
7	南充	CINRAD/SC
8	达州	CINRAD/SC

续表

序号	雷达站点	雷达型号
9	遵义	CINRAD/CD
10	怀化	CINRAD/CB
11	恩施	CINRAD/SB
12	宜昌	CINRAD/SA

1.2 分类结果初步分析

2015—2016年重庆发生了12次区域性暴雨过程，各暴雨过程的开始与结束时间以及对应的研究区域见表2。这里将研究区域划分为西部、东北部、东南部，并分别用A、B、C表示，其对应范围见图1b。这样划分的主要原因有3点：一是每次暴雨过程的主要降雨具有区域性特点，暴雨过程对应的研究区域即为主要的降水区，可以减少6 min间隔的高频次资料处理；二是研究区域均在闪电定位仪探测网覆盖范围内，确保在研究区域内闪电资料的全覆盖，并以此作为检验资料；三是根据重庆范围内的地形地貌及气候特点，西部主要位于四川盆地的东南部，主要以丘陵为主，东北部与东南部主要是山区地带，且东北部与东南部在气候特点上也存在明显的差异。三个研究区对应的暴雨过程均为4次，应用FLM将12次暴雨过程逐6 min间隔的拼图产品区分为层状云与对流云降水，得到6 min间隔的降水分类产品。

图2是暴雨1在2015年6月30日03时的分类结果，其中蓝色表示层状云降水，红色表示对流性降水。图2a是应用FLM直接分类的结果，可以看出存在较多孤立的对流性回波。采用3×3格点中值滤波处理后得到图2b，可以看出滤波处理后去掉了部分孤立对流点，对流性降水回波的整体形态更符合实际情况。图3是对应时刻的雷达拼图组合反射率因子及滤波后的分类降水产品与地闪的叠加。从雷达回波(图3a)看，降水回波范围较大，回波强度变化范围较大，其中小于25 dBZ一般是层状云降水回波，大于45 dBZ一般认为是对流性降水回波，具有回波强度强、回波密实、回波发展高度较高、水平梯度较大、更容易产生雷电等特点。25~45 dBZ为混合性降水回波，在较均匀的层状云降水回波中镶嵌有较强的对流性回波。从分类结果与地闪的叠加图(图3b)看，对流性降水与强回波具有较好的一致性，在研究区域内，大部分地闪在对流性降水区及其附近。图4是对应时刻的4个识别量，对应的对流云降水均表现出较大值的特点，也符合对流云降水的特点，同时也进一步说明降水分类结果具有合理性。

表2 选取的12次暴雨过程的研究时段及对应区域

序号	暴雨过程时段/年月日时—月日时	简称	研究区域
1	2015062908—070120	暴雨1	A
2	2015071408—071514	暴雨2	C
3	2015072120—072220	暴雨3	A
4	2015081614—081908	暴雨4	A
5	2015091020—091208	暴雨5	A
6	2016050614—050808	暴雨6	C
7	2016053120—060202	暴雨7	B
8	2016061818—062014	暴雨8	C
9	2016062318—062508	暴雨9	B
10	2016063002—070110	暴雨10	B
11	2016071319—071508	暴雨11	B
12	2016071814—072008	暴雨12	C

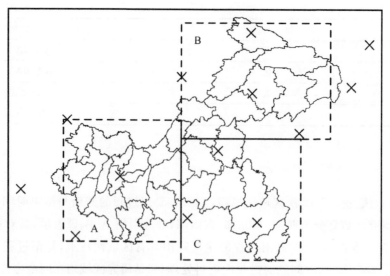

图 1 研究区域及 13 部闪电定位仪分布示意图

虚线框代表研究区域,西部、东北部、东南部分别用 A、B、C 表示,×表示闪电定位仪

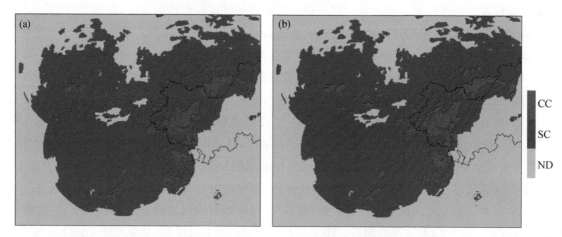

图 2 暴雨 1 在 2015 年 6 月 30 日 03 时层状云与对流云降水分类结果

(a) 3×3 中值滤波前;(b) 3×3 中值滤波后

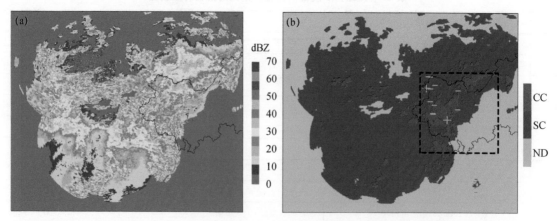

图 3 暴雨 1 在 2015 年 6 月 30 日 03 时反射率因子与降水分类(虚线框代表研究区域)

(a) 雷达拼图组合反射率因子;(b) 层状云与对流云降水分类结果与地闪叠加

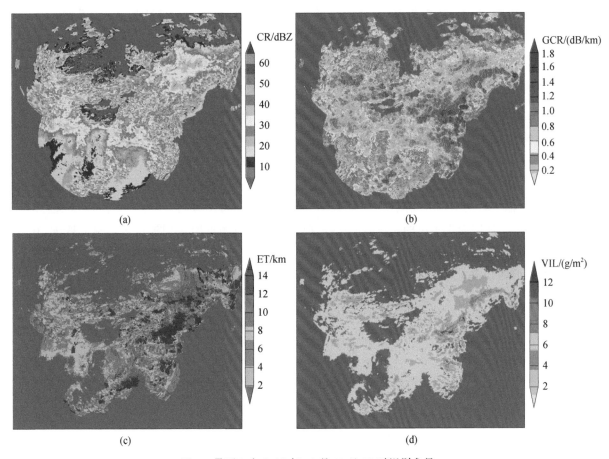

图4 暴雨1在2015年06月30日03时识别参量
(a)组合反射率因子;(b)组合反射率因子水平梯度;(c)回波顶高;(d)垂直累积液态水含量

2 分类效果定量检验

2.1 检验方法

前面简单分析了层状云与对流性降水分类结果,利用雷达回波强度与地闪资料,从宏观上初步分析了层状云与对流云分类识别效果,但没有给出定量的检验效果。这里尝试定量方法来评估识别效果,首先利用ADTD地闪资料作为检验效果的实况资料。ADTD是中国科学院空间科学与应用研究中心研制的闪电定位系统,各子站探测范围250 km,按150 km的基线距离及一定几何形状组成探测网,其探测效率高于85%(曾金全 等,2016),时间精度为0.1 μm,网内探测定位精度小于300 m(支树林 等,2018)。本文中ADTD资料来自于13部闪电定位仪组成的探测网,完全覆盖研究区域,其空间分布见图1。ADTD观测的是二维地闪信息,假定对流云降水产生闪电,且闪电与地面接通,即云地闪,且云地闪被ADTD观测定位,这样就可以用ADTD观测到的地闪来检验对流云的识别效果。在实际定位中,由于定位方法、电磁干扰、地形地貌及数据传输等影响,一部分数据存在失真的情况,参照曾金全等(2016)的处理方式,剔除电流幅值或陡度为0的闪电。从ADTD的探测原理可以知:ADTD主要探测到的是云地之间的闪电。因而实际情况中,对流云降水产生云间、云内及云对空气的闪电就不能被探测到。在业务中有这样的经验,能听到雷声,而ADTD系统上没有观测到相应的闪电。鉴于此,采用适当宽松的模糊时间、空间匹配方法来定量检验对流云降水的分类效果。实际上,国内研究人员已采用模糊方式检验或评估不能完全确定的目标(李佰平 等,2016;马申佳 等,2018)。表 3 是对流云降水识别与 ADTD 地闪观测列联表,正确率 R_r(保留一位小数)表达式为

$$R_r = \frac{N_r}{N_t} \times 100\%$$

$$N_t = N_r + N_w \tag{3}$$

式中,N_t 表示与雷达观测匹配的 6 min 地闪发生次数;N_r 表示正确次数;N_w 表示错误次数。如,在 6 min 内研究区域内观测到 10 次地闪,其中 6 次地闪的匹配区有对流云降水,则对该次识别效果评定正确率为 60.0%。地闪与对流云采用点面的空间匹配方法,观测到的地闪对应一定范围内有对流云即为正确,否则错误。设置地闪周围 2、5、10、20 km 半径范围内作为与地闪的匹配面区域,在时间匹配上,地闪处理成与拼图产品一致的 6 min 间隔,设置时间起点分别为雷达拼图资料时间点的前 6 min、前 3 min、0 min 及后 3 min。地闪与识别出的对流云的空间、时间匹配示意图见图 5。图 5a 表示空间匹配,在以地闪为中心半径 R 范围内有识别出的对流云即为正确。图 5b 表示时间匹配,地闪与雷达拼图均为 6 min 间隔,地闪起始时间相对于雷达拼图起始时间分别为 -6、-3、0 及 3 min,并分别用 A、B、C 及 D 表示。

表 3 对流云降水识别与 ADTD 地闪观测列联表

对流云降水识别/N_t	ADTD 地闪观测 N_t		
	对	N_r	$N_t = N_r + N_w$
	错	N_w	

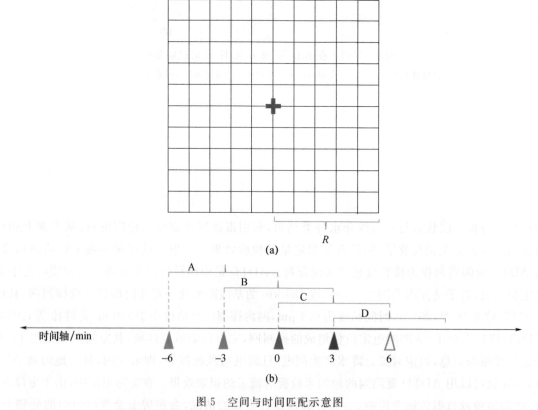

图 5 空间与时间匹配示意图

(a)空间匹配,✚ 代表地闪位置,R 表示地闪匹配的空间半径;(b)时间匹配,▲ 代表雷达拼图产品起始时刻,其后的红色直线表示拼图产品的时间长度 6 min,△ 表示地闪资料相对于拼图产品时间的起始时刻,地闪数据的时长与雷达拼图产品一致,均为 6 min,并根据其相对位置,依次用 A、B、C、D 表示

2.2 检验结果

图6是暴雨1在半径2 km的空间匹配与A方式的时间匹配的逐6 min检验结果序列图,在研究区域内观测到地闪共4369次,其中识别出对流性降水对应的有2883次,正确率为66.0%(表4、表5)。从逐6 min检验结果看,6 min的地闪频次变化很大,最大达到177次,对应有151次评定为正确,正确率为85.3%。单从正确率看,有的时次达到100%,有的为0,正确率100%与0对应的地闪频次一般较低,多数在几次之内。在A方式的时间匹配下,不同半径(2、5、10和20 km)空间匹配的逐6 min检验结果序列见图7。从图中可以看出,随着匹配半径的增大,正确率也增大,分别为66.0%、74.0%、81.8%与88.4%(表4、表5)。图8是在5 km半径的空间匹配方式下,不同时间匹配方式的检验序列。从图中可以看出,检验结果变化很小,A、B、C和D时间匹配方式逐6 min的正确率非常一致,分别为74.0%、74.3%、74.5%和74.7%(表4、表5)。表4、表5是在不同的时空匹配方式下的检验结果,考虑随机因素可能带来的影响,给出了去掉1次及≤3次地闪后的检验结果。从表4中可以看出,无论是否去掉1次及≤3次地闪的情况,A、B、C和D 4种时间匹配下的2、5、10和20 km匹配半径的平均正确率略有增大,但增大都不明显,在1%以内。在去掉1次及≤3次地闪时,正确率略有提高,约为1%,对检验结果影响较小。在未去掉少频次地闪时,平均正确率分别为77.6%、77.9%、78.0%和78.2%,去掉1次地闪时,平均正确率分别为78.4%、78.7%、78.8%和79.0%,去掉≤3次地闪时,平均正确率分别为79.5%、79.8%、79.9%和80.1%。表5是2、5、10及20 km 4种匹配半径下的A、B、C和D时间匹配的检验结果,同时也给出去掉1次及≤3次地闪时的检验结果。从表中可以看出,随着匹配半径的增大,正确率明显提高,在未去掉少频次地闪时,2 km半径的空间匹配在4种时间匹配方式下的平均正确率为66.4%,5、10和20 km分别为74.4%、82.2%和88.8%。当去掉1次及≤3次地闪时,正确率依次提高1%左右,对检验结果影响较小,检验结果较稳定。由此可见,当匹配半径增大时,实际上是放宽了评定正确率的条件,正确率提高明显,符合实际情况,而在时间匹配变化不大的情况下(6 min以内),对流云的移动、变化较小,其地闪的空间时空变化不大,因而评定的正确率变化不大,也符合实际情况。从此次暴雨过程对流云降水分类的定量评分结果看,总体评分较高,在不同的时空匹配方式下的评分结果也符合实际情况,说明对流云与层状云降水分类效果较好,同时也是对对流云降水识别定量检验的一次尝试。

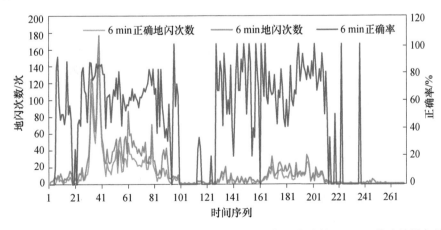

图6 暴雨1采用半径2 km的空间匹配及时间匹配为A方式的逐6 min检验结果序列

表6、表7是12次暴雨过程的检验结果,由于去掉1次及≤3次地闪时对检验结果影响较小,表中没有给出这两种情况的检验结果。从表中可以看出,其检验结果与以上分析的暴雨1趋势完全一致,并且在12次暴雨过程的平均情况中,正确率高于暴雨1。12次暴雨过程在时间匹配方式为相对于拼图产品前6 min、前3 min、0 min和后3 min时,4种空间匹配方式的平均正确率分别为84.5%、84.7%、84.6%和

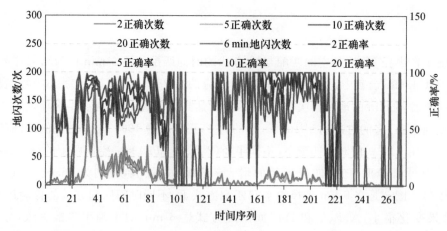

图 7　暴雨 1 分别采用半径 2、5、10 和 20 km 的空间匹配及时间匹配为 A 方式的逐 6 min 检验结果序列

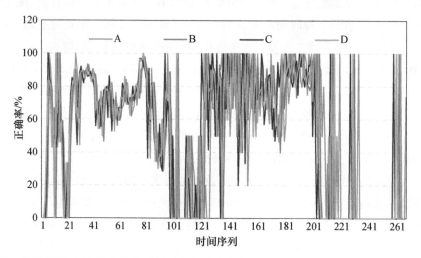

图 8　暴雨 1 分别采用 A、B、C 和 D 的时间匹配方式及半径 5 km 的空间匹配的逐 6 min 的正确率序列

84.2%。在 2、5、10 和 20 km 的空间半径匹配时,4 种时间匹配方式的平均正确率分别为 75.1%、82.2%、87.8% 与 92.8%。

表 4　暴雨 1 不同时间匹配方式在不同空间匹配半径下的对流性
降水分类定量检验结果(无处理/去掉 1 次地闪/去掉≤3 次地闪)

时间匹配方式	时间匹配长度/min	空间匹配半径/km	对流性降水对应的地闪次数	地闪总次数	正确率/%	平均值/%
A	6	2	2883/2879/2857	4369/4315/4212	66.0/66.7/67.8	77.6/78.4/79.5
		5	3232/3227/3197		74.0/74.8/75.9	
		10	3573/3565/3529		81.8/82.6/83.8	
		20	3864/3852/3804		88.4/89.3/90.3	
B	6	2	2879/2876/2861	4346/4288/4201	66.2/67.1/68.1	77.9/78.7/79.8
		5	3227/3221/3203		74.3/75.1/76.2	
		10	3568/3556/3533		82.1/82.9/84.1	
		20	3861/3845/3818		88.8/89.7/90.9	

续表

时间匹配方式	时间匹配长度/min	空间匹配半径/km	对流性降水对应的地闪次数	地闪总次数	正确率/%	平均值/%
C	6	2	2896/2891/2872	4347/4294/4198	66.6/67.3/68.4	78.0/78.8/79.9
		5	3238/3233/3207		74.5/75.3/76.4	
		10	3573/3565/3531		82.2/83.0/84.1	
		20	3859/3848/3805		88.8/89.6/90.6	
D	6	2	2886/2880/2861	4336/4279/4190	66.6/67.3/68.3	78.2/79.0/80.1
		5	3237/3230/3209		74.7/75.5/76.6	
		10	3576/3564/3540		82.5/83.3/84.5	
		20	3858/3841/3811		89.0/89.8/91.0	

表5 同表4，但为不同空间匹配半径方式在不同的时间匹配下的结果

空间匹配半径/km	时间匹配长度/min	时间匹配方式	对流性降水对应的地闪次数	地闪总次数	正确率/%	平均值/%
2	6	A	2883/2879/2857	4369/4315/4212	66.0/66.7/67.8	66.4/67.1/68.2
		B	2879/2876/2861	4346/4288/4201	66.2/67.1/68.1	
		C	2896/2891/2872	4347/4294/4198	66.6/67.3/68.4	
		D	2886/2880/2861	4336/4279/4190	66.6/67.3/68.3	
5	6	A	3232/3227/3197	4369/4315/4212	74.0/74.8/75.9	74.4/75.2/76.3
		B	3227/3221/3203	4346/4288/4201	74.3/75.1/76.2	
		C	3238/3233/3207	4347/4294/4198	74.5/75.3/76.4	
		D	3237/3230/3209	4336/4279/4190	74.7/75.5/76.6	
10	6	A	3573/3565/3529	4369/4315/4212	81.8/82.6/83.8	82.2/83.0/84.1
		B	3568/3556/3533	4346/4288/4201	82.1/82.9/84.1	
		C	3573/3565/3531	4347/4294/4198	82.2/83.0/84.1	
		D	3576/3564/3540	4336/4279/4190	82.5/83.3/84.5	
20	6	A	3864/3852/3804	4369/4315/4212	88.4/89.3/90.3	88.8/89.6/90.7
		B	3861/3845/3818	4346/4288/4201	88.8/89.7/90.9	
		C	3859/3848/3805	4347/4294/4198	88.8/89.6/90.6	
		D	3858/3841/3811	4336/4279/4190	89.0/89.8/91.0	

表6 暴雨1~12不同时间匹配方式在不同空间匹配方式下的对流性降水分类正确率

时间匹配方式	时间匹配长度/min	空间匹配半径/km	对流性降水对应的地闪次数	地闪总次数	正确率/%	平均值/%
A	6	2	30018	39756	75.5	84.5
		5	32635		82.1	
		10	34882		87.7	
		20	36864		92.7	

续表

时间匹配方式	时间匹配长度/min	空间匹配半径/km	对流性降水对应的地闪次数	地闪总次数	正确率/%	平均值/%
B	6	2	30119	39736	75.8	84.7
B	6	5	32758	39736	82.4	84.7
B	6	10	34888	39736	87.8	84.7
B	6	20	36861	39736	92.8	84.7
C	6	2	29868	39697	75.2	84.6
C	6	5	32730	39697	82.4	84.6
C	6	10	34844	39697	87.8	84.6
C	6	20	36867	39697	92.9	84.6
D	6	2	29230	39630	73.8	84.2
D	6	5	32507	39630	82.0	84.2
D	6	10	34854	39630	87.9	84.2
D	6	20	36809	39630	92.9	84.2

表7 同表6，但为不同空间匹配方式在不同时间匹配方式下的结果

空间匹配半径/km	时间匹配长度/min	时间匹配方式	对流性降水对应的地闪次数	地闪总次数	正确率/%	平均值/%
2	6	A	30018	39756	75.5	75.1
2	6	B	30119	39736	75.8	75.1
2	6	C	29868	39697	75.2	75.1
2	6	D	29230	39630	73.8	75.1
5	6	A	32635	39756	82.1	82.2
5	6	B	32758	39736	82.4	82.2
5	6	C	32730	39697	82.4	82.2
5	6	D	32507	39630	82.0	82.2
10	6	A	34882	39756	87.7	87.8
10	6	B	34888	39736	87.8	87.8
10	6	C	34844	39697	87.8	87.8
10	6	D	34854	39630	87.9	87.8
20	6	A	36864	39756	92.7	92.8
20	6	B	36861	39736	92.8	92.8
20	6	C	36867	39697	92.9	92.8
20	6	D	36809	39630	92.9	92.8

3 结论与讨论

利用SWAN雷达拼图产品资料，采用模糊逻辑法（FLM）实现暴雨过程中的对流云与层状云降水分类。选取发生在重庆的12次暴雨过程作为试验对象，并利用ADTD地闪观测资料对逐6 min的对流性降水分类产品进行定量检验。得到如下结论：

（1）根据对流云与层状云降水的雷达回波及其产品所表现出的不同特征，选取组合反射率因子(CR)、组合反射率因子水平梯度(GCR)、回波顶高(ET)及垂直累积液态水含量(VIL)4个识别量，采用模糊逻辑法(FLM)实现暴雨过程中基于SWAN逐6 min拼图产品生成逐6 min的对流云与层状云降水分类产品。

（2）采用ADTD地闪探测资料，假设对流云降水产生闪电并被ADTD定位观测到，以ADTD地闪作为实况数据，采用较宽松的模糊时间与空间的匹配方式对流云降水分类产品进行定量检验。结果显示：空间匹配半径分别为2、5、10、20 km时，随着匹配半径的增大，正确率明显提高，分别为75.1%、82.2%、87.8%与92.8%。而6 min闪电相对于6 min对流性降水识别产品提前6、3、0 min及滞后3 min 4种时间匹配方式，其正确率变化很小，分别为84.5%、84.7%、84.6%及84.2%。

FLM的应用很广，本文的对流云与层状云降水分类应用关键点主要有三点：第一是识别量的选取。选取的基本原则是识别量在对流云与层状云降水应表现出不同的特点。第二是识别量门限值的确定。理想情况下，识别量门限值应该通过大量的样本统计，通过识别量的概率密度分布特征来确定，然而统计本身需要基于确定的对流云与层状云降水样本，实际情况是没有对流云与层状云降水的统一区分标准，也就是对流云与层状云降水统计样本本身存在一定的主观性。在实际应用中，识别量门限值一般通过典型的个例统计或结合经验来确定，这也正是FLM的优势，不需要识别量的具体值，仅需要较宽松的分级区间。第三是识别量的权重分配。权重应该由识别量与识别对象的相关性确定，识别量与识别对象相关性越强其权重应该越大，反之亦然。实际应用中定量描述识别量与识别对象较难，一般也由经验给出权重，本文没有对识别量的权重进行分析与研究。

本文以ADTD地闪资料作为对流云降水的实况观测，采用模糊时空匹配方式，对基于SWAN拼图产品识别出的对流云降水进行定量检验，检验本身是基于对流性降水的物理过程，具有清楚的物理意义，从检验结果分析看也具有合理性，这是对对流云降水分类效果定量检验的一次探索。

参考文献

白慧卿,方宗义,吴蓉璋,等,1998.基于人工神经网络的GMS云图四类云系的识别[J].应用气象学报,9(4):402-409.
曹俊武,刘黎平,葛润生,2005.模糊逻辑法在双线偏振雷达识别降水粒子相态中的研究[J].大气科学,29(5):827-836.
方德贤,李红斌,董新宁,等,2016.风暴分类识别技术在人工防雹中的应用[J].气象,42(9):1124-1134.
勾亚彬,刘黎平,杨杰,等,2014.基于雷达组网拼图的定量降水估测算法业务应用及效果评估[J].气象学报,72(4):731-748.
黄钰,阮征,罗秀明,等,2015.垂直探测雷达的降水云分类方法在北京地区的应用[J].高原气象,34(3):815-824.
李佰平,戴建华,张欣,等,2016.三类强对流天气临近预报的模糊检验试验与对比[J].气象,42(2):129-143.
李国翠,刘黎平,连志鸾,等,2014.利用雷达回波三维拼图资料识别雷暴大风统计研究[J].气象学报,72(1):168-181.
刘黎平,吴林林,杨引明,2007.基于模糊逻辑的分步式超折射地物回波识别方法的建立和效果分析[J].气象学报,65(2):252-260.
马申佳,陈超辉,智协飞,等,2018.基于时空不确定性的对流尺度集合预报效果评估检验[J].气象学报,76(4):578-589.
郄秀书,刘冬霞,孙竹玲,2014.闪电气象学研究进展[J].气象学报,72(5):1054-1068.
师春香,瞿建华,2002.用神经网络方法对NOAA-AVHRR资料进行云客观分类[J].气象学报,60(2):250-255.
王静,程明虎,2007.用神经网络方法对雷达资料进行降水类型的分类[J].气象,33(7):55-59.
肖艳姣,刘黎平,2006.新一代天气雷达网资料的三维格点化及拼图方法研究[J].气象学报,64(5):647-657.
肖艳姣,刘黎平,2007.三维雷达反射率资料用于层状云和对流云的识别研究[J].大气科学,31(4):645-654.
杨吉,郑媛媛,夏文梅,等,2015.雷达拼图资料上中尺度对流系统的跟踪与预报[J].气象,41(6):738-744.
曾金全,杨超,王颖波,等,2016.基于统计分布特征的闪电强度等级划分[J].暴雨灾害,35(6):585-589.
支树林,李婕,陈娟,2018.江西不同类型强对流天气的地闪统计特征及与雷达回波特征对比分析[J].气象,44(2):

222-232.

仲凌志,刘黎平,顾松山,2007. 层状云和对流云的雷达识别及在估测雨量中的应用[J]. 高原气象,26(3):593-602.

ADLER R F,NEGRI A J,1988. A satellite infrared technique to estimate tropical convective and stratiform rainfall[J]. J Appl Meteor,27(1):30-51.

BALDWIN M E,KAIN J S,LAKSHMIVARAHAN S,2005. Development of an automated classification procedure for rainfall systems[J]. Mon Wea Rev,133(4):844-862.

BAUM B A,TOVINKERE V,TITLOW J,et al,1997. Automated cloud classification of global AVHRR data using a fuzzy logic approach[J]. J Appl Meteor,36(11):1519-1540.

BIGGERSTAFF M I,LISTEMAA S A,2000. An improved scheme for convective/stratiform echo classification using radar reflectivity[J]. J Appl Meteor,39(12):2129-2150.

CHEN M X,WANG Y C,GAO F,et al,2012. Diurnal variations in convective storm activity over contiguous North China during the warm season based on radar mosaic climatology[J]. J Geophys Res,117(D20):D20115.

CHEN X C,ZHAO K,XUE M,2014. Spatial and temporal characteristics of warm season convection over Pearl River Delta region,China,based on 3 years of operational radar data[J]. J Geophys Res,119(22):12447-12465.

CHURCHILL D D,HOUZE R A JR,1984. Development and structure of winter monsoon cloud clusters on 10 December 1978[J]. J Atmos Sci,41(6):933-960.

HOUZE R A JR,1973. A climatological study of vertical transports by cumulus-scale convection[J]. J Atmos Sci,30(6):1112-1123.

RAO T N,KIRANKUMAR N V P,RADHAKRISHNA B,et al,2008a. Classification of tropical precipitating systems using wind profiler spectral moments. Part Ⅰ: algorithm description and validation[J]. J Atmos Oceanic Technol,25(6):884-897.

RAO T N,KIRANKUMAR N V P,RADHAKRISHNA B,et al,2008b. Classification of tropical precipitating systems using wind profiler spectral moments. Part Ⅱ: statistical characteristics of rainfall systems and sensitivity analysis[J]. J Atmos Oceanic Technol,25(6):898-908.

RAO T N,RAO D N,MOHAN K,et al,2011. Classification of tropical precipitating systems and associated Z-R relationships[J]. J Geophys Res,106(D16):17699-17711.

STEINER M,HOUZE R A JR,YUTER S E,1995. Climatological characterization of three-dimensional storm structure from operational radar and rain gauge data[J]. J Appl Meteor,34(9):1978-2007.

WILLIAMS C R,ECKLUND W L,GAGE K S,1995. Classification of precipitating clouds in the tropics using 915-MHz wind profilers[J]. J Atmos Oceanic Technol,12(5):996-1012.

ZADEH L A,1968. Fuzzy algorithms[J]. Inf Control,12(2):94-102.

基于SCMOC的贵州最高气温预报方法研究

李 刚[1]　杨秀庄[1]　刘彦华[1]　陈贞宏[2]　余 清[1]　吴昌航[1]

(1. 贵州省气象台,贵阳,550002；2. 贵州省安顺市气象局,安顺,561000)

摘　要：为探究最高气温预报准确率偏低这一现象,采用2013—2018年SCMOC精细化指导预报资料及气象站观测资料,对贵州省85站24～72 h最高气温预报展开研究。通过建立横向预报模型(F1)、纵向预报模型(F2)以及横向与纵向预报相整合的预报模型(Fzh),对贵州日最高气温进行试验。结果表明,无论是平均均方根误差还是预报准确率,各模型预报效果均有不同程度改进,3种客观订正预报中Fzh表现最优；相对于SCMOC,F1在春、夏季的预报优于秋、冬季,且在贵州省北部地区改进较南部明显,F2在4个季节均有改进,总体较平稳；Fzh预报结果明显优于F1和F2,均方根误差得到明显改善,平均RMSE下降1.0～2.0 ℃,准确率平均提高11%～13%。

关键词：最高气温；SCMOC；均方根误差；预报准确率

Forecast of Maximum Temperature Based on Refined Guidance SCMOC Data in Guizhou Province

LI Gang[1]　YANG Xiuzhuang[1]　LIU Yanhua[1]　CHEN Zhenhong[2]
YU Qing[1]　WU Changhang[1]

(1. Guizhou Meteorological Observatory, Guiyang, 550002;
2. Anshun Meteorological Bureau of Gui Zhou Province, Anshun, 561000)

Abstract: Due to the low accuracy of the maximum temperature in Guizhou Province, the refined guidance forecast SCMOC data and meteorological station observations data from 2013 to 2018 were used to study the maximum temperature forecast for 24-72 hour of 85 stations in Guizhou. The horizontal prediction model(F1), the vertical forecast model(F2) and the combination model with F1 and F2(Fzh) were established to forecast daily maximum temperature in Guizhou. The results show that Fzh took a best performance among three models, and for these three models, both mean root-mean-square error (RMSE) and accuracy rate were improved to varying degrees. Compared with SCMOC, the forecast results of F1 in spring and summer were better than that in autumn and winter, and the improvement in the northern areas of Guizhou Province was more obvious than southern areas. The forecast results of F2

---

① 本文发表于《干旱气象》2020年第3期。
资助项目：国家预报员专项"深秋初冬时节静止锋减弱北抬对贵州气温的差异性分析"(CMAYBY2016-065)；国家气象关键技术集成与应用(面上)项目"贵州气温的多模式集合预报研究与应用"(CMAGJ2014M45)；2019年贵州省气象台业务项目"贵州省实况格点偏差订正及研究"及公益性行业(气象)科研专项(201106009)"地形复杂地区的MOS预报效果改进方法研究"。
第一作者：李刚(1983—)，男，贵州黔西县人，硕士，高级工程师，主要从事短期预报、数值预报释用研究。E-mail:lg0857@163.com。

were improved all the year round, and the overall improvement was relatively stable. The forecast accuracy rate of Fzh was obviously better than F1 and F2, and RMSE decreased significantly, the mean RMSE decreased 1.0 to 2.0 ℃, the accuracy rate increased by 11% to 13%.

Key words: maximum temperature; SCMOC; root mean square errors; forecast accuracy rate

引 言

随着气象科技的不断进步和社会经济的发展,对外公众预报服务及政府决策气象服务要求也越来越高。尽管当前数值预报技术发展迅速,其精细化程度及预报准确率都有改进和提高,但与观测值之间仍存在误差[1,2]。为提高天气预报精度和预报准确率,更好地进行公众及政府决策气象服务,有必要对模式直接输出结果进行解释应用和客观技术订正[3]。近年来,针对数值预报后处理的客观订正研究较多,特别是在气温预报方面取得不少有意义的研究成果[4-7]。在当前天气预报服务工作中,气温预报一直是很复杂的科学问题[8,9],其影响因子较多,且不同天气形势下影响气温的因子各有不同。为此,在实际气温预报业务中,多数气象台站皆是在数值模式预报产品基础上,利用多年历史累积观测资料,通过分析不同因子对气温的影响程度而建立不同的预报模型得出最终预报结论[10-13],或是利用不同的后处理技术对数值预报进行订正[14-16]。这些预报方法无论是主观经验分析还是客观统计回归分析等,在一定程度上都有效延长了温度预报时效,改善了预报准确率。

多元回归分析是目前气象统计分析中最常用也是最基本的方法之一[17,18],其理论严谨成熟,适用于气温等气象要素的变化特性(在一定时间和空间范围内变化可视为线性变化)。韦淑侠[19]利用多因子线性回归MOS统计方法对青海省51个观测站24～240 h预报间隔24 h的日最低气温、最高气温进行研究,结果表明当天气形势变化较平稳时,MOS方法制作的气温预报结果不但可用,而且相当稳定。在气温客观订正预报中,除了MOS方法和单个预报模式输出结果的应用外,多模式集合预报也取得较好地发展。该方式较好地发挥并集成了各预报模式优势的同时给出较单模式更优的预报结果[20-22],大幅度改进了预报效果。

近年来,关于贵州气温预报已有一些研究成果,主要是某一技术方法下的某种或多种预报资料、预报因子的研究应用,而在同一预报资料中采用多种技术方法进行整合却尚未开展。基于此,本文根据贵州特殊地理位置及现行业务运行实际情况,针对近年来贵州最高气温预报在全国一直处于排名倒数状态(来源于中央气象台业务网 http://10.1.64.146/npt/product/iframe/42250),利用中国气象局每日下发2次的城镇精细化预报指导产品(SCMOC),一方面运用精细化预报产品在同一预报时刻的不同要素(如最高气温、最低气温、降水量、相对湿度、最低云量及总云量等)进行影响因子选取,建立横向预报模型[3];另一方面利用精细化指导预报提供的最高气温预报与一定时段的观测序列建立纵向预报模型[11,23];最后再对横向预报与纵向预报进行整合集成,得出新的最高气温预报结果,旨在检验SCMOC预报产品的误差订正能力。此外,选用SCMOC站点预报资料进行研究,一方面因为它是当前预报业务应用参考产品,其预报结果有较好的参考价值;另一方面因为,国家气象局及贵州省当前预报质量考核与评估仍然以站点数据为基础,直接应用下发的站点指导预报,相对网格预报插值到站点的预报值误差更小。

1 资料与方法

1.1 资料

选取2013—2018年中国气象局每日下发贵州省逐日20:00起报的24～72 h精细化预报指导产

品,包括日最高气温(T_{max})、最低气温(T_{min})及逐日 14:00 的最低云量(Lcc)、总云量(Tcc)、降水量(Rain)、相对湿度(RH)等,预报间隔为 24 h。利用 2013—2018 年贵州省 85 个国家地面观测站逐日 20:00 至次日 20:00 地面日最高气温(来源于国家局统一建设布置的数据环境 CIMISS)进行预报模型的计算和检验。

1.2 预报方法

1.2.1 一元线性回归模型

在纵向预报方程建立中,应用精细化指导预报产品的最高气温与所对应观测站点的逐日最高气温建立回归模型,对未来时刻气温预报进行求解和判断,以下简称"纵向预报"(F2),具体公式如下:

$$Y_{it} = aX_{it} + b \tag{1}$$

式中:X_{it} 指 t 时刻第 i 个站点或格点预报值;Y_{it} 指 t 时刻第 i 个站点的观测值;回归系数 a、b 由时间序列 $t=1,2,\cdots,t-1$ 的观测和预报值经过训练计算而确定[4,21]。为尽可能减小计算误差和提高预报准确率,其训练期长度主要选取 15～30 d 中计算误差最小的天数作为该站点或格点上的计算训练期。

1.2.2 多元线性回归模型

通过对同一预报时效上的不同要素进行相关性分析,选取相关性较为显著的影响因子进行预报模型的建立,以下简称"横向预报"(F1),具体公式如下:

$$y = \sum_{j=1}^{m} b_j x_j + b_0 + \varepsilon \tag{2}$$

式中:y 是预报结果;x_j 是第 j 个预报因子;b_0,b_j 是回归系数($j=1,2,\cdots,m$),由最小二乘法计算而得;ε 是服从正态分布的随机误差[24]。

1.2.3 预报整合模型

为进一步改进预报准确率,将纵向预报 F2 和横向预报 F1 的预报结果进行整合集成,以下简称"整合预报"(Fzh),具体公式如下:

$$y_p = \sum_{k=1}^{m} a_k y_k \tag{3}$$

式中:y_p 为整合集成后预报模型的预报结果;y_k 为第 k 个预报模型的预报值;a_k 为第 k 个预报值的系数,主要通过一定时间序列绝对误差的倒数取其平均值最小化而求解,其计算训练期与公式(1)一致。

1.2.4 滑动系数的应用及气候偏差订正

以往预报模型对训练期的分析计算中,其系数不变,使得预报时间越向后其误差越来越大。为使预报结果进一步改进和误差最优化,在预报模型中应用滑动系数及气候偏差订正[16,22],即在保持使用训练期长度不变的情况下,通过每天增加新的实况和预报值录入,舍弃第一天的观测及对应预报值,从而达到模型系数的滑动更新。而气候偏差订正主要利用短期内预报模型与气候态的系统偏差对预报结果进行订正,具体订正如下:

$$\tilde{F}_l = F_l - (\bar{F}_l - \bar{O}) \tag{4}$$

式中:$\tilde{F}_l$ 为订正后的预报值;F_l 为第 l 个模型的预报值;$\bar{F}_l$ 为第 l 个模型的预报值平均,时间长度与公式(1)的训练期长度保持一致;$\bar{O}$ 为气候平均值。

1.3 检验方法

1.3.1 预报准确率

根据中国气象局对各省、地(市)气象部门现行业务的考核标准,以预报值与观测值的绝对误差≤2.0 ℃作为温度预报准确率检验的阈值。即绝对误差小于等于2.0 ℃为正确,否则为错误,其计算公式如下:

$$TT = \frac{N_r}{N_f} \tag{5}$$

式中:TT 为预报准确率;N_r 为预报正确的站(次)数;N_f 为预报的总站(次)数。

1.3.2 均方根误差

对于某种预报方法,通过不同的检验手段从不同角度对其进行评估,除上述准确率外,还采用均方根误差对预报结果进行检验。具体计算如下:

$$RMSE = \left[\frac{1}{n}\sum_{s=1}^{n}(F_s - O_s)^2\right]^{\frac{1}{2}} \tag{6}$$

式中:RMSE 为均方根误差;F_s 为预报值;O_s 为观测值;s 为资料长度($s=1,2,3,\cdots,n$)。

2 结果分析

2.1 预报因子选取

选取影响日最高气温预报的诸多因子进行分析,如日最高气温、最低气温、每日 14:00 的低云量、总云量、相对湿度和降水量等。为在预报模型中得到较好的影响因子而建立最优化的预报模型,有必要对这些因子进行分析和筛选。表1列出 2013—2018 年贵州省实测日最高气温与精细化指导预报产品 24~72 h 预报中逐 24 h 各预报要素的相关系数。可以看出,实测日最高气温与精细化城镇预报指导产品提供的最高气温、最低气温呈正相关性,相关系数较高(均在 0.88 以上),且通过 99.9% 的显著性检验;与 14:00 低云量和总云量均呈负相关,相关系数分别在 −0.27 和 −0.14 以下,且分别通过 98.0% 和 90.0% 的显著性检验;与 14:00 相对湿度、降水量也呈负相关,但相关系数均较低(绝对值均在 0.1 以下),且未通过 90.0% 的显著性检验,即相对湿度和降水量对最高气温的预报影响较小,在预报模型中作剔除处理。

表1 2013—2018 年贵州省实测日最高气温与精细化指导预报产品各要素在不同预报时效中的相关系数

预报时效	T_{max}	T_{min}	14:00 Lcc	14:00 Tcc	14:00 Rain	14:00 RH
24 h	0.93***	0.88***	−0.27**	−0.14*	−0.06	−0.08
48 h	0.92***	0.88***	−0.32**	−0.16*	−0.07	−0.07
72 h	0.91***	0.88***	−0.34**	−0.17*	−0.07	−0.07

注:*、**、*** 分别表示通过 0.1、0.02、0.001 的显著检验。

2.2 结果对比

图1为 2013—2018 年精细化指导预报(SCMOC)、横向预报(F1)、纵向预报(F2)和整合预报(Fzh)在 24~72 h 逐 24 h 预报的平均均方根误差和预报准确率。可以看出,相对 SCMOC 预报结果,横向预报、纵向预报及整合预报的平均均方根误差(RMSE)均有不同程度的减小,且整合预报改进效果最明显,在

24～72 h预报中,各时效平均RMSE减小约1.0 ℃,其中72 h预报改进最大(平均RMSE减小1.1 ℃)。不同预报模型的平均准确率显示,在24～72 h预报中,横向预报、纵向预报及整合预报在准确率上均有不同的改进,且效果最好的同样是整合预报,在整个预报时效中平均准确率在原有基础上提高了10%左右,其中改进最大的为72 h预报,达11%。

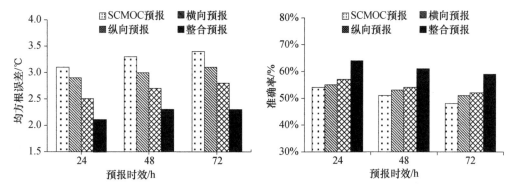

图1 2013—2018年贵州省SCMOC预报、横向预报(F1)、纵向预报(F2)和整合预报(Fzh)最高气温的平均均方根误差(a)和准确率(b)

表2、表3分别列出2013—2018年精细化指导预报(SCMOC)、横向预报(F1)、纵向预报(F2)和整合预报(Fzh)的地面最高气温在不同预报时效、不同阈值下的平均均方根误差和平均预报准确率所占预报区域百分率。由表2可见,在24～72 h内各预报随着各阈值的增大其平均均方根误差所占区域百分率均明显增加,且各阈值下SCMOC、F1、F2和Fzh预报所占比例也依次增加,其中占比最大、效果最明显的为整合预报(Fzh)。在SCMOC提供的24 h预报中,阈值小于2.0 ℃的平均均方根误差所占区域比例为0,经过整合后预报可达24%,而在阈值小于2.5 ℃时整合预报Fzh平均均方根误差的区域占比可达78%;SCMOC 72 h预报通过整合后,在阈值小于2.5 ℃时占全省的比例从0增至38%。因此相对于同时次、同阈值的F1与F2,Fzh改进最明显,特别是相对于精细化指导预报(所占百分比为0),预报效果得到大幅度改善。

由表3可以看出,在不同阈值下,Fzh较其他3类预报表现好,所占区域百分率明显增加。在阈值大于40%时,72 h内SCMOC、F1、F2和Fzh预报模型均几乎覆盖全省范围,表现较好;阈值大于50%时,SCMOC在24～72 h预报中所占全省预报区域从96%降至13%,通过整合预报后Fzh结果所覆盖区域均在99%以上,其中改进最显著的为72 h预报,占全省区域比率从13%增加至99%;在阈值大于60%及以上时,SCMOC几乎没有任何预报能力(所占区域几乎小于1%),但通过客观订正后,F1、F2和Fzh平均预报准确率均有不同程度改进,其中效果最显著的为24 h预报,Fzh预报占全省比率从1%增加至72%,而准确率大于70%的区域占全省比率从0增加至13%,较其他3类预报表现出更优的预报性能。

表2 2013—2018年SCMOC预报、横向预报(F1)、纵向预报(F2)和整合预报(Fzh)的地面最高气温在不同预报时效、不同阈值下的平均均方根误差占贵州省的百分比

单位:%

平均RMSE阈值	24 h时效				48 h时效				72 h时效			
	SCMOC	F1	F2	Fzh	SCMOC	F1	F2	Fzh	SCMOC	F1	F2	Fzh
<2.0 ℃的区域	0	0	0	24	0	0	0	12	0	0	0	5
<2.5 ℃的区域	0	27	41	78	0	29	7	45	0	22	0	38
<3.0 ℃的区域	28	45	99	100	5	36	99	99	1	35	98	96
<3.5 ℃的区域	96	93	100	100	76	78	100	100	59	72	100	100

表3 2013—2018年SCMOC预报、横向预报(F1)、纵向预报(F2)和整合预报(Fzh)的
地面最高气温在不同预报时效、不同阈值下的平均预报准确率占贵州省的百分比

单位：%

预报准确率阈值	24 h时效				48 h时效				72 h时效			
	SCMOC	F1	F2	Fzh	SCMOC	F1	F2	Fzh	SCMOC	F1	F2	Fzh
>40%的区域	100	98	100	100	100	89	100	100	100	87	100	100
>50%的区域	96	61	99	100	51	52	95	99	13	41	56	99
>60%的区域	1	32	7	72	0	32	1	45	0	31	0	34
>70%的区域	0	0	0	13	0	1	0	4	0	0	0	1

为进一步了解各客观预报的性能，以72 h预报为例，图2、图3分别列出了2013—2018年精细化指导预报(SCMOC)和其他3类客观预报(F1、F2和Fzh)的平均均方根误差(RMSE)及平均预报准确率的空间分布。可以看出，F1、F2和Fzh预报的平均RMSE在全省范围内均有明显减小。其中，Fzh预报改进最明显，相对SCMOC预报，全省平均RMSE下降1.0~1.5 ℃；其次为F2预报。另外，还可以看出，F1对贵州中部以北部地区改进极为明显，平均RMSE由最高4.2 ℃下降至最低2.7 ℃，改善幅度达1.5 ℃；南部地区改进略差，平均RMSE下降幅度在0.5~1.0 ℃，在西南部的预报却不如SCMOC。F2较

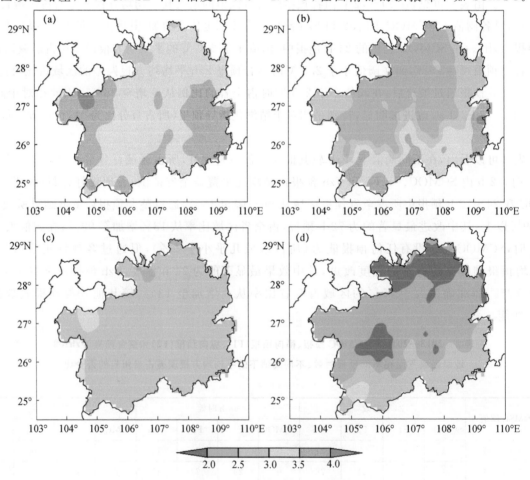

图2 2013—2018年贵州省SCMOC预报(a)、横向预报F1(b)、纵向预报F2(b)和整合预报Fzh(d)
在72 h的最高气温的平均均方根误差空间分布(单位：℃)

SCMOC 改进较为平稳,全省平均 RMSE 下降 1.0 ℃ 左右。对横向预报与纵向预报进行集成整合后,Fzh 相对于 SCMOC 全省范围内预报效果明显提升,对北部地区改进幅度突破 2.0 ℃ 以上,对 F1 在西南部预报不足的情况也得到较大改善。图 3 为预报准确率的空间分布,其表现形势与图 2 的平均 RMSE 比较类似,即 F1 较 SCMOC 中、北部较明显,F2 较 SCMOC 整体较平稳。其中效果最显著的仍为 Fzh,Fzh 全省平均准确率在原有基础上(SCMOC)平均提升 11%,在中部以北提升 15%~22%,在中部以南提升 5%~9%。

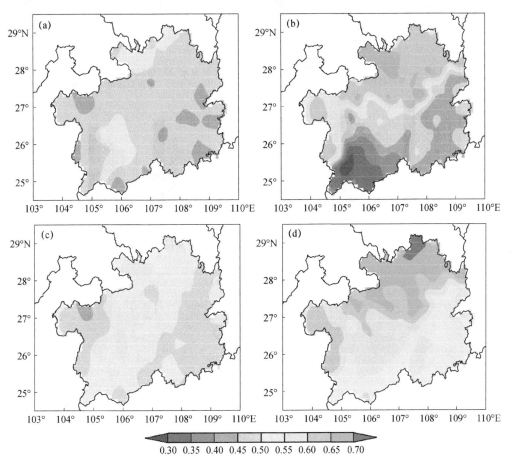

图 3　2013—2018 年贵州省 SCMOC 预报(a)、横向预报 F1(b)、纵向预报 F2(b)和
整合预报 Fzh(d)在 72 h 最高气温的准确率空间分布

为详细了解各客观预报的改进能力,图 4 展示了 2013—2018 年贵州省不同季节、不同时效下的平均预报准确率。可以看出,通过客观方法订正后,各季节在各预报时效上准确率均有不同程度提高,但由于季节差异,SCMOC 预报在春、夏季总体预报准确率低于秋、冬季。F1 在春季和夏季相对于 SCMOC 各预报时效上改进较为突出,尤其在春季预报准确率超过 F2,但在秋季各预报时效及冬季 24 h 预报时,其订正效果不如 SCMOC 预报。相对于 SCMOC 各预报时效,F2 在四个季节中均有改进,且较稳定。3 种客观预报模型中,改进最明显的为整合预报 Fzh,在 24~72 h 预报中,各季节的平均准确率在原有基础上(SCMOC)均提升 10% 左右,其中,提升最高为夏季 72 h 预报,其平均准确率提升达 12%。

图 5 为 2013—2018 年贵州省夏季 SCMOC 预报、横向预报 F1、纵向预报 F2 和整合预报 Fzh 在 72 h 最高气温的平均准确率空间分布。可以看出,3 类客观订正预报相对于 SCMOC 均有改进,且改进效果最好的仍为整合预报,相对于 SCMOC 全省平均准确率提升 12% 以上。其中,改进最明显的是贵州省北部地区,预报准确率由 42% 提升至 74%,改进幅度达 32%。

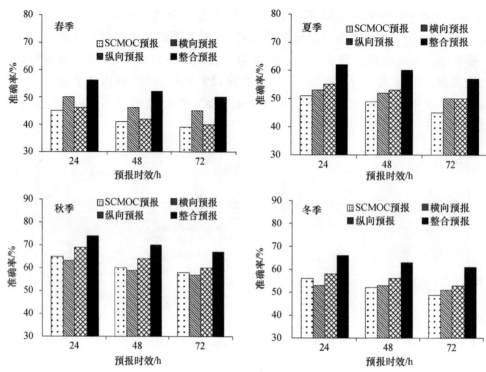

图 4　2013—2018 年贵州省不同季节 SCMOC 预报、横向预报 F1、纵向预报 F2 和整合预报 Fzh 的平均预报准确率

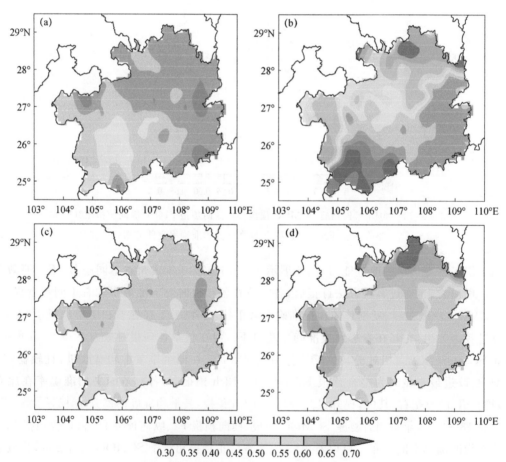

图 5　2013—2018 年贵州省夏季 SCMOC 预报(a)、横向预报 F1(b)、纵向预报 F2(b)和整合预报 Fzh(d)在 72 h 最高气温的平均准确率空间分布

3 结论与讨论

(1)横向预报(F1)的最高气温在24～72 h预报中较SCMOC预报均有明显改进,特别是在贵州省中北部地区改进较显著,平均RMSE减小幅度达1.5 ℃。

(2)纵向预报(F2)的改进效果相对于F1在贵州省中北部地区略差一些,但全省总体优于F1,相对于SCMOC,平均RMSE和平均预报准确率均有改进,且其性能在空间、时间上表现相对较稳定。

(3)整合预报(Fzh)相对于SCMOC、F1和F2,在24～72 h预报中各阈值上,其平均预报准确率和平均均方根误差的改善幅度最明显,全省平均RMSE下降1.0～2.0 ℃,准确率平均提高11%,其中在贵州省中部以北地区为15%～22%。

由于季节和空间的差异,不同订正方法均有各自优势,在不同季节和区域上表现出不同的改进效果和稳定程度。如在季节上,F1订正效果整体优于F2,但秋季F1却不如F2;在空间分布上,F1在贵州省的中北部优于F2,但南部地区F1又不如F2。预报集成整合弥补了不同订正方法在不同季节和空间上的差异,Fzh展现出更好的预报效果,大幅度提升预报准确率。

参考文献

[1] 薛志磊,张书余. 气温预报方法研究及其应用进展综述[J]. 干旱气象,2012,30(3):451-458.
[2] 关吉平,张立凤,张铭. 集合预报研究现状与展望[J]. 气象科学,2006,26(2):228-235.
[3] 王丹,黄少妮,高红燕,等. 递减平均法对陕西SCMOC精细化温度预报的订正效果[J]. 干旱气象,2016,34(3):575-583.
[4] 智协飞,林春泽,白永清,等. 北半球中纬度地区地面气温的超级集合预报[J]. 气象科学,2009,29(5):578-583.
[5] 白永清,林春泽,陈正洪,等. 基于LAPS分析的WRF模式逐时气温精细化预报释用[J]. 气象,2013,39(4):460-465.
[6] 贾丽红,张云惠,何耀龙,等. 基于多模式的新疆最高(低)气温预报误差订正及集成方法研究[J]. 干旱气象,2018,36(2):310-318.
[7] 李佰平,智协飞. ECMWF模式地面气温预报的四种误差订正方法的比较研究[J]. 气象,2012,38(8):897-902.
[8] 王丹,高红燕,张宏芳,等. 一种逐时气温预报方法[J]. 干旱气象,2015,33(1):89-97.
[9] 张弛,王东海,巩远发. 基于WRF/CALMET的近地面精细化风场的动力模拟试验研究[J]. 气象,2015,41(1):34-44.
[10] 李玲萍,尚可政,钱莉,等. 最优子集回归在夏季高温极值预报中的应用[J]. 兰州大学学报(自然科学版),2010,46(6):54-58.
[11] 罗菊英,周建山,闫永财. 基于数值预报及上级指导产品的本地气温MOS预报方法[J]. 气象科技,2014,42(3):443-450.
[12] 吴乃庚,曾沁,刘段灵,等. 日极端气温主客观预报能力评估及多模式集成网格释用[J]. 气象,2017,43(5):581-590.
[13] 翟宇梅,赵瑞星,高建春,等. 遗忘因子自适应最小二乘算法及其在气温预报中的应用[J]. 气象,2014,40(7):881-885.
[14] 马清,龚建东,李莉,等. 超级集合预报的误差订正与集成研究[J]. 气象,2008,34(3):42-48.
[15] 吴启树,韩美,郭弘,等. MOS温度预报中最优训练期方案[J]. 应用气象学报,2016,27(4):426-434.
[16] 李刚,甘文强,谢清霞. 贵州最高气温预报方法研究[J]. 中国农学通报,2016,32(23):165-170.
[17] 孟祥翼. 基于特定因子的河南省干热风客观预报方法[J]. 气象科技,2017,45(6):1049-1057.
[18] 王在文,郑祚芳,陈敏,等. 支持向量机非线性回归方法的气象要素预报[J]. 应用气象学报,2012,23(5):562-570.
[19] 韦淑侠. 青海省气温精细化预报方法研究[J]. 青海科技,2009,1(3):46-48.
[20] KRISHNAMURTI T N,KISHTAWAL C M,LAROW T E,et al. Improved weather and seasonal climate forecasts from multimodel superensemble[J]. Science,1999,285(5433):1548-1550.

[21] ZHI X, QI H, BAI Y, et al. A comparison of three kinds of multimodel ensemble forecast techniques based on the TIGGE data[J]. Acta Meteorologica Sinica, 2012, 26(1): 41-51.
[22] 李刚, 吴春燕, 肖若. 地面气温的概率预报试验[J]. 气象科技, 2015, 43(1): 97-102.
[23] 陈豫英, 陈晓光, 马筛艳, 等. 精细化MOS相对湿度预报方法研究[J]. 气象科技, 2006, 34(2): 143-146.
[24] 张连成, 胡列群, 李帅, 等. 基于GIS的新疆地区两种气温插值方法对比研究[J]. 干旱气象, 2017, 35(2): 330-336.

陕西省精细化网格预报业务系统技术方法[①]

王建鹏[1]　薛春芳[2]　潘留杰[1]　胡　皓[1]　戴昌明[1]　王　丹[3]

(1. 陕西省气象台,西安,710014;2. 陕西省气象局,西安,710014;3. 陕西省气象服务中心,西安,710014)

摘　要:精细化网格预报不仅是目前中国气象局主推的预报业务,而且是未来天气预报的发展方向。本文详细阐述了陕西省精细化网格预报业务系统中数据产品的技术方法。主要内容包括4个方面:①建立陕西网格预报技术框架,提出"动态交叉最优要素预报"(DCOEF)的方法来建立基础网格预报场。②提出"站点订正值向格点场传递"的格点连续性要素订正方法,交叉检验表明,该方法在格点场上24 h最低、最高温度<2 ℃的准确率较模式降尺度数据分别提高34%和23%。此外,该方法在背景场协同、主观站点预报和客观格点预报要素值融合一致方面有较好的应用价值。③基于"偏差订正"方法订正格点降水,结果表明,通过计算预报偏差BIAS来"消空"小雨频率、"补漏"暴雨频率,ECMWF降水预报24 h小雨、暴雨TS评分较原模式分别提高2.5%和4.82%。④提出"反向离差数据归一化"算法,处理因客观方法或主观订正后数据在时间序列上的矛盾问题。该方法不改变原模式对要素的预报趋势,同时使要素在时间上协同一致,很好地解决了网格要素预报的时间协同性问题。

关键词:网格预报;动态交叉最优要素预报;偏差订正;温度站点逼近;要素协同;降尺度

Operation System of Fine Grid Forecast in Shaanxi Province:Technical Methods

WANG Jianpeng[1]　XUE Chunfang[2]　PAN Liujie[1]　HU Hao[1]　DAI Changming[1]　WANG Dan[3]

(1. Shaanxi Meteorological Observatory,Xi'an,710014;2. Shaanxi Meteorological Bureau, Xi'an,710014;3. Shaanxi Meteorological Service Centre,Xi'an,710014)

Abstract:Fine grid forecast is the main service of China Meteorological Administration,and also the future development direction of weather forecast. This system improves the spatial resolution(0.025°× 0.025°),and at the same time, meteorological elements such as precipitation and temperature forecast quality. This article described the technical methods in the data products of this system,from four aspects:(1)established the technical framework for grid forecast, using the Dynamic Cross Optimal Elements Forecast(DCOEF) method to establish the background field of grid forecast,which means comparing different model's elements forecast result and selecting that with higher forecast quality in past 15 days as the base field for forecasters. (2)proposed the methods of "station-revised value transmitting to the grid field" for consecutive elements correction. The cross test shows that the accurate rate of 24-hour

---

[①] 本文发表于《气象科技》2018年第5期。
资助项目:中国气象局预报员专项(CMAYBY2018-074,CMAYBY2018-075);气象预报业务关键技术发展专项(YBGJXM2018;03-13)。
第一作者:王建鹏,男,1972年生,高级工程师,主要从事天气预报与研究工作。E-mail:xawjp@163.com。

minimum and maximum temperature (<2 ℃) are improved by 34% and 23%, respectively, by this method compared to the model downscaling data, and also, the method has better application value in the combination of the background field collaborative and subjective station forecast and objective grid element forecasts. (3) based on the Bias Correction method to correct grid precipitation; the results show that through calculating forecast bias to decrease light rain frequency and increase rainstrom frequency, the 24-hour TS (Threat Score) improved by 2.5% and 4.82%, respectively, compared to the original model. (4) proposed the reverse deviation data normalization algorithm to deal the inconsistent problem of the objective or subjective correction data in the time series, which does not change the elements forecast trends of original models, and at the same time, the elements are coordinated in time, so to solve the problem of time coordination of grid elements.

Key words: grid forecast; dynamic cross optimal elements forecast; bias correction; temperature station approximation; coordinated element; downscaling

引　言

高分辨率网格要素预报是目前中国气象局的主推业务和未来天气预报的发展方向。高时空分辨率的网格产品可以增强气象预报的服务能力，做到时间、空间预报的无缝隙衔接[1,2]。这种时空无缝隙的网格预报的准确率一方面取决于高分辨率数值模式预报能力的提高，另一方面依赖于对模式产品的合理释用，因此，在模式预报性能基本稳定的前提下，加强对高分辨率模式预报产品的解释应用能力就成为提高网格要素预报能力的关键问题[3,4]。

针对各种模式不同气象要素站点预报的释用方法，中国气象学者做了大量深入和卓有成效的研究工作。在温度预报方面，刘建国等[5]分析了基于多模式集合和BMA((Bayesian Model Averaging)方法的气温概率预报；李佰平等[6]归纳了一元线性回归、多元线性回归、单时效消除偏差和多时效消除偏差4种订正方法在温度释用中的优缺点和适用性；吴启树等[7]研究了MOS(Model Output Statistics Method)方法温度预报中的最优训练期方案；王倩等[8]采用平均法、双权重平均法、滑动平均法和滑动双权重平均法分别对Grapes-Rulfs系统2 m温度预报产品进行偏差订正，并比较4种方法的订正效果；翟宇梅等[9]基于遗忘因子的线性自适应最小二乘建模算法，进行最高气温和最低气温预报试验，结果表明适当地选择遗忘因子有助于提高温度的预报准确率。降水预报释用方面发展了频率订正[10]、偏差订正[1]、区域建模[11]、贝叶斯降水订正[12]等一系列客观方法。其他要素释用方法还包括风速的客观订正和预报[13,14]以及基于数值模式的能见度产品预报方法[15]。

系统建设是现代化气象业务中的重要环节[16-20]。本文作为陕西省精细化网格预报业务系统的客观方法部分，主要介绍系统技术思路、总体框架和一些主要网格要素产品的释用方法。

1　技术思路和总体架构

发展高分辨率中尺度快速循环同化模式是实现精细化网格预报最关键的核心技术和有效途径，然而在实际业务中往往面临诸多问题。一是由于目前对大气运动的中小尺度物理过程并不十分清楚，模式中描述中小尺度物理过程的参数化方案仍然需要改进，加上快速循环同化中的资料质量控制及同化算法对模式输出结果有重要影响，因此一些区域中尺度模式的预报性能较世界上先进的全球模式还存在差距。二是面对不断增多的中尺度和全球模式预报产品，当其预报结论出现冲突时，如何实现客观判断，提高要素预报质量,是网格预报业务系统必须面对的问题。

以往面对海量模式数据,气象工作者提出了多模式集成的方法。多模式集成对气候预测、不同情景下的未来气候预估有非常重要的应用价值,实际计算时可采用等权重集成、比例权重和动态权重集成3种方法。等权重集成预报性能在多数情况下仅代表模式预报的平均水平,是一种相对朴素的方法。而比例权重和动态权重在天气预报业务中存在一些问题:首先,集成后的要素预报整体效果理论上可能低于集成前最优模式的预报效果;其次,模式预报往往具有系统性误差,这种系统性误差对数值预报释用具有关键性作用,但集成混淆了不同模式的系统性误差,使得后期释用更加困难;最后,与大尺度的气候预估不同,小尺度异常在天气预报中非常重要,而集成方法通常使得预报要素趋于平均态,譬如对降水来说,集合之后往往对小雨空报偏多,对暴雨预报频次预报则显著减少[21]。

考虑到平均集成和权重集成在要素预报方法方面可能存在的不足,陕西省精细化网格预报系统采用模式动态交叉最优要素预报 DCOEF(Dynamic Cross Optimal Element Forecast)方法。有关 DCOEF 的详细技术在第 2 节中详细介绍。图 1 给出了系统详细的技术方法架构图,可以看到整个系统从原始模式产品出发,经过数据降尺度处理、客观方法订正、动态检验形成一次最优预报网格预报产品后,再引入中央气象台优秀指导预报 SCMOC 预报产品,经过二次取优、客观订正、要素协同,最终形成一套全要素网格预报产品。

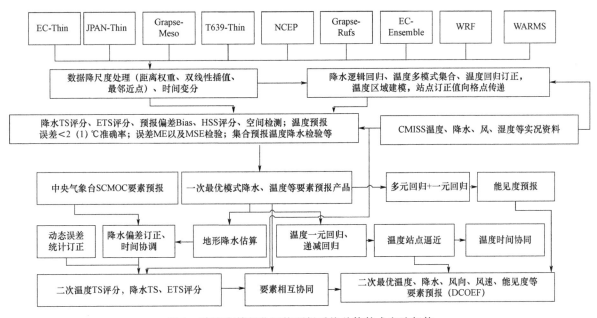

图 1 陕西省精细化网格预报系统总体技术方法架构

2 技术方法

2.1 动态交叉最优要素预报

与简单集成不同,陕西省精细化网格预报系统使用了 DCOEF 算法。DCOEF 是建立在对大量模式要素预报实时动态检验的基础上的。具体方法包括:①面向业务评价体系的传统降水检验 TS(Threat Score)评分、ETS(Equitable Threat Score)评分、预报偏差 BIAS 和温度预报的平均误差 ME、平均绝对误差 MAE 以及预报误差<1 ℃和 2 ℃的准确率,这些指标是动态交叉选优的标准。②适合高分辨率网格预报的邻域检验[22]。由于气象站点稀疏,当允许半径内观测到事件发生时,即认为预报正确,其计算指标

包括传统预报评分和邻域法特有的 FSS(Fractions Skill Score)评分。③面向对象的 MODE 检验[23]，目前主要针对降水预报。MODE 检验可以从降水场中提取降水对象，进而给出降水强度、空间位移、空间形态相对于观测的定量描述，不仅为预报员提供降水落区的准确信息，而且为不同类型降水落区的系统性误差订正提供可能。④集合预报和概率预报检验，计算集合预报融合统计量确定性预报评分和不同量级降水的 Brier 评分。

DCOEF 的基本出发点是：不同模式在不同时段对不同的要素预报性能存在较大的差异，某个模式在特定的时段内对一个特定要素的预报有较好的表现，但并不总是预报最好的，为此对所有现有模式地面预报要素进行 15 d 动态检验，选择前 15 d 预报评分最好的模式结果作为基础背景场的要素预报。但这也存在两个问题：①与集成方法相同，动态选择要素混淆了不同模式的系统性误差，不利于整体的再次释用；②由于最优要素可能来自于不同的模式，其要素的基础值不一致，动态选优后的要素值可能存在跳跃。针对上述第 1 个问题，首先将消除模式系统性误差工作前移，即需要关注每一个模式输出的每一个相关要素，先加入客观方法，然后与原有模式输出进行比较选优。针对第 2 个问题，发展一套要素时间协同、空间协同的相互协同技术，包含两个方面：①来自不同模式的同一要素在数值上可能存在跳跃，本文后续章节中提出的"反向离差数据归一化"算法能够较好地处理这一问题。②不同要素之间相互协同，主要解决的是不同要素相互冲突的问题，其基本原则是定义一系列的规则，通过大量的判断来处理不协调数据，本文不作介绍。

2.2 空间降尺度方法

双线性插值、数值守恒插值、有限元单元片回归法、反距离权重法是数据插值的通用方法[24]。图 2 给出了将温度场格点化为 $0.025°×0.025°$ 时所使用其余 3 种方法的对比。为了清楚分析插值效果，选择分辨率相对较粗的日本模式，预报原始分辨率为 $0.5°×0.5°$，资料时段为 2017 年 3 月 11 日 12:00 UTC 起报的未来 24 h 的 2 m 温度。从模式直接输出(图 2a)来看，模式预报出了温度的局地变化，但关中平原、秦岭山脉等大地形对温度的影响表现不清楚。数值守恒插值(图 2 c)法能够较好地保留插值前的数据信息，空间数值积分和插值前的数据最为吻合，但显著缺陷在于，插值后仅增加了格点数，但原格点内的数值基本没有改变，格点之间数据不连续，新数据在原始数据分辨率格点内存在"一个格点范围内趋同，两个格点边界跳跃"的现象，不适用于精细化网格预报。双线性插值(图 2b)和有限元单元片回归法(图 2 d)均能较好地刻画地形对温度的影响。比较发现，单个格点上双线性插值和有限元单元片回归法的温度数值差异最大可达 0.43 ℃。有限元单元片回归法插值后，空间场内的最大值和最小值均高于双线性插值法，且双线性插值法对空间内的最大值和最小值具有约束性，不会超出原空间场数值的范围。此外，如图 2b 和图 2 d 中黑色方框显示，双线性插值法能更好地保留原场(图 2a)中的数据信息，因此，后期系统在数据高分辨率格点化方面统一使用双线性插值法。

2.3 连续性要素格点订正

网格预报场通常可以分为时空上连续和非连续两种气象要素预报场。温度是具有代表性的连续要素预报场，同时也最受公众关注，为此，以温度为例阐述连续性网格要素的订正方法，其他连续性要素可以采用相同的方法订正。对温度来说，任意空间格点的温度观测值无法准确获取，给高分辨率格点温度释用订正带来了较大的困难，为此，陕西省精细化网格预报系统提出了"站点订正值向格点场传递"的方法[25]。通过常规方法获得较好的站点预报后，将误差返回格点场，进而改进格点要素的预报准确率。具体思路为：在获得一个较好的站点订正温度值后，取模式高分辨率格点化后最邻近站点的格点值作为模式的站点温度预报，同时计算订正温度和预报值的差值，并将误差分配到整个空间，获得一个误差场，最后将误差场叠加到原空间分辨率0.025°的模式预报场上。值得注意的是,在计算过程中要保证站点温度预报值不变。

中央气象台站点指导预报SCMOC有很好的站点温度预报效果,为此,利用中央气象台站点温度预报来订正最低、最高格点温度场订正个例。如图3所示,原ECMWF(European Centre for Medium-Range Weather Forecasts)模式预报的最高(图3a)和最低温度(图3d)较好地表现出温度的纬向变化和地形特征,对秦岭山脉、关中盆地和秦岭南部汉中、安康等小盆地的地形对温度的影响都有很好的表现。

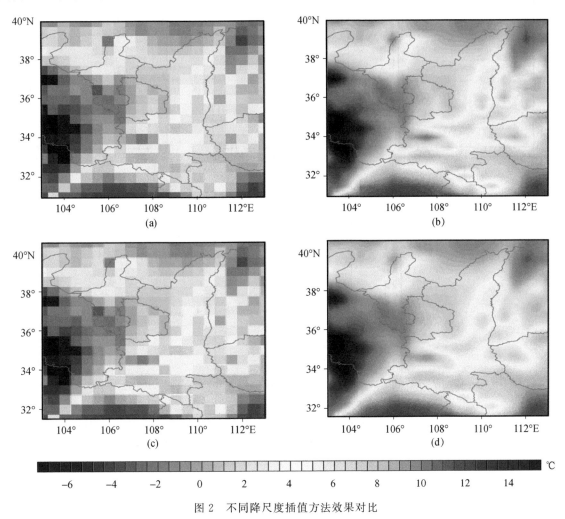

图2 不同降尺度插值方法效果对比
(a)原始数据;(b)双线性插值;(c)数值守恒插值;(d)有限元单元片回归法

与SCMOC相比,主要差异在于ECMWF的最高温度(图3b)预报在秦岭南部显著偏低,在关中北部整体偏高,偏低、偏高幅度最大分别达-8 ℃和1 ℃;最低温度(图3e)的空间形态和最高温度(图3b)整体一致,在秦岭南部偏低,在关中北部偏高,但偏低、偏高幅度和最高温度显著不同,分别为-1 ℃和7 ℃。因此,假如误差小于1 ℃认为预报正确,则ECMWF最高温度主要表现为偏低,最低温度则相反,相对SCMOC偏高。利用SCMOC误差场订正后,最高温度(图3c)在秦岭南部整体升高,最低温度在关中北部明显下降,但订正后,整体不改变ECMWF原温度预报场的空间形态和原模式预报对地形的刻画特征。

统计2016年1—12月的数据,将中央气象台SCMOC温度指导预报的98个县级站的订正差值传递到ECMWF模式预报场后的客观订正结果。结果表明,1289个乡镇站的整体温度预报表现,订正后最高温度小于1 ℃的准确率除在144 h略有下降外,其余时段一致提高;小于2 ℃的准确率一致高于原模式,平均增幅达 20% 以上。最低温度无论是小于 1 ℃，还是小于 2 ℃ 的 TS 准确率均高于 ECMWF 预报，其中小于 1 ℃ 和 2 ℃ 的准确率平均增幅分别达到 0.22 和 0.34。从相对误差 ME 来看，整个预报时效上 ME 都是减小的，特别前 72 h 订正后的 ME 误差减小非常显著，最高和最低温度的 ME 平均分别从原来的 −3.0 ℃ 和 −1.7 ℃ 减小为 0.52 ℃ 和 0.12 ℃。同样，订正后的绝对误差也有不同程度的减小。

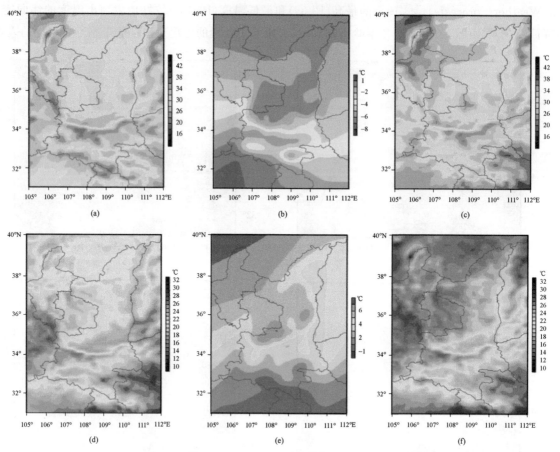

图 3　2016 年 7 月 26 日 12:00 UTC 对未来 24 h 最高温度、最低温度预报的 SCMOC 订正个例
(a)ECMWF 预报最高温度；(b)最高温度订正误差；(c)订正后最高温度；
(d)ECMWF 预报最低温度；(e)最低温度订正误差；(f)订正后最低温度

2.4　格点降水订正

降水是非连续性网格预报中最具有代表性的要素，事实上，风速也在空间上不连续，可以采用类似的方法。对降水来说，陕西省精细化格点预报系统主要采用偏差订正的方法。预报偏差 BIAS 定义为预报事件发生的次数与观测事件发生的次数的比率。BIAS>1 表示预报事件高于观测事件的发生频率；BIAS<1 则相反；理想情况下，BIAS=1。有研究表明，降水预报偏差不仅直接和业务考核指标 TS、ETS 评分相关，而且降水偏差对格点降水订正有很好的指示意义[1]。预报偏差 BIAS>1，反映了降水预报频次较观测偏多；反之偏少；BIAS=1 为理想值。调整 BIAS 可以改变预报评分，考虑到空报和漏报的情况，BIAS=1 并不一定能有最好的预报评分。为此，通过调整降水阈值，进而改进 BIAS，可能会使得 TS、ETS 评分有较好的表现。

图 4 给出了利用 2013—2015 年 5—10 月 ECMWF 集合预报系统降水集合平均调整不同量级降水阈值统计评分的结果。计算时分别为小雨以 0.1 mm 步长，假定模式预报从 0.1~1.0 mm 来预报观测大于 0 mm 的降水发生；大雨以 1 mm 为步长，假定模式预报从 15~25 mm 来预报观测大于 25 mm 降水的发

生。结果显示,随着预报阈值从 0.1~0.9 mm,预报偏差逐渐接近 1,小雨预报频率偏多的情况得到显著改善(图 4a)。但当预报阈值调整到 1.0 mm 时,预报偏差突然变大,比调整前更偏大,而且 TS 评分、ETS 评分也降低,说明阈值调整到大于等于 1.0 mm 已经不适合了。随着阈值的调整,模式的 TS 评分(图 4c)、ETS 评分(图 4e)指数均大幅度提高,除了 1.0 mm 的阈值外,其他阈值的 TS 评分均大于调整前,且当预报阈值为 0.5 和 0.6 mm 时,TS 评分达到最高,超过了控制预报,48 h TS 评分达到 0.58,较调整前提高了 0.06。与 TS 评分不同的是,当预报阈值从 0.1 mm 调整到 0.9 mm 时,ETS 评分随着阈值增大逐渐提高;当预报阈值为 0.9 mm 时,达到最优。在预报的前 8 天,当集合平均的预报阈值调整为 0.5~0.9 mm 时,它们的 ETS 评分都显著优于控制预报。与小雨情况类似,调整后的大雨预报评分也有一定提高,且模式预报阈值为 17~20 mm 时,评分达到最优,在前 7 天,超过控制预报,在第 10~15 天,控制预报仍然表现较好。实际计算时,调整降水阈值,动态计算前 15 天的降水预报表现,从而获得最优订正阈值。2016 年 1—12 月业务运行实际表明,采用偏差订正后,ECMWF 降水预报 24 h 小雨、暴雨 TS 评分较原模式分别提高 2.5% 和 4.82%。

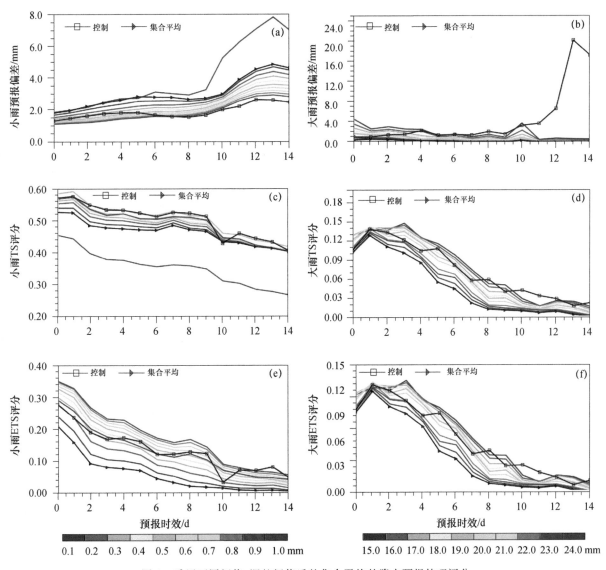

图 4 采用不同阈值,调整阈值后的集合平均的降水预报技巧评分

2.5 同一要素协同

客观订正或动态交叉取优后的预报要素尽管有较好的预报评分,但可能出现要素在时间上不协调的情况,诸如订正后的最高温度低于原模式时间变分后的定时温度、偏差订正后的24 h总降水量与逐小时降水累加值不匹配、来自不同模式的要素在时间序列上不协调等一系列问题,因此,在技术方法的最后一部分阐述同一要素在时间序列及空间分布上的协同处理。有关不同要素之间的相互协同,软件模块中定义了一系列的判定规则,不作为客观方法介绍。

2.5.1 同一要素时间协同

离差数据标准化处理方法可以对某个数据序列进行变换,使其结果落到[0,1]区间,同时其数据序列之间的相互关系保持不变。为此,业务系统建设中反向使用离差数据标准化方法来对数据进行协同。对一个数据序列 x 进行离差标准化,有公式:

$$x_i^* = \frac{x_i - \min(x)}{\max(x) - \min(x)} \tag{1}$$

式中,x_i^* 表示对数据序列 x 中第 i 个数据标准化处理后的值,$\max(x)$ 和 $\min(x)$ 分别表示数据序列中的最大值和最小值。假如客观方法订正了数据序列中的最大值、最小值或序列中任意值,则与 x_i 对应的协同值 y_i 的计算公式为

$$y_i = x_i^* \times [\max(x) - \min(x)] + \min(x) \tag{2}$$

图5给出了以温度和降水为例,对要素进行协同的结果。可以看出,不同情况下调整最高或最低温度后(图5a~e),该方法能够在保持原有数据变化趋势的同时,使得数据协同一致。此外,值得注意的是,如果仅订正了某一或几个时刻的温度预报值,但不影响最高和最低温度,则无须协同,即使协同也将使数据序与初始序列重合,失去订正意义。降水协同与温度略有差异,数据标准化处理时仍然采用公式(1),反算则用序列降水总量值代替序列中最大降水值 $\max(x)$。该方法也适用于前期降水实况已知,保持数据已有趋势前提下的降水实时滚动订正,计算分为3步:①将前期降水实况纳入数据序列,对数据进行标准化;②用已知降水实况和未来降水预报序列计算降水累计值;③用累计值代替最大值采用公式(2)对降水进行回算。图5f为降水主观订正个例,假定预报员修改了某个点的24 h总降水量,可以看出,协同后每个时段的降水量很好地保持了数据序列的原有趋势。

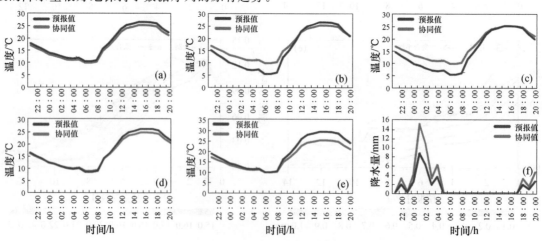

图5 同一个要素时间协同的个例

(a)最高温度和最低温度同时变低;(b)最高温度变低,最低温度变高;
(c)最高温度不变,最低温度变高;(d)最高温度变低,最低温度变高;
(e)最高温度变低,最低温度不变;(f)24 h总降水量增加,逐小时降水量分配

2.5.2 DCOEF 取优空间协同

来自不同模式的预报要素在时间上不连续,在空间上原始分辨率存在差异,可能导致取优后要素场在相邻时段存在跳跃。极端情况下,假如当前时次要素取优来自 ECMWF,下一时次来自 NCEP,再下一个时次又来自 ECMWF,则可能出现连续跳跃。事实上,由于统计检验的实况总是来自站点,采用前文所述温度订正方法可以很好地解决这个问题。图 6 给出了 2017 年 7 月 1 日 00:00 UTC 起报湿度场取优的时空协同个例,可以看出,ECMWF 预报未来 3 h 湿度场在研究区有一条西南—东北向的湿度大值带(图 6a),平均湿度达 70% 以上,随着时间的推移,湿度大值带逐渐向东移动(图 6b、c)。就整个 ECMWF 模式自身预报来看,湿度场在时间系数上比较连续,较好地反映了系统时间演变。假如极端情况,03:00 UTC(图 6b)和 09:00 UTC(图 6c)的检验结果 ECMWF 模式较优,而 06:00 UTC(图 6d)检验结果 NCEP 预报较好,如果简单取 06:00 UTC 的 NCEP 湿度作为最终预报结果,湿度在研究区西南和东北角区域出现不连续现象。采用类似温度订正时的站点订正差值向格点传递的方法,取 NCEP 在研究区 432 个站点上的湿度预报值,协同 06:00 UTC 的湿度预报,结果如图 6e 所示,可以看出协同后,较好地连续了 ECMWF 在时空上的预报特征,同时预报值也有了较好的订正(图 6f),订正湿度与原 ECMWF 预报最大差值达 20% 以上。

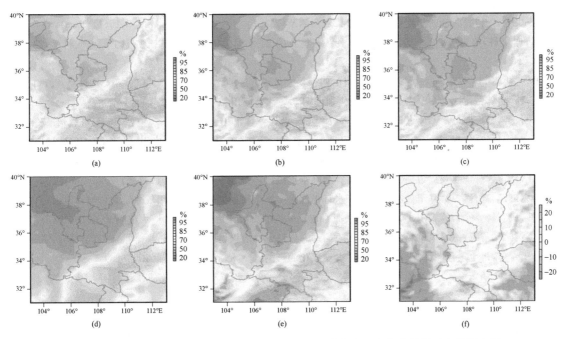

图 6 2017 年 7 月 1 日 00:00 UTC 起报的 DCOEF 动态交叉取优背景场时空协同个例
(a)03 时湿度场预报;(b)06 时湿度场预报;(c)09 时湿度场预报;(d)NCEP 06 时预报;
(e)以 NCEP 站点值协同 06 时预报;(f)协同值与 ECMWF 预报差值

3 结论和讨论

精细化网格预报是中国气象局的主推业务和未来的发展方向。本文基于陕西省精细化网格预报系统的客观方法,主要阐述了使用的技术思路和订正协同方法,以期为网格预报发展提供参考。主要结论如下:

(1)基于不同数值模式预报产品,采用动态交叉取优 DCOEF 的方法建立最优网格预报背景场,形成网格预报场的基础产品。

(2)不同降尺度插值方法结果差异较大。结果比较表明,双线性插值能较好地刻画地形的影响,同时对空间内的最大值和最小值具有较好的约束性,在模式降尺度处理方面有较好的适用性。

(3)提出"站点订正差值向格点场传递"的连续要素格点预报订正方法。该方法在提高格点温度预报准确率的同时不改变原模式预报场要素的空间结构。此外,"站点订正差值向格点场传递"在要素的空间协同方面也有较好的应用价值。

(4)基于"偏差订正"方法订正格点降水,结果表明,通过计算预报偏差 BIAS 来"消空"小雨频率、"补漏"暴雨频率,24 h 小雨、暴雨 TS 评分较原模式分别提高 2.5% 和 4.82%。

(5)采用"反向离差数据归一化"算法,处理因客观方法或主观订正后数据在时间序列上的矛盾问题。该方法不改变原模式对要素的预报趋势,同时使得要素在时间上协同一致,很好地解决了格点要素预报的时间协同性问题。

参考文献

[1] 张宏芳,潘留杰,杨新. ECMWF、日本高分辨率模式降水预报能力的对比分析[J]. 气象,2014,40(4):424-432.
[2] 潘留杰,张宏芳,王建鹏. 数值天气预报检验方法研究进展[J]. 地球科学进展,2014,29(3):327-335.
[3] 潘留杰,薛春芳,张宏芳,等. 三种高分辨率格点降水预报检验方法的对比[J]. 气候与环境研究,2017,22(1):45-58.
[4] 卫捷,张庆云,陶诗言. 2004年夏季短期气候集成预测及检验[J]. 气候与环境研究,2005,10(1):19-31.
[5] 刘建国,谢正辉,赵琳娜,等. 基于 TIGGE 多模式集合的 24 小时气温 BMA 概率预报[J]. 大气科学,2013,37(1):43-53.
[6] 李佰平,智协飞. ECMWF 模式地面气温预报的四种误差订正方法的比较研究[J]. 气象,2012,38(8):897-902.
[7] 吴启树,韩美,郭弘,等. MOS 温度预报中最优训练期方案[J]. 应用气象学报,2016(4):426-434.
[8] 王婧,徐枝芳,范广洲,等. GRAPES_RAFS 系统 2 m 温度偏差订正方法研究[J]. 气象,2015,41(6):719-726.
[9] 翟宇梅,赵瑞星,高建春,等. 遗忘因子自适应最小二乘算法及其在气温预报中的应用[J]. 气象,2014,40(7):881-885.
[10] 李俊,杜钧,陈超君. "频率匹配法"在集合降水预报中的应用研究[J]. 气象,2015,41(6):674-684.
[11] 赵翠光,赵声蓉. 华北及周边地区夏季分区客观降水预报[J]. 应用气象学报,2011,22(5):558-566.
[12] 梁莉,赵琳娜,齐丹,等. 基于贝叶斯原理降水订正的水文概率预报试验[J]. 应用气象学报,2013,24(4):416-424.
[13] 罗连升,段春峰,杨玮,等. MRI-CGCM 模式对东亚夏季风的模拟评估及订正[J]. 大气科学,2016,40(6):1320-1332.
[14] 吴息,黄林宏,周海,等. 风电场风速数值预报的动态修订方法的探讨[J]. 大气科学学报,2014,37(5):665-670.
[15] 孟蕾,周奇越,牛生杰,等. 降水对雾中能见度参数化的影响[J]. 大气科学学报,2010,33(6):731-737.
[16] 熊劦,邓卫华,胡佳军,等. 基于 CIMISS 的区域灾害性天气实时监测与报警系统的设计与实现[J]. 气象科技,2017,45(3):453-459.
[17] 刘立明,常飚. 基于 MARS 的模式数据精细化服务平台设计[J]. 气象科技,2017,45(2):240-246.
[18] 李建,郑伟才,邓闯,等. 基于移动互联网的浙江台风信息发布系统研发与应用[J]. 气象科技,2017,45(2):254-260.
[19] 朱传林,王学良,范宏飞,等. 闪电数据三维可视化统计分析系统设计与实现[J]. 气象科技,2017,45(1):59-63.
[20] 张宏芳,李建科,陈小婷,等. 基于百度地图的精细化格点预报显示[J]. 气象科技,2017,45(2):261-268.
[21] 张宏芳,潘留杰,卢珊,等. ECMWF 集合预报系统对秦岭周边地区降水确定性预报的性能分析[J]. 气候与环境研究,2017,22(5):551-562.
[22] 潘留杰,张宏芳,陈小婷,等. 基于邻域法的高分辨率模式降水的预报能力分析[J]. 热带气象学报,2015,31(5):632-642.
[23] 薛春芳,潘留杰. 基于 MODE 方法的日本细网格模式降水预报的诊断分析[J]. 高原气象,2016,35(2):406-418.
[24] 周璞,江志红. 自组织映射神经网络(SOM)降尺度方法对江淮流域逐日降水量的模拟评估[J]. 气候与环境研究,2016,21(5):512-524.
[25] 潘留杰,薛春芳,王建鹏,等. 一个简单的格点温度预报订正方法[J]. 气象,2017,43(12):1584-1593.

基于小波分析的客观预报方法在智能网格高低温预报中的应用

刘新伟[1]　段伯隆[1,*]　黄武斌[1]　段明铿[2]　李蓉[1]　狄潇泓[1]　魏素娟[3]

(1. 兰州中心气象台,兰州,730020; 2. 南京信息工程大学气象灾害预报预警与评估协同创新中心/气象灾害教育部重点实验室/气候与环境变化国际合作联合实验室,南京,210044; 3. 甘肃省生态环境科学设计研究院,兰州,730020)

摘　要:基于2017—2018年中国气象局高分辨率数值预报产品、甘肃实时城镇预报产品和国家级地面观测站数据,利用小波分析、滑动训练、最优融合等技术,研发甘肃省智能网格高低温客观订正产品。检验分析表明:城镇预报产品、滑动训练订正产品、最优融合产品3种订正产品对CMA预报均有订正能力,3种客观订正产品的最高气温订正能力强于最低气温订正能力;滑动训练法与最优融合法产生的高低温订正产品,在系统误差明显地区(甘南、陇南等)的预报结果要好于模式客观预报,而高低温城镇预报产品在气温局地性强或者模式客观预报能力差的区域有优势;最优融合预报方法生成的高低温产品预报能力略高于滑动训练订正产品且与现有预报员制作城镇预报产品基本持平,初步具备了替代主观预报的能力。

关键词:智能网格;小波分析;温度订正

Application of Objective Prediction Method Based on Wavelet Analysis in Intelligent Grid High and Low Temperature Prediction

LIU Xinwei[1]　DUAN Bolong[1,*]　HUANG Wubin[1]　DUAN Mingkeng[2]　LI Rong[1]　DI Xiaohong[1]　WEI Sujuan[3]

(1. Lanzhou Central Meteorological Observatory, Lanzhou, 730020; 2. Collaborative Innovation Center on Forecast and Evaluation of Meteorological Disaster (CIC-FEMD)/Key Laboratory of Meteorological Disasters, Ministry of Education (KLME)/Joint International Reasearch Laboratory of Climate and Environment Change(ILCEC), Nanjing University of Information Science & Technology, Nanjing, 210044; 3. Gansu Academy of Environmental Sciences, Lanzhou, 730020)

Abstract: Based on the 2017-2018 high-resolution numerical forecast products of China Meteorological

---

① 本文发表于《大气科学学报》2020年第3期。
资助项目:国家重点研发计划项目(2017YFC1502002);中国气象局预报员专项项目(CMAYBY2019-122);甘肃对流性暴雨预报预警关键技术创新团队(GSQXCXTD-2020-01);甘肃省气象局十人计划(GSMArc2019-04)。
第一作者:刘新伟。E-mail:liunavip666@163.com。

Administration(CMA), real-time urban forecast products of Gansu Province and data of national ground-based observation stations, the intelligent grid high and low temperature objective correction products of Gansu Provinceare developed by using wavelet analysis, sliding training, optimal fusion and other technologies. The test results show that the three correction products(urban forecast products, sliding training correction products and optimal fusion products)have the ability to correct CMA forecast, and the maximum temperature correction ability of the three objective correction products is stronger than the minimum temperature correction ability. The prediction results of the high and low temperature correction products produced by the sliding training method and the optimal fusion method are better than those of the model objective prediction in the areas with obvious systematic errors(Gannan, Longnan, etc.), while the high and low temperature urban prediction products have advantages in the areas with strong temperature localization or poor model objective prediction ability. The prediction ability of the high and low temperature products generated by the optimal fusion prediction method is slightly higher than that of the sliding training correction products, and is basically the same as that of the urban prediction products produced by the existing forecasters, which initially has the ability to replace the subjective prediction.

Key words: intelligent grid; wavelet analysis; temperature correction

  天气预报的精细化发展一方面得益于天气预报业务的现代化进程,高性能计算机的提升和预测预报技术的发展为精细化预报提供了硬件基础和理论支撑;另一方面还得益于气象服务的精细化需求,社会经济水平的发展和环境生态发展对天气预报提出更高、更精细化的要求(矫梅燕,2010)。目前,天气预报的能力水平已经较高,能够满足大部分天气预报的需求,在此基础上发展更为精细化的天气预报是对现有天气预报的补充,更是服务新型社会经济的重要途径(矫梅燕,2007)。2016年以来,中国气象局提出建立时间分辨率为3 h(10 d)、空间分辨率为5 km的高时空分辨率的天气预报系统,并在全国推广试验,同时开展智能网格预报服务。高时空分辨率的精细化智能网格预报区别于以往的人工主观预报,是完全基于客观定量化的预报。由于智能网格预报服务产品的服务对象面向城镇,因此智能网格预报业务的开展一方面要考虑与对外服务产品的衔接,另一方面更要在产品准确率方面与城镇预报相比较,这使得气象要素的客观预报和预报方法的研发更加困难。

  在数值预报产品释用方面,诸多气象工作者开展了大量研究并取得丰硕的成果,这些研究成果在实际预报业务中得到了很好的应用(赵声蓉 等,2012;Krishnamurti et al.,2016;赵滨 等,2018;蔡凝昊 等,2019)。目前,针对不同的气象要素预报方法,诸多学者提出诸如MOS方法、卡尔曼滤波方法、相似方法和神经网络等不同方法的应用研究,并在各级台站的数值预报中发挥出重要的作用。其中,针对气温和降水要素的预报最为广泛(Bowler et al.,2008;智协飞 等,2019a,2019b),如潘留杰等(2017)根据ECMWF高分辨率温度预报产品和中央气象台SCMOC最高、最低温度指导预报产品,提出基于回归方法的"站点订正差值向格点传递"的格点温度预报订正方法,并针对研究区域内98个县级气象站温度的开展订正预报研究,在不改变原有ECMWF温度预报的空间形态的基础上很好地纠正了模式格点温度预报的系统性误差;李佰平等(2012)则比较了一元线性回归、多元线性回归、单时效消除偏差和多时效消除偏差4种不同的方法对ECMWF温度预报误差的表征能力,优选出适用于温度的误差订正方法;吴启树等(2016)提出准对称混合滑动训练期方法并对ECMWF模式预报产品进行订正,订正后预报效果明显提高。除此以外,吴乃庚等(2017)研究了日极端气温的主客观预报方法,戴翼等(2019)发展了基于一元线性回归和克里金插值方法的北京地区智能网格温度客观预报方法。这些方法的研究在提升温度预报水平方面开展了有益的尝试,也为模式温度的精准释用提供了技术支撑和理论依据,在加强气象台站温度的预报服务能力方面具有重要的科学意义和实用价值。上述研究也发现,大多数学者在开展模式产品释用方面的研究都不同程度地依赖于 ECMWF 数值模式预报产品,针对中国自己的本地数值模式产品的应用研究反而较少,而后者恰恰是彰显本地数值预报产品和提升数值预报能力的关键,因此亟须相应的理论研究和技术研发。

本文结合实际业务工作,在中国气象局智能网格高低温指导预报产品的基础上,开展基于小波分析的高低温客观订正方法研究。订正后的预报产品对甘肃省城镇预报站点的预报准确率有了明显的改进,基本具备了客观预报代替预报员主观预报的能力,为智能网格高低温预报业务提供了关键技术支撑。

1 资料和方法

1.1 资料

资料为中国气象局(China Meteorological Administration,CMA)智能网格指导预报产品、甘肃省逐日城镇预报产品和甘肃省 78 个国家气象观测站(图 1)的地面最高、最低气温数据,选取 2017 年 1 月 1 日至 12 月 31 日的预报及观测产品作为训练样本,选取 2018 年 1 月 1 日至 12 月 31 日的数据作为测试样本,预报数据为每日 08:00(北京时,下同)起报,预报时效 72 h,时间间隔为 24 h。

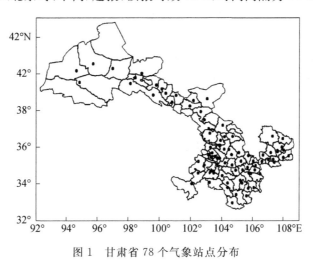

图 1 甘肃省 78 个气象站点分布

1.2 方法

1.2.1 小波分析法

利用小波变换法分析各个站点最高、最低气温的周期,采用 Morlet 小波函数(王蕊 等,2009),如式(1)所示。其中 c 为常数(设定为 6.2)、i 表示虚数。

$$\Psi(t) = e^{ict} e^{-t^2/2} \tag{1}$$

Morlet 小波函数为复数,其实部表示不同特征时间尺度信号在不同时间上的分布和位相两方面的信息。不同时间尺度下的小波系数,可以反映系统在该时间尺度下的变化特征:正的小波系数对应偏多期,负的小波系数对应偏少期,小波系数绝对值越大,表明该时间尺度的周期特征越显著。

1.2.2 滑动训练方法

(1)根据中国气象局的业务要求,智能网格预报产品与城镇站点的预报产品需采取邻域法来对应,即选择距离站点最邻近的格点预报作为站点预报值。若存在多个距离相等的格点,则取东北角格点。据此,本文分别计算得到 24、48、72 h 智能网格预报产品,并计算与对应日期观测数据的差值,得到甘肃省 78 个城镇考核站的最高、最低气温差值时间序列。

(2)针对 3 个预报时效的最高、最低气温差值时间序列,采用 Morlet 小波方法,分析每个城镇站不同预报时效的最高、最低气温的周期特征。

(3)对 78 个城镇站的最高、最低气温周期进行滑动训练,以兰州站最高气温(滑动周期为 t)为例进行说明。当计算 24 h 滑动订正时,将所选日期向前滑动 t d,把前 t d 预报与观测的差值进行平均,所得的平均值称为 24 h 最高气温预报的滑动订正偏差,将订正偏差与指导预报相加,可得出兰州站 24 h 最高气温的订正值。

1.2.3 最优融合方法

考虑到滑动训练方法对个别站点进行订正后天气预报准确率反而低于 CMA 指导产品,因此在本文中将 CMA 指导产品及滑动订正产品进行融合。分别计算前 t d($t=1\sim30$ d)CMA 指导产品与滑动训练订正产品的预报准确率,对比结果发现,1～30 d 的预报准确率相差在 0.5 以内很难区分最优天数,进而对 30 d 的预报准确率进行滤波分析(图略),最终发现 24 h 的高低温产品取 9 d 作为最优融合订正周期订正效果较为显著,48 h 的高低温产品取 13 d 作为最优融合订正周期订正效果较为显著,72 h 的高低温产品取 10 d 作为最优融合订正周期订正效果较为显著。最终结合甘肃本地预报经验,选取 15 d 作为两种产品预报准确率对比天数,当前时刻向前取 15 d CMA 指导产品及滑动订正产品的预报准确率进行对比,保留 15 d 平均预报准确率高的高低温产品,即得到最优融合产品。

1.2.4 检验方法

采用高低温预报准确率、技巧评分和平均绝对误差三个指标对预报结果进行检验,各指标定义如下:
(1)高低温预报准确率

$$F_2 = \frac{n_2}{n} \times 100\% \tag{2}$$

式中,n_2 表示不超过 2 ℃的样本量;n 表示样本总量;F_2 表示温度预报值与观测值误差不超过 2 ℃的百分率。

(2)技巧评分

$$\text{BIAS} = \frac{1}{N}\sum_{i=1}^{N} |\text{obs}_i - \text{pre}_i| \tag{3}$$

式中,obs_i 为第 i 个点的观测值;pre_i 为第 i 个点的预报值;BIAS 表示绝对误差。

(3)平均绝对误差

$$T_{ss} = (T_{\text{MAEN}} - T_{\text{MAEF}})/T_{\text{MAEN}} \times 100\% \tag{4}$$

式中,T_{MAEN} 表示 CMA 指导产品预报的最高、最低温度平均偏差;T_{MAEF} 表示滑动训练法或最优融合法的最高、低温度平均绝对误差。

2 计算及检验

2.1 滑动训练周期计算

文中分别计算 24、48、72 h 甘肃省 78 个国家级考核站最高、最低气温的滑动订正周期,为简要说明,以甘肃省兰州站作为代表站进行说明。对该站进行小波分析,结果如图 2 所示,小波变换系数实部时频分布和功率谱分布均表明该站最高温的小波周期为 21 d,且在 0～90 d 更为显著。同样方法可以分别给出甘肃省 13 个市(州)主要国家级气象站的最高、最低气温的主要周期,见表 1。结果表明,72 h 周期(平均高温周期达到 29.50 d)要大于 24 和 48 h 周期(平均高温周期分别为 21.14 和 21.00 d),其中最高温的周期要大于最低温的周期(24、48、72 h 的平均低温周期分别为 14.43、20.07、28.57 d)。

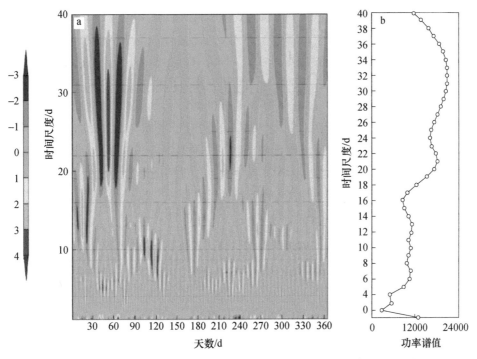

图 2 兰州站 24 h 最高气温差值的小波周期(a)和相应的功率谱随时间变化(b)

表 1 甘肃省 13 个市(州)主要国家级气象站最高、最低气温(℃)的主要周期

站名	24 h		48 h		72 h	
	最高气温	最低气温	最高气温	最低气温	最高气温	最低气温
敦煌	22	14	22	23	22	21
酒泉	30	13	23	20	24	23
张掖	29	18	20	21	26	25
金昌	27	22	24	24	24	24
武威	15	23	24	24	25	24
白银	14	10	12	13	25	24
兰州	21	17	28	27	30	28
临夏	15	13	30	28	32	31
定西	13	12	14	15	32	31
武都	22	14	14	15	40	39
合作	14	10	10	10	36	38
天水	21	6	20	21	30	31
平凉	30	25	27	15	37	32
庆阳	23	5	26	25	30	29

此外,利用小波分析方法得出甘肃省 78 个国家站不同预报时次的最高、最低温度滑动周期,整体来

讲滑动周期集中在 3~30 d,个别站点的滑动周期超过 30 d;24 h 以内的滑动周期多集中在 3~20 d,而 48、72 h 滑动周期更多地集中在 15~30 d,最长的玛曲站高低温滑动周期分别达到 40 和 39 d。

2.2 预报准确率检验

针对 2018 年 1 月 1 日—12 月 31 日甘肃高低温预报开展城镇预报产品、滑动训练订正产品、最优融合产品与 CMA 指导预报产品预报能力的对比研究,分析 24、48、72 h 预报产品的预报能力,对比结果见表 2。结果表明:①城镇预报产品、滑动训练订正产品、最优融合产品 3 种订正产品绝对误差明显低于 CMA 指导预报产品,意味着 3 种预报方法对 CMA 预报均有订正能力。值得注意的是,客观方法预报产品和 CMA 指导预报产品对甘肃城镇站点最低气温的预报能力均高于最高气温的预报。②甘肃高低温的预报能力存在差距,随着时效的延长,预报准确率逐渐下降,72 h 内最高气温的预报准确率为 63%~73%,而最低气温的预报准确率为 72%~77%;主客观方法对于最低气温的预报能力均明显高于最高气温的预报能力。③由预报员主观制作的城镇预报产品与两种客观方法订正产品对比来看,最高气温的城镇预报产品准确率要仍然高于客观预报,24 h 主观预报与客观预报的准确率相差不到 1%,48、72 h 的准确率差距在 3%~4%;最低气温的主观预报准确率与客观预报基本持平,尤其是 48 h 的准确率均低于客观预报。最大差距为 1.5%。订正技巧的结果也证明了上述的观点。

表 2 3 种产品 24、48、72 h 高低温平均绝对误差、预报准确率及订正技巧

预报时效	方法	高温绝对误差	低温绝对误差	高温准确率	低温准确率	高温订正技巧	低温订正技巧
24 h	CMA	1.76	1.67	65.80	68.00	/	/
	城镇预报	1.53	1.36	73.40	77.50	13.07	18.56
	滑动训练	1.55	1.38	72.50	77.00	11.93	17.37
	最优融合	1.54	1.36	72.90	77.70	12.50	18.56
48 h	CMA	2.10	1.78	57.90	64.60	/	/
	城镇预报	1.80	1.49	67.20	73.20	14.29	16.29
	滑动训练	1.93	1.50	63.30	73.70	8.10	15.73
	最优融合	1.90	1.46	64.10	74.70	9.25	17.98
72 h	CMA	2.26	1.84	54.80	62.50	/	/
	城镇预报	1.96	1.55	63.50	71.80	13.27	15.76
	滑动训练	2.10	1.57	59.10	71.10	7.08	14.67
	最优融合	2.08	1.55	59.70	71.50	7.96	15.76

2.3 空间误差检验

对比 CMA 高低温指导预报产品、城镇预报产品、滑动训练订正产品、最优融合产品不同预报时效内的高低温预报准确率空间分布,24、48、72 h 的预报准确率随时效增加而降低,但时空分布情况大致相同,以 24 h 结果为例(图 3、图 4)表明。

(1)CMA 高低温指导预报产品对甘肃最低气温的预报准确率高于最高气温的预报准确率。其中,最高气温的预报在河西地区要好于河东地区,而最低温则相反。CMA 指导预报产品在祁连山区东部的古浪、永登和甘岷山区的舟曲、岷县、宕昌等地最高、最低温度的预报能力均较低。

(2)高低温城镇预报产品对 CMA 指导预报产品有一定的订正能力。长期来看,最高气温在武威、兰州、白银、天水、陇南、平凉、庆阳以及酒泉西部等地区的预报得到有效订正,而最低气温则在酒泉、武威、甘南、陇南、天水及庆阳部分地区得到有效订正。

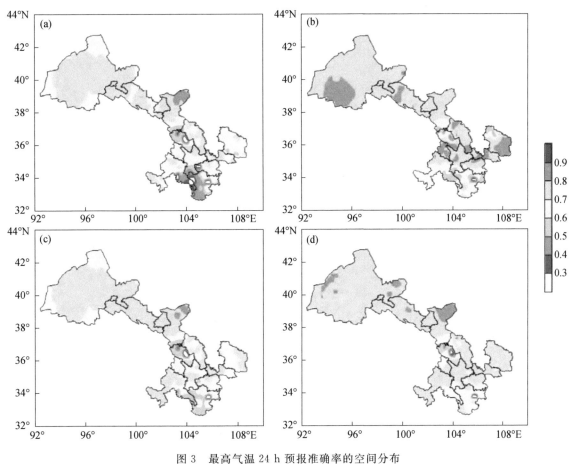

图 3 最高气温 24 h 预报准确率的空间分布
(a)CMA;(b)城镇预报;(c)滑动训练订正产品;(d)最优融合产品

(3)滑动训练方法订正后的高低温产品与 CMA 指导预报高低温产品相比,最高气温的订正能力主要体现在河东地区,如甘南、陇南交界地区;最低气温的订正能力在河东大部地区、酒泉西部得到体现,特别地在甘南、陇南交界区预报准确率提升 30% 以上。

(4)高低温最优融合产品与 CMA 指导预报产品相比,最高、最低气温的预报准确率均有明显的提升。其中,在河东地区(甘南、陇南交界区)站点预报准确率提升超过 10%。

(5)对比高低温最优融合产品与城镇预报产品可以发现,针对最高气温的预报,城镇预报在酒泉西部、临夏、兰州、定西、白银、平凉、庆阳等地的预报准确率要高于最优融合预报,在甘南、陇南、武威北部以及庆阳北部等地区,最优融合预报的预报准确率则要高于城镇预报的准确率;另外对于最低气温的预报,城镇预报产品在酒泉西部、庆阳北部的预报准确率要高于最优融合预报,在定西、甘南、陇南地区则相反。

综合来讲,城镇预报产品在某些局地性的高低温变化中具有比模式客观预报产品更高的预报能力,这些地区的模式客观预报产品仍然需要研发更先进的方法进行客观定量化预报,而对像甘南、陇南这些模式客观预报系统性误差较大的地区,客观最优融合预报方法的订正能力彰显出更大的优势。

3 讨论和结论

本文针对智能网格预报业务开展以来所面对的客观化预报的问题,提出基于中国气象局现行的 CMA 指导预报产品的智能网格预报(城镇站点预报)方法的研究,利用小波分析的方法,通过滑动训练、最优融合等技术,研发出甘肃省最高、最低气温客观订正方法,并与现有城镇站点预报进行对比。结果表

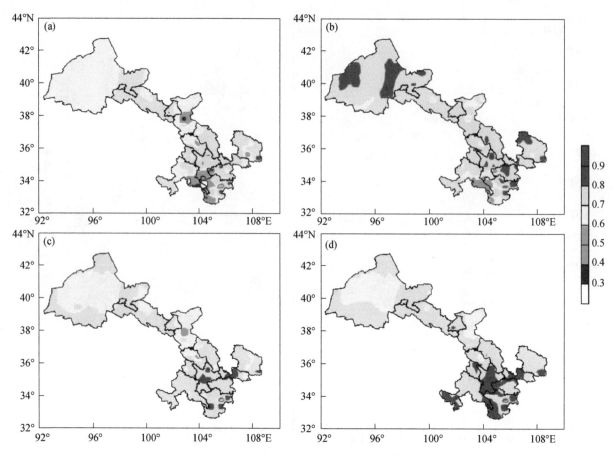

图 4 最低气温 24 h 预报准确率的空间分布
(a)CMA;(b)城镇预报;(c)滑动训练订正产品;(d)最优融合产品

明前者有效地提升了客观预报的准确率,主要结论如下:

(1)城镇预报产品、滑动训练订正产品、最优融合产品 3 种订正产品绝对误差明显低于 CMA 指导预报产品,意味着 3 种预报方法对 CMA 预报均有订正能力,且客观方法预报产品和 CMA 指导预报产品对甘肃城镇站点最低气温的预报能力均高于最高气温的预报。

(2)高低温模式预报产品对甘肃最低温的预报能力要高于最高温,经过客观方法订正后,72 h 内最高温的预报准确率为 63%～73%,而最低温的预报准确率为 72%～77%。

(3)空间误差检验表明,两种订正技术以及预报员城镇预报均对 CMA 高低温指导产品有明显的订正能力,最优融合技术的优势主要体现在甘南、陇南等模式客观预报误差较大、系统误差明显的区域,预报员的优势集中在气温局地性特点强或者模式客观预报能力差的区域。

(4)与预报员主观制作的高低温城镇预报相比,基于小波分析的最优融合预报方法最低气温的预报准确率已经达到主观订正水平,最高气温的预报准确率也只是略低于主观预报,气温要素的客观预报能力已经初步具备了替代主观预报的能力。

本文只是针对业务中的国家考核站高低温产品进行了客观订正与检验评估,对于更加需要客观订正方法的乡镇站点预报、格点预报还在研发中。

参考文献

蔡凝昊,俞剑蔚,2019. 基于数值模式误差分析的气温预报方法[J]. 大气科学学报,42(6):864-873.

戴翼,何娜,付宗钰,等,2019.北京智能网格温度客观预报方法(BJTM)及预报效果检验[J].干旱气象,37(2):339-344,350.

矫梅燕,2007.关于提高天气预报准确率的几个问题[J].气象,33(11):3-8.

矫梅燕,2010.天气业务的现代化发展[J].气象,36(7):1-4.

李佰平,智协飞,2012.ECMWF模式地面气温预报的四种误差订正方法的比较研究[J].气象,38(8):897-902.

潘留杰,薛春芳,王建鹏,等,2017.一个简单的格点温度预报订正方法[J].气象,43(12):1584-1593.

王蕊,王盘兴,吴洪宝,等,2009.小波功率谱Monte Carlo显著性检验的一个简易方案[J].南京气象学院报,32(1):140-144.

吴乃庚,曾沁,刘段灵,等,2017.日极端气温的主客观预报能力评估及多模式集成网格释用[J].气象,43(5):581-590.

吴启树,韩美,郭弘,等,2016.MOS温度预报中最优训练期方案[J].应用气象学报,27(4):426-434.

赵滨,张博,2018.一种2 m温度误差订正方法再复杂地形区数值预报中的应用[J].大气科学学报,41(5):657-667.

赵声蓉,赵翠光,赵瑞霞,等,2012.中国精细化客观气象要素预报进展[J].气象科技进展,2(5):12-21.

智协飞,黄闻,2019a.基于卡尔曼滤波的中国区域气温和降水的多模式集成预报[J].大气科学学报,42(2):197-206.

智协飞,吴佩,俞剑蔚,等,2019b.GFS模式地形高度偏差对地面2 m气温预报的影响[J].大气科学学报,42(5):652-659.

BOWLER N E,ARRIBAS A,MYLNE K R,2008. The benefits of multianalysis and poor man's ensembles[J]. Mon Wea Rev,136(11):4113-4129.

KRISHNAMURTI T N,KUMAR V,SIMON A,et al,2016. A review of multimodel superensemble forecasting for weather,seasonal climate,and hurricanes[J]. Reviews of Geophysics,54(2):336-377.